QUANTUM DYNAMICS

Applications in Biological and Materials Systems

QUANTUM DYNAMICS

Applications in Biological
and Materials Systems

Eric R. Bittner

CRC Press
Taylor & Francis Group
Boca Raton London New York

CRC Press is an imprint of the
Taylor & Francis Group, an **informa** business

CRC Press
Taylor & Francis Group
6000 Broken Sound Parkway NW, Suite 300
Boca Raton, FL 33487-2742

First issued in paperback 2019

© 2010 by Taylor and Francis Group, LLC
CRC Press is an imprint of Taylor & Francis Group, an Informa business

No claim to original U.S. Government works

ISBN-13: 978-1-4200-8053-7 (hbk)
ISBN-13: 978-0-367-38543-9 (pbk)

Library of Congress Cataloging-in-Publication Data

Bittner, Eric R.
    Quantum dynamics : applications in biological and materials systems / Eric R. Bittner.
      p. cm.
    Includes bibliographical references and index.
    ISBN 978-1-4200-8053-7 (hardcover : alk. paper)
    1. Quantum theory--Textbooks. I. Title.

QC174.12.B54 2010
530.12--dc22
                                              2009019261

Visit the Taylor & Francis Web site at
http://www.taylorandfrancis.com

and the CRC Press Web site at
http://www.crcpress.com

Contents

Preface

Why do we need another book on quantum mechanics? Go to any university library and you're bound to find hundreds of textbooks on this subject. A number are truly outstanding and nearly everyone has a favorite. In my case, I very much like Cohen-Tannoudji's two-volume text, although Merzbach, Feynman, and Hibbs and Landau and Lifshitz hold their own places of honor on my bookshelf. So, why go through the bother and effort of trying to say something new? The majority of leading texts focus upon the solution of Schrödinger's equation for a handful of solvable problems. Some will venture into the realm of scattering theory and most will have a good presentation of second quantization. However, the discussion of time-dependent quantum dynamics is typically limited to the spread of a Gaussian wave packet, the dispersionless evolution of a Gaussian in a parabolic potential, and time-dependent perturbation theory leading to Fermi's golden rule.

This book grew out of the need to fill a glaring gap between the standard quantum mechanics textbooks and more specialized texts. It has evolved out of a series of lecture notes for a course on this topic that I have presented intermittently over the past decade, and it has grown out of my own attempts to study the underlying physics of quantum relaxation dynamics as applied to chemical systems. For certain, this book draws from a variety of deep wells.

One significant focus of modern chemical physics is the experimental detection of quantum dynamical processes that occur in chemical systems, typically in a condensed phase environment. With the rapid advance of multiphonon spectroscopies, we are beginning to probe some of nature's most important processes, such as the light-harvesting mechanism in photosynthetic systems or the mechanism of photodamage to DNA. We have also turned these tools to study similar ultrafast processes in nanoscale materials that may eventually be used for artificial photosynthetic systems, electronic switches, or light sources. Understanding these systems requires an in-depth knowledge of time-dependent quantum mechanics beyond what is presented in a typical graduate-level course.

Regarding scope and level, I deliberately chose not to include much detail on solving the standard models for the harmonic oscillator, hydrogen atom, quantized angular momentum, and so forth. These appear in all standard textbooks and I saw little need to rework these models here. A truly comprehensive text would fill at least two complete bookshelves. However, I do rely upon such models for bases and approximations, and I summarize the essential features (eigenstates, spectrum, and so on) as needed. I assume that the reader is familiar with the essential theory of quantum mechanics as presented in a typical undergraduate-level physical chemistry course, and we have used this material in our first-year graduate quantum chemistry course at the University of Houston. My assumption is that students are acquainted with the notion of quantization and its role in molecular spectroscopy. Applications and codes for further illustration can be found on the accompanying Web site (http://k2.chem.uh.edu/quantum_dynamics). A solutions manual is also available for download.

Much of this was committed to text over the course of my sabbatical at Cambridge in 2007, and I wish to thank all the students, postdocs, and colleagues who helped track down typos, clarify sessions, provide figures, and so on. I thank the editors at CRC/Taylor & Francis for keeping me on target to complete this. I also thank the postdocs and graduate students in my group for contributing figures, proofreading, and working problems.

Eric R. Bittner
Cambridge, U.K. & Houston, Texas

About the Author

Eric Bittner is currently the John and Rebecca Moores Distinguished Professor of chemical physics at the University of Houston. He received his PhD from the University of Chicago in 1994 and was a National Science Foundation Postdoctoral Fellow at the University of Texas at Austin and at Stanford University before moving to the University of Houston in 1997. His accolades include an NSF Career Award and a Guggenheim Fellowship. He has also held visiting appointments at the University of Cambridge, the Ecôle Normale Supérieure, Paris, and at Los Alamos National Lab. His research is focused in the areas of quantum dynamics as applied to organic polymer semiconductors, object linking and embedding directory services (OLEDS), solar cells, and energy transport in biological systems.

1 Survey of Classical Mechanics

Quantum mechanics is in many ways the cumulation of many hundreds of years of work and thought about how mechanical things move and behave. Since ancient times, scientists have wondered about the structure of matter and have tried to develop a generalized and underlying theory that governs how matter moves at all length scales.

For ordinary objects, the rules of motion are very simple. By ordinary, I mean objects that are more or less on the same length and mass scale as you and I, say (conservatively) 10^{-7}m to 10^6 m and 10^{-25}g to 10^8g moving at less than 20% of the speed of light. On other words, almost everything you can see and touch and hold obeys what are called *classical* laws of motion. The term *classical* means that that the basic principles of this class of motion have their foundation in antiquity. Classical mechanics is an extremely well-developed area of physics. While you may think that because classical mechanics has been studied extensively for hundreds of years there really is little new development in this field, it remains a vital and extremely active area of research. Why? Because the majority of universe "lives" in a dimensional realm where classical mechanics is extremely valid. Classical mechanics is the workhorse for atomistic simulations of fluids, proteins, and polymers. It provides the basis for understanding chaotic systems. It also provides a useful foundation of many of the concepts in quantum mechanics.

Quantum mechanics provides a description of how matter behaves at very small length and mass scales, that is, the realm of atoms, molecules, and below. It has been developed over the past century to explain a series of experiments on atomic systems that could not be explained using purely classical treatments. The advent of quantum mechanics forced us to look beyond the classical theories. However, it was not a drastic and complete departure. At some point, the two theories must correspond so that classical mechanics is the limiting behavior of quantum mechanics for macroscopic objects. Consequently, many of the concepts we will study in quantum mechanics have direct analogs to classical mechanics: momentum, angular momentum, time, potential energy, kinetic energy, and action.

Much as classical music is cast in a particular style, classical mechanics is based upon the principle that the motion of a body can be reduced to the motion of a point particle with a given mass m, position x, and velocity v. In this chapter, we will review some of the concepts of classical mechanics which are necessary for studying quantum mechanics. We will cast these in forms whereby we can move easily back and forth between classical and quantum mechanics. We will first discuss Newtonian motion and cast this into the Lagrangian form. We will then discuss the principle of least action and Hamiltonian dynamics and the concept of phase space.

1.1 NEWTON'S EQUATIONS OF MOTION

1.1.1 NEWTON'S POSTULATES

Why do things move? Why does an apple fall from a tree? This is usually the first sort of problem we face in trying to study the motion and dynamics of particles and develop laws of nature that are independent of a particular situation.

We understand the concept of *force*. We all have pushed, pulled, or thrown something. Those actions require an action or force from the muscles in our body. Newton proposed a set of basic rules or postulates which he thought could describe the rules that all objects obey under the influence of any kind of force.

Postulate 1.1
Law of Inertia: A free particle always moves without acceleration.

That is, a particle that is not under the influence of an outside force moves along a straight line at constant speed, or remains at rest.

Postulate 1.2
Law of Motion: The rate of change of an object's momentum is equal to the force acting upon it.

$$\frac{d\vec{p}}{dt} = \vec{F} \tag{1.1}$$

This is equivalent to $\vec{F} = m\vec{a}$ where $\vec{a} = d\vec{v}/dt$ is the acceleration. Note that in Newton's first postulate, we assume that the mass does not change with time.

Postulate 1.3
Law of Action: For every action, there is an equal and opposite reaction.

$$\vec{F}_{12} = -\vec{F}_{21} \tag{1.2}$$

This is to say that if particle 1 pushes on particle 2 with force F, then particle 2 pushes on particle 1 with a force $-F$. In SI units, the unit of force is the Newton, $1N = 1kg \cdot m \cdot s^{-2}$.

Newton's *Principia* set the theoretical basis of mathematical mechanics and analysis of physical bodies. The equation that force equals mass times acceleration is the fundamental equation of classical mechanics. Stated mathematically,

$$m\ddot{x} = f(x) \tag{1.3}$$

The dots refer to differentiation with respect to time. We will use this notion for time derivatives. We may also use x' or dx/dt as well. So,

$$\ddot{x} = \frac{d^2x}{dt^2} \tag{1.4}$$

For now we are limiting ourselves to one particle moving in one dimension. For motion in more dimensions, we need to introduce vector components. In Cartesian coordinates, Newton's equations are

$$m\ddot{x} = f_x(x, y, z) \tag{1.5}$$
$$m\ddot{y} = f_y(x, y, z) \tag{1.6}$$
$$m\ddot{z} = f_z(x, y, z) \tag{1.7}$$

where the force vector $\vec{f}(x, y, z)$ has components in all three dimensions and varies with location. We can also define a position vector $\vec{x} = (x, y, z)$ and velocity vector $\vec{v} = (\dot{x}, \dot{y}, \dot{z})$. We can also replace the second-order differential equation with two first-order equations

$$\dot{x} = v_x \tag{1.8}$$
$$\dot{v}_x = f_x/m \tag{1.9}$$

These, along with the initial conditions $x(0)$ and $v(0)$, are all that are needed to solve for the motion of a particle with mass m given a force f. We could have chosen two endpoints as well and asked, What path must the particle take to get from one point to the next? Let us consider some elementary solutions.

First, the case in which $f = 0$ and $\ddot{x} = 0$. Thus, $v = \dot{x} = const$. So, unless there is an applied force, the velocity of a particle will remain unchanged.

Second, we consider the case of a linear force $f = -kx$. This is restoring force for a spring and such force laws are termed *Hooke's law* and k is termed the *force constant*. Our equations are

$$\dot{x} = v_x \tag{1.10}$$
$$\dot{v}_x = -k/mx \tag{1.11}$$

or $\ddot{x} = -(k/m)x$. So we want some function which is its own second derivative multiplied by some number. The cosine and sine functions have this property, so let us try

$$x(t) = A\cos(at) + B\sin(bt) \tag{1.12}$$

Taking time derivatives,

$$\dot{x}(t) = -aA\sin(at) + bB\cos(bt) \tag{1.13}$$
$$\ddot{x}(t) = -a^2 A\cos(at) - b^2 B\sin(bt) \tag{1.14}$$

So we get the required result if $a = b = \sqrt{k/m}$, leaving A and B undetermined. Thus, we need two initial conditions to specify these coefficients. Let us pick $x(0) = x_o$ and $v(0) = 0$. Thus, $x(0) = A = x_o$ and $B = 0$. Notice that the term $\sqrt{k/m}$ has units of angular frequency,

$$\omega = \sqrt{\frac{k}{m}} \tag{1.15}$$

So, our equations of motion are

$$x(t) = x_o \cos(\omega t) \tag{1.16}$$
$$v(t) = -x_o \omega \sin(\omega t) \tag{1.17}$$

Let us now consider a two-dimensional example where we have a particle launched upwards at some initial velocity and we wish to predict where it will land. We shall neglect frictional forces.

The equations of motion in each direction are as follows. In the vertical direction,

$$m\ddot{y} = -mg \tag{1.18}$$

where g is the gravitational constant and the force $-mg$ is the attractive force due to gravity. In x, we have

$$m\ddot{x} = 0 \tag{1.19}$$

since there are no net forces acting in the x direction. Hence, we can solve the x equation immediately since $\dot{v}_x = 0$ and thus, $x(t) = v_x(0)t + x_o = v_o t \cos(\phi)$. For the y equation, denote $v_y = \dot{y}$,

$$m\frac{d}{dt}v_y = -mg \tag{1.20}$$

Integrating, $v_y = -gt + const$. Evaluating this at $t = 0$, $v_y(0) = v_o \sin(\phi) = const$. Thus,

$$v_y(t) = -gt + v_o \sin(\phi) \tag{1.21}$$

This we can integrate as

$$\int dy = \int (-gt + v_o \sin(\phi))dt \tag{1.22}$$

that is,

$$y = v_o \sin(\phi)t - \frac{g}{2}t^2 \tag{1.23}$$

So the trajectory in y is parabolic. To determine the point of impact, we seek the roots of the equation

$$\left(v_o \sin(\phi)t - \frac{g}{2}t^2\right) = 0 \tag{1.24}$$

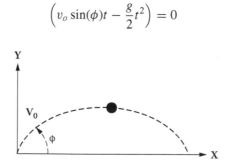

Either $t = 0$ or

$$t_I = \frac{2}{g}v_o \sin(\phi) \tag{1.25}$$

We can now ask this question: What angle do we need to point our cannon to hit a target X meters away? In time t_I the cannon ball will travel a distance $x = v_o \cos(\phi)t_I$. Substituting our expression for the impact time:

$$X = v_o^2 \cos(\phi)\frac{2}{g}\sin(\phi) = \frac{v_o^2 \sin(2\phi)}{g} \tag{1.26}$$

Thus,

$$\sin(2\phi) = \frac{g}{v_o^2}X \tag{1.27}$$

One can also see that the maximum range is obtained when $\phi = \pi/4$.

1.2 LAGRANGIAN MECHANICS

1.2.1 THE PRINCIPLE OF LEAST ACTION

The most general form of the law governing the motion of a mass is the principle of least action or Hamilton's principle. The basic idea is that every mechanical system is described by a single function of coordinate, velocity, and time: $L(x, \dot{x}, t)$ and that the motion of the particle is such that certain conditions are satisfied. That condition is that the time integral of this function

$$S = \int_{t_o}^{t_f} L(x, \dot{x}, t)\, dt \tag{1.28}$$

takes the least possible value given a path that starts at x_o at the initial time and ends at x_f at the final time.

Let us take $x(t)$ to be a function for which S is minimized. This means that S must increase for any variation about this path, $x(t) + \delta x(t)$. Since the endpoints are specified, $\delta x(0) = \delta x(t) = 0$ and the change in S upon replacement of $x(t)$ with $x(t) + \delta x(t)$ is

$$\delta S = \int_{t_o}^{t_f} L(x + \delta x, \dot{x} + \delta \dot{x}, t)dt - \int_{t_o}^{t_f} L(x, \dot{x}, t)dt = 0 \tag{1.29}$$

This is zero because S is a minimum. Now, we can expand the integrand in the first term

$$L(x + \delta x, \dot{x} + \delta \dot{x}, t) = L(x, \dot{x}, t) + \left(\frac{\partial L}{\partial x}\delta x + \frac{\partial L}{\partial \dot{x}}\delta \dot{x}\right) \tag{1.30}$$

Thus, we have

$$\int_{t_o}^{t_f} \left(\frac{\partial L}{\partial x}\delta x + \frac{\partial L}{\partial \dot{x}}\delta \dot{x}\right) dt = 0 \tag{1.31}$$

Since $\delta \dot{x} = d\delta x/dt$ and integrating the second term by parts

$$\delta S = \left[\frac{\partial L}{\partial \dot{x}}\delta x\right]_{t_o}^{t_f} + \int_{t_o}^{t_f} \left(\frac{\partial L}{\partial x} - \frac{d}{dt}\frac{\partial L}{\partial \dot{x}}\right)\delta x\, dt = 0 \tag{1.32}$$

The surface term vanishes because of the condition imposed above. This leaves the integral. It too must vanish and the only way for this to happen is if the integrand itself vanishes. Thus we have

$$\frac{\partial L}{\partial x} - \frac{d}{dt}\frac{\partial L}{\partial \dot{x}} = 0 \tag{1.33}$$

L is known as the Lagrangian. Before moving on, we consider the case of a free particle. The Lagrangian in this case must be independent of the position of the particle since a freely moving particle defines an inertial frame. Since space is isotropic, L must depend upon only the magnitude of v and not its direction. Hence,

$$L = L(v^2) \tag{1.34}$$

Since L is independent of x, $\partial L/\partial x = 0$, so the Lagrange equation is

$$\frac{d}{dt}\frac{\partial L}{\partial v} = 0 \tag{1.35}$$

So, $\partial L/\partial v = const$, which leads us to conclude that L is quadratic in v. In fact,

$$L = \frac{1}{m}v^2 \tag{1.36}$$

which is the *kinetic energy* for a particle

$$T = \frac{1}{2}mv^2 = \frac{1}{2}m\dot{x}^2 \tag{1.37}$$

For a particle moving in a potential field V, the Lagrangian is given by

$$L = T - V \tag{1.38}$$

L has units of energy and gives the difference between the energy of motion and the energy of location.

This leads to the equations of motion:

$$\frac{d}{dt}\frac{\partial L}{\partial v} = \frac{\partial L}{\partial x} \tag{1.39}$$

Substituting $L = T - V$ yields

$$m\dot{v} = -\frac{\partial V}{\partial x} \tag{1.40}$$

which is identical to Newton's equations given above once we identify the force as the minus of the derivative of the potential. For the free particle, $v = const$. Thus,

$$S = \int_{t_o}^{t_f} \frac{m}{2}v^2 dt = \frac{m}{2}v^2(t_f - t_o) \tag{1.41}$$

You may be wondering at this point why we needed a new function and derived all this from some minimization principle. The reason is that for some systems we have constraints on the type of motion they can undertake. For example, there may be bonds, hinges, and other mechanical hindrances that limit the range of motion a given particle can take. The Lagrangian formalism provides a mechanism for incorporating these extra effects in a consistent and correct way. In fact we will use this principle later in deriving a variational solution to the Schrödinger equation by constraining the wave function solutions to be orthonormal.

Lastly, it is interesting to note that $v^2 = (dl/d)^2 = (dl)^2/(dt)^2$ is the square of the element of an arc in a given coordinate system. Thus, within the Lagrangian formalism it is easy to convert from one coordinate system to another. For example, in Cartesian coordinates: $dl^2 = dx^2 + dy^2 + dz^2$. Thus, $v^2 = \dot{x}^2 + \dot{y}^2 + \dot{z}^2$. In cylindrical coordinates, $dl = dr^2 + r^2 d\phi^2 + dz^2$, we have the Lagrangian

$$L = \frac{1}{2}m(\dot{r}^2 + r^2\dot{\phi}^2 + \dot{z}^2) \tag{1.42}$$

and for spherical coordinates, $dl^2 = dr^2 + r^2 d\theta^2 + r^2 \sin^2\theta d\phi^2$; hence,

$$L = \frac{1}{2}m(\dot{r}^2 + r^2\dot{\theta}^2 + r^2 \sin^2\theta \dot{\phi}^2) \tag{1.43}$$

1.2.2 EXAMPLE: THREE-DIMENSIONAL HARMONIC OSCILLATOR IN SPHERICAL COORDINATES

Here we take the potential energy to be a function of r alone (isotropic)

$$V(r) = kr^2/2 \tag{1.44}$$

Thus, the Lagrangian in Cartesian coordinates is

$$L = \frac{m}{2}(\dot{x}^2 + \dot{y}^2 + \dot{z}^2) + \frac{k}{2}r^2 \tag{1.45}$$

Since $r^2 = x^2 + y^2 + z^2$, we could easily solve this problem in Cartesian space since

$$L = \frac{m}{2}(\dot{x}^2 + \dot{y}^2 + \dot{z}^2) + \frac{k}{2}(x^2 + y^2 + z^2) \tag{1.46}$$

$$= \left(\frac{m}{2}\dot{x}^2 + \frac{k}{2}x^2\right) + \left(\frac{m}{2}\dot{y}^2 + \frac{k}{2}y^2\right) + \left(\frac{m}{2}\dot{z}^2 + \frac{k}{2}z^2\right) \tag{1.47}$$

and we see that the system is separable into three independent oscillators. To convert to spherical polar coordinates, we use

$$x = r\sin(\phi)\cos(\theta) \tag{1.48}$$

$$y = r\sin(\phi)\sin(\theta) \tag{1.49}$$

$$z = r\cos(\theta) \tag{1.50}$$

and the arc length given above

$$L = \frac{m}{2}(\dot{r}^2 + r^2\dot{\theta}^2 + r^2 \sin^2\theta \dot{\phi}^2) - \frac{k}{2}r^2 \tag{1.51}$$

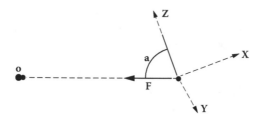

FIGURE 1.1 Vector diagram for motion in central forces. The particle's motion is along the Z axis, which lies in the plane of the page.

The equations of motion are

$$\frac{d}{dt}\frac{\partial L}{\partial \dot{\phi}} - \frac{\partial L}{\partial \phi} = \frac{d}{dt} mr^2 \sin^2 \theta \dot{\phi} = 0 \tag{1.52}$$

$$\frac{d}{dt}\frac{\partial L}{\partial \dot{\theta}} - \frac{\partial L}{\partial \theta} = \frac{d}{dt}(mr^2 \dot{\theta}) - mr^2 \sin \theta \cos \theta \dot{\phi} = 0 \tag{1.53}$$

$$\frac{d}{dt}\frac{\partial L}{\partial \dot{r}} - \frac{\partial L}{\partial r} = \frac{d}{dt}(m\dot{r}) - mr\dot{\theta}^2 - mr \sin^2 \theta \dot{\phi}^2 + kr = 0 \tag{1.54}$$

We now prove that the motion of a particle in a central force field lies in a plane containing the origin. The force acting on the particle at any given time is in a direction toward the origin. Now, place an arbitrary Cartesian frame centered about the particle with the z axis parallel to the direction of motion as sketched in Figure 1.1. Note that the y axis is perpendicular to the plane of the page, and hence, there is no force component in that direction. Consequently, the motion of the particle is constrained to lie in the zx plane, that is the plane of the page, and there is no force component that will take the particle out of this plane.

Let us make a change of coordinates by rotating the original frame to a new one whereby the new z' is perpendicular to the plane containing the initial position and velocity vectors. In Figure 1.1, this new z' axis would be perpendicular to the page and would contain the y axis we placed on the moving particle. In terms of these new coordinates, the Lagrangian will have the same form as previously since our initial choice of axis was arbitrary. However, now we have some additional constraints. Because the motion is now constrained to lie in the $x'y'$ plane, $\theta' = \pi/2$ is a constant, and $\dot{\theta} = 0$. Thus $\cos(\pi/2) = 0$ and $\sin(\pi/2) = 1$ in the previous equations. From the equations for ϕ we find

$$\frac{d}{dt} mr^2 \dot{\phi} = 0 \tag{1.55}$$

or

$$mr^2 \dot{\phi} = const = p_\phi \tag{1.56}$$

This we can put into the equation for r

$$\frac{d}{dt}(m\dot{r}) - mr\dot{\phi}^2 + kr = 0 \tag{1.57}$$

$$\frac{d}{dt}(m\dot{r}) - \frac{p_\phi^2}{mr^3} + kr = 0 \tag{1.58}$$

where we notice that $-p_\phi^2/mr^3$ is the centrifugal force. Taking the last equation, multiplying by $\dot{r}$, and then integrating with respect to time gives

$$\dot{r}^2 = -\frac{p_\phi^2}{m^2 r^2} - kr^2 + b \tag{1.59}$$

that is,

$$\dot{r} = \sqrt{-\frac{p_\phi^2}{m^2 r^2} - kr^2 + b} \tag{1.60}$$

Integrating once again with respect to time,

$$t - t_o = \int \frac{r\,dr}{\dot{r}} \tag{1.61}$$

$$= \int \frac{r\,dr}{\sqrt{-\frac{p_\phi^2}{m^2} - kr^4 + br^2}} \tag{1.62}$$

$$= \frac{1}{2} \int \frac{dx}{\sqrt{a + bx + cx^2}} \tag{1.63}$$

where $x = r^2$, $a = -p_\phi^2/m^2$, b is the constant of integration, and $c = -k$. This is a standard integral and we can evaluate it to find

$$r^2 = \frac{1}{2\omega}(b + A\sin(\omega(t - t_o))) \tag{1.64}$$

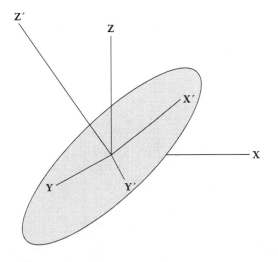

where

$$A = \sqrt{b^2 - \frac{\omega^2 p_\phi^2}{m^2}} \tag{1.65}$$

What we see then is that r follows an elliptical path in a plane determined by the initial velocity.

This example also illustrates another important point that has tremendous impact on molecular quantum mechanics, namely, that the angular momentum about the axis of rotation is conserved. We can choose any axis we want. In order to avoid confusion, let us define χ as the angular rotation about the body-fixed Z' axis and ϕ as angular rotation about the original Z axis. So our conservation equations are

$$mr^2 \dot{\chi} = p_\chi \tag{1.66}$$

about the Z' axis and

$$mr^2 \sin\theta \dot{\phi} = p_\phi \tag{1.67}$$

for some arbitrary fixed Z axis. The angle θ will also have an angular momentum associated with it, $p_\theta = mr^2\dot{\theta}$, but we do not have an associated conservation principle for this term since it varies with ϕ. We can connect p_χ with p_θ and p_ϕ about the other axis via

$$p_\chi d\chi = p_\theta d\theta + p_\phi d\phi \tag{1.68}$$

Consequently,

$$mr^2 \dot{\chi}^2 d\chi = mr^2(\dot{\phi}\sin\theta d\phi + \dot{\theta}d\theta) \tag{1.69}$$

Here we see that the the angular momentum vector remains fixed in space in the absence of any external forces. Once an object starts spinning, its axis of rotation remains pointing in a given direction unless something acts upon it (torque); in essence, in classical mechanics we can fully specify L_x, L_y, and L_z as constants of the motion since $d\vec{L}/dt = 0$. In a later chapter, we will cover the quantum mechanics of rotations in much more detail. In the quantum case, we will find that one cannot make such a precise specification of the angular momentum vector for systems with low angular momentum. We will, however, recover the classical limit in the end as we consider the limit of large angular momenta.

1.3   CONSERVATION LAWS

We just encountered one extremely important concept in mechanics, namely, that some quantities are conserved if there is an underlying symmetry. Next, we consider a conservation law arising from the homogeneity of time. For a closed dynamical system, the Lagrangian does not explicitly depend upon time. Thus we can write

$$\frac{dL}{dt} = \frac{\partial L}{\partial x}\dot{x} + \frac{\partial L}{\partial \dot{x}}\ddot{x} \tag{1.70}$$

Replacing $\partial L / \partial x$ with Lagrange's equation, we obtain

$$\frac{dL}{dt} = \dot{x}\frac{d}{dt}\left(\frac{\partial L}{\partial \dot{x}}\right) + \frac{\partial L}{\partial \dot{x}}\ddot{x} \qquad (1.71)$$

$$= \frac{d}{dt}\left(\dot{x}\frac{\partial L}{\partial \dot{x}}\right) \qquad (1.72)$$

Now, rearranging this a bit,

$$\frac{d}{dt}\left(\dot{x}\frac{\partial L}{\partial \dot{x}} - L\right) = 0 \qquad (1.73)$$

So, we can take the quantity in the parenthesis to be a constant, and

$$E = \left(\dot{x}\frac{\partial L}{\partial \dot{x}} - L\right) = const \qquad (1.74)$$

is an integral of the motion. This is the *energy* of the system. L can be written in the form $L = T - V$ where T is a quadratic function of the velocities, and using Euler's theorem on homogeneous functions:

$$\dot{x}\frac{\partial L}{\partial \dot{x}} = \dot{x}\frac{\partial T}{\partial \dot{x}} = 2T \qquad (1.75)$$

This gives

$$E = T + V \qquad (1.76)$$

which says that the energy of the system can be written as the sum of two different terms: the kinetic energy or energy of motion and the potential energy or the energy of location.

One can also prove that linear momentum is conserved when space is homogeneous. That is, when we can translate our system, some arbitrary amount ε and our dynamical quantities must remain unchanged. We will prove this in the problem sets.

1.3.1 CONSERVATIVE FORCES

A conservative force has nothing to do with its particular political bend. In a loose sense, it is a force in which the total energy is conserved. More precisely, a conservative force acts in such a way that the potential energy of an object does not depend upon the path taken by the object. Recall that work is force times the distance moved. More precisely, work is an integral of the force along a given line or trajectory. In one dimension

$$W = \int_a^b F(x)dx \qquad (1.77)$$

where a and b are beginning and end of the path. In multiple dimensions, we have to extend this concept so that the integral is taken along some arbitrary path.

Suppose we have a curve, C, connecting two points either on a plane or in a volume. This curve may twist and bend, but it is fixed at the two endpoints and our integral must be taken along C from one endpoint to the other. First, let us cut C into N short straight segments of length Δs_i so that the segments $\{\Delta s_1 \cdots \Delta s_N\}$ make up a piecewise continuous approximation for C. The work performed along any one of the segments can be approximated as

$$W_i = \Delta s_i F(x_i, y_i, z_i) \tag{1.78}$$

Consequently, the total work in moving along C is approximately

$$W \approx \sum_i^N \Delta s_i F(x_i, y_i, z_i) \tag{1.79}$$

Taking $\Delta_s \to 0$ and $N \to \infty$, we can write the work performed in moving along path C as

$$W = \lim_{\Delta s_i \to \infty} \sum_i^N \Delta s_i F(x_i, y_i, z_i) = \int_C F(s)ds \tag{1.80}$$

Now, suppose the force can be written as the gradient of some scaler potential function

$$F = \nabla G \tag{1.81}$$

and that our curve C can be parametrized via a single variable t. For example, t could be the length traveled along C or the time. Thus,

$$\frac{dG}{dt} = \nabla G \frac{ds}{dt} = F(s(t)) \frac{ds}{dt} \tag{1.82}$$

Inserting this into the work integral,

$$W = \int_C F(s)ds = \int_C F(s(t)) \frac{ds}{dt} dt = \int_C \frac{dG}{dt} dt = G(a) - G(b) \tag{1.83}$$

where a and b are the two endpoints. As you can see, the integral now depends only upon the two endpoints and does not depend upon the particular details of path C.

Suppose an object starts at point A and moves about some arbitrary closed path P such that after some time it is again at point A. It may still be moving, but the net *work* done on the object is exactly zero. That is, for a conservative force

$$W = \oint \vec{F}(s) \cdot d\vec{s} = 0 \tag{1.84}$$

Although most forces encountered in molecular systems are conservative, many are not, particularly those that depend upon velocity. For such forces, the three criteria are not mathematically equivalent. For example, a magnetic force will satisfy the first requirement, but its curl is not defined and it cannot be written as the gradient of a potential. However, the magnetic force $F = q\vec{v} \times \vec{B}$ can be counted as conservative since the force acts perpendicular to the velocity vector $\vec{v}$ and as such the work is always zero. Nonconservative forces often arise when we neglect or exclude various degrees of freedom. For example, for Brownian motion, the Brownian particle feels a random kick and a viscous drag. These forces arise from the microscopic motion of the surrounding atoms and molecules in the liquid. If we were to treat their motions explicitly, the force acting on the Brownian particle would be conservative. Treating the forces and interactions statistically makes for a far simpler description at the cost of introducing a nonconservative force.

Example: Let us take for an example a force given by $F(x, y) = (x + y)$ and let us compute the work along three different paths. First, a path C_1 from the origin to $(1, 1)$; second, along a path C_2 from $(0, 0)$ to $(1, 0)$ then to $(1, 1)$; and finally along a curved parabolic path C_3 given by $y = x^2$ from the origin to $(1, 1)$. Along C_1, we take s as the distance traveled along C_1 so that $x = s/\sqrt{2}$ and $y = s/\sqrt{2}$. Thus,

$$W_1 = \int_{C_1} (x + y)ds = \sqrt{2} \int_0^{\sqrt{2}} s\,ds = \sqrt{2} \tag{1.85}$$

Moving on to C_2, it is easier to break this into two segments. Along the segment from $(0,0)$ to $(1,0)$, $x = s$ and $y = 0$. Thus,

$$W_2^{(1)} = \int_0^1 s\,ds = \frac{1}{2} \tag{1.86}$$

Along the next segment from $(1, 0)$ to $(1, 1)$, $x = 1$ and $y = s$, so we integrate

$$W_2^{(2)} = \int_0^1 (1 + s)ds = \frac{3}{2} \tag{1.87}$$

then add $W_2 = W_2^{(1)} + W_2^{(2)} = 2$. Finally, along the parabolic path, let $x = s$ and $y = s^2$ and we integrate

$$W_3 = \int_0^1 (s + s^2)ds = \frac{5}{6} \tag{1.88}$$

Clearly, we are not dealing with a conservative force in this case! In fact, in most cases, line integrals depend upon the path taken.

1.4 HAMILTONIAN DYNAMICS

Hamiltonian dynamics is a further generalization of classical dynamics and provides a crucial link with quantum mechanics. Hamilton's function, H, is written in terms of the particle's position and momentum, $H = H(p, q)$. It is related to the Lagrangian via

$$H = \dot{x}p - L(x, \dot{x}) \tag{1.89}$$

Taking the derivative of H with respect to x,

$$\frac{\partial H}{\partial x} = -\frac{\partial L}{\partial x} = -\dot{p} \tag{1.90}$$

Differentiation with respect to p gives

$$\frac{\partial H}{\partial p} = \dot{q} \tag{1.91}$$

These last two equations give the conservation conditions in the Hamiltonian formalism. If H is independent of the position of the particle, then the generalized momentum, p, is constant in time. If the potential energy is independent of time, the Hamiltonian gives the total energy of the system,

$$H = T + V \tag{1.92}$$

It is often easier and more convenient to express Newton's equations of motion as two first-order differential equations rather than a single second-order differential equation. Both are equally valid. However, it is far easier to obtain equations of motion in other coordinate systems than the x, y, z Cartesian coordinates we work with as a more general set of equations. For this, we define a more general quantity for the energy of a system,

$$H = T(v_1, v_2, \ldots, v_N) + V(q_1, q_2, \ldots q_N) \tag{1.93}$$

where T is the kinetic energy that depends upon the velocities of the N particles in the system and V is the potential energy describing the interaction between all the particles and any external forces. V is the energy of position whereas T is the energy of motion. For a single particle moving in three dimensions,

$$T = \frac{1}{2}m\left(v_x^2 + v_y^2 + v_z^2\right) \tag{1.94}$$

If we write the momentum as $p_x = mv_x$, then

$$T = \frac{1}{2m}\left(p_x^2 + p_y^2 + p_z^2\right) \tag{1.95}$$

Notice that we can also define the momentum as the velocity derivative of T:

$$p_x = \frac{\partial T}{\partial v_x} \tag{1.96}$$

This defines a generalized momentum such that q_x is the conjugate coordinate to p_x and (q_x, p_x) are a pair of conjugate variables. This relation between T and p_x is important since we can define the canonical momentum in any coordinate frame. In the Cartesian frame, $p_x = mv_x$. However, in other frames, this will not be so simple. We can also define the following relations:

$$\frac{\partial H}{\partial p_i} = \frac{\partial T}{\partial p_i} = \frac{p_i}{m} = \frac{\partial q_i}{\partial t} \tag{1.97}$$

$$\frac{\partial H}{\partial q_i} = \frac{\partial V}{\partial q_i} = -F_i = -\frac{\partial(mv_i)}{\partial t} \tag{1.98}$$

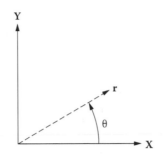

where i now denotes a general coordinate (not necessarily x, y, z). In short, we can write the following equations of motion:

$$\frac{\partial H}{\partial p_i} = \frac{\partial q_i}{\partial t} \tag{1.99}$$

$$-\frac{\partial H}{\partial q_i} = \frac{\partial p_i}{\partial t} \tag{1.100}$$

These hold in any coordinate frame and are termed *Hamilton's equations*.

Example: Hamilton's Equations in Polar Coordinates

Let us consider the transformation between polar and two-dimensional Cartesian coordinates, x and y.

$$x = r \cos\theta \quad \text{and} \quad y = r \sin\theta \tag{1.101}$$

Our Hamiltonian in x, y coordinates is

$$H = \frac{m}{2}\left(v_x^2 + v_y^2\right) + V(x, y) \tag{1.102}$$

Thus,

$$v_x = \frac{dx}{dt} = v_r \cos\theta - v_\theta r \sin\theta \tag{1.103}$$

$$v_y = \frac{dy}{dt} = v_r \sin\theta + v_\theta r \cos\theta \tag{1.104}$$

$$v^2 = v_x^2 + v_y^2 = v_r^2 + v_\theta^2 r^2 \tag{1.105}$$

Thus,

$$H = \frac{m}{2}\left(v_r^2 + v_\theta^2 r^2\right) + V(r, \theta) \tag{1.106}$$

We can now proceed to write this in terms of the conjugate variables,

$$p_r = \frac{\partial T}{\partial v_r} = m v_r \tag{1.107}$$

$$p_\theta = \frac{\partial T}{\partial v_\theta} = m v_\theta r^2 \tag{1.108}$$

Note that p_θ is the *angular momentum* of the system. Thus, we can write

$$H = \frac{1}{2m}\left(p_r^2 + \frac{p_\theta^2}{r^2}\right) + V(r,\theta) \tag{1.109}$$

Next, consider the case where $V(r,\theta)$ has no angular dependence. Thus,

$$\frac{\partial p_r}{\partial t} = -\frac{\partial H}{\partial r} = \frac{p_\theta^2}{mr^3} - \frac{\partial V}{\partial r} \tag{1.110}$$

$$\frac{\partial p_\theta}{\partial t} = -\frac{\partial H}{\partial \theta} = -\frac{\partial V}{\partial \theta} = 0 \tag{1.111}$$

$$\frac{\partial r}{\partial t} = \frac{\partial H}{\partial p_r} = \frac{p_r}{m} \tag{1.112}$$

$$\frac{\partial \theta}{\partial t} = \frac{\partial H}{\partial p_\theta} = \frac{p_\theta}{mr^2} \tag{1.113}$$

Notice that p_θ does not change in time; that is, the angular momentum is a constant of the motion. The radial force we obtain from $\partial p_r/\partial t = F_r$ is

$$F_r = \frac{p_\theta^2}{mr^3} - \frac{\partial V}{\partial r} \tag{1.114}$$

The first term is constant (since $p_\theta = const$) and represents the radial force produced by the angular momentum. It always points *outward* toward larger values of r and is termed the *centrifugal force*. The second term is the force due to the attraction between the moving object and the origin. It could be the gravitational forces, the Coulombic force between charged particles, and so forth. Using the expression for p_θ (Equation 1.111), we can also write the force equation as

$$F_r = \frac{(mv_\theta r^2)^2}{mr^3} - \frac{\partial V}{\partial r} = mv_\theta^2 r - \frac{\partial V}{\partial r} \tag{1.115}$$

If the two forces counterbalance each other, then the net force is $F_r = 0$ and we have

$$mv_\theta^2 r = \frac{\partial V}{\partial r} \tag{1.116}$$

Since $v_\theta = \dot{\theta} = const$, $\theta = \omega t + const$. Where ω is the angular velocity and using $v_\theta = \omega$, we can write

$$m\omega^2 r = \frac{\partial V}{\partial r} \tag{1.117}$$

Finally, we note that the linear velocity is related to the angular velocity by $\omega = vr$,

$$\frac{mv^2}{r} = \frac{\partial V}{\partial r} \tag{1.118}$$

Hence we have a relation between the kinetic energy T and the potential energy V for a centro-symmetric system:

$$mv^2 = 2T = r\frac{\partial V}{\partial r} \tag{1.119}$$

This relation is extremely useful in deriving the classical orbital motion for Coulomb-bound charges as in the hydrogen atom or for planetary motion.

1.4.1 PHASE PLANE ANALYSIS

Often we cannot determine the closed-form solution to a given problem and we need to turn to more approximate methods or even graphical methods. Here, we will look at an extremely useful way to analyze a system of equations by plotting their time derivatives.

First, let us look at the oscillator we just studied. We can define a vector $s = (\dot{x}, \dot{v}) = (v, -k/mx)$ and plot the vector field. Figure 1.2 shows how to do this in Mathematica. The superimposed curve is one trajectory and the arrows give the "flow" of trajectories on the phase plane.

We can examine more complex behavior using this procedure. For example, the simple pendulum obeys the equation $\ddot{x} = -\omega^2 \sin x$. This can be reduced to two first-order equations: $\dot{x} = v$ and $\dot{v} = -\omega^2 \sin(x)$.

We can approximate the motion of the pendulum for small displacements by expanding the pendulum's force about $x = 0$:

$$-\omega^2 \sin(x) = -\omega^2 \left(x - \frac{x^3}{6} + \cdots \right) \tag{1.120}$$

For small x, the cubic term is very small, and we have

$$\dot{v} = -\omega^2 x = -\frac{k}{m}x \tag{1.121}$$

which is the equation for harmonic motion. So, for small initial displacements, we see that the pendulum oscillates back and forth with an angular frequency ω. For large initial displacements, $x_o = \pi$, or if we impart some initial velocity on the system $v_o > 1$, the pendulum does not oscillate back and forth but undergoes librational motion (spinning!) in one direction or the other.

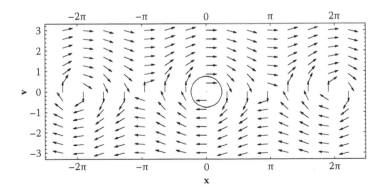

FIGURE 1.2 Tangent field for simple pendulum with $\omega = 1$. The superimposed curve is a linear approximation to the pendulum motion.

1.4.2 INTERACTION BETWEEN A CHARGED PARTICLE AND AN ELECTROMAGNETIC FIELD

We consider here a free particle with mass m and charge e in an electromagnetic field. The Hamiltonian is

$$H = p_x \dot{x} + p_y \dot{y} + p_z \dot{z} - L \tag{1.122}$$

$$= \dot{x}\frac{\partial L}{\partial \dot{x}} + \dot{y}\frac{\partial L}{\partial \dot{y}} + \dot{z}\frac{\partial L}{\partial \dot{z}} - L \tag{1.123}$$

Our goal is to write this Hamiltonian in terms of momenta and coordinates.

For a charged particle in a field, the force acting on the particle is the Lorenz force. Here it is useful to introduce a vector and scalar potential and to work in centimeter-gram-second (cgs) units

$$\vec{F} = \frac{e}{c}\vec{v} \times (\vec{\nabla} \times \vec{A}) - \frac{e}{c}\frac{\partial \vec{A}}{\partial t} - e\vec{\nabla}\phi \tag{1.124}$$

The force in the x direction is given by

$$F_x = \frac{d}{dt}m\dot{x} = \frac{e}{c}\left(\dot{y}\frac{\partial A_y}{\partial x} + \dot{z}\frac{\partial A_z}{\partial x}\right)$$

$$- \frac{e}{c}\left(\dot{y}\frac{\partial A_x}{\partial y} + \dot{z}\frac{\partial A_x}{\partial z} + \frac{\partial A_x}{\partial t}\right) - e\frac{\partial \phi}{\partial x} \tag{1.125}$$

with the remaining components given by cyclic permutation. Since

$$\frac{dA_x}{dt} = \frac{\partial A_x}{\partial t} + \dot{x}\frac{\partial A_x}{\partial x} + \dot{y}\frac{\partial A_x}{\partial y} + \dot{z}\frac{\partial A_x}{\partial z} \tag{1.126}$$

with the force in x given by

$$F_x = \frac{e}{c}\left(\dot{x}\frac{\partial A_x}{\partial x} + \dot{y}\frac{\partial A_x}{\partial y} + \dot{z}\frac{\partial A_x}{\partial z}\right) - \frac{e}{c}\dot{\vec{v}} \cdot \vec{A} - e\phi \tag{1.127}$$

and we find that the Lagrangian is

$$L = \frac{1}{2}m\dot{x}^2 + \frac{1}{2}m\dot{y}^2 + \frac{1}{2}m\dot{z}^2 + \frac{e}{c}\vec{v} \cdot \vec{A} - e\phi \tag{1.128}$$

where ϕ is a velocity independent and static potential.

Continuing on, the Hamiltonian is

$$H = \frac{m}{2}(\dot{x}^2 + \dot{y}^2 + \dot{z}^2) + e\phi \tag{1.129}$$

$$= \frac{1}{2m}((m\dot{x})^2 + (m\dot{y})^2 + (m\dot{z})^2) + e\phi \tag{1.130}$$

The velocities, $m\dot{x}$, are derived from the Lagrangian via the canonical relation

$$p = \frac{\partial L}{\partial \dot{x}} \tag{1.131}$$

From this we find,

$$m\dot{x} = p_x - \frac{e}{c}A_x \qquad (1.132)$$

$$m\dot{y} = p_y - \frac{e}{c}A_y \qquad (1.133)$$

$$m\dot{z} = p_z - \frac{e}{c}A_z \qquad (1.134)$$

and the resulting Hamiltonian is

$$H = \frac{1}{2m}\left[\left(p_x - \frac{e}{c}A_x\right)^2 + \left(p_y - \frac{e}{c}A_y\right)^2 + \left(p_z - \frac{e}{c}A_z\right)^2\right] + e\phi \qquad (1.135)$$

We see here an important concept relating the velocity and the momentum. In the absence of a vector potential, the velocity and the momentum are parallel. However, when a vector potential is included, the actual velocity of a particle is no longer parallel to its momentum and is in fact deflected by the vector potential.

1.4.3 TIME DEPENDENCE OF A DYNAMICAL VARIABLE

One of the important applications of Hamiltonian mechanics is in the dynamical evolution of a variable that depends upon p and q, $G(p,q)$. The total derivative of G is

$$\frac{dG}{dt} = \frac{\partial G}{\partial t} + \frac{\partial G}{\partial q}\dot{q} + \frac{\partial G}{\partial p}\dot{p} \qquad (1.136)$$

From Hamilton's equations, we have the canonical definitions

$$\dot{q} = \frac{\partial H}{\partial p}, \quad \dot{p} = -\frac{\partial H}{\partial q} \qquad (1.137)$$

Thus,

$$\frac{dG}{dt} = \frac{\partial G}{\partial t} + \frac{\partial G}{\partial q}\frac{\partial H}{\partial p} - \frac{\partial G}{\partial p}\frac{\partial H}{\partial q} \qquad (1.138)$$

$$\frac{dG}{dt} = \frac{\partial G}{\partial t} + \{G, H\} \qquad (1.139)$$

where $\{A, B\}$ is called the Poisson bracket of two dynamical quantities, G and H:

$$\{G, H\}, = \frac{\partial G}{\partial q}\frac{\partial H}{\partial p} - \frac{\partial G}{\partial p}\frac{\partial H}{\partial q} \qquad (1.140)$$

We can also define a linear operator $\mathcal{L}$ as generating the Poisson bracket with the Hamiltonian:

$$\mathcal{L}G = \frac{1}{i}\{H, G\} \qquad (1.141)$$

so that if G does not depend explicitly upon time,

$$G(t) = \exp(i\mathcal{L}t)G(0) \tag{1.142}$$

where $\exp(i\mathcal{L}t)$ is the propagator that carried $G(0)$ to $G(t)$.

Also, note that if $\{G, H\} = 0$, then $dG/dt = 0$ so that G is a constant of the motion. This too, along with the construction of the Poisson bracket, has considerable importance in the realm of quantum mechanics.

1.4.4 VIRIAL THEOREM

Finally, we turn our attention to a concept that has played an important role in both quantum and classical mechanics. Consider a function G that is a product of linear momenta and coordinate,

$$G = pq \tag{1.143}$$

The time derivative is simply

$$\frac{G}{dt} = q\dot{p} + p\dot{q} \tag{1.144}$$

Now, let us take a *time average* of both sides of this last equation:

$$\left\langle \frac{d}{dt} pq \right\rangle = \lim_{T \to \infty} \frac{1}{T} \int_0^T \left(\frac{d}{dt} pq \right) dt \tag{1.145}$$

$$= \lim_{T \to \infty} \frac{1}{T} \int_0^T d(pq) \tag{1.146}$$

$$= \lim_{T \to \infty} \frac{1}{T} ((pq)_T - (pq)_0) \tag{1.147}$$

If the trajectories of the system are bounded, both p and q are periodic in time and are therefore finite. Thus, the average must vanish as $T \to \infty$ giving

$$\langle p\dot{q} + q\dot{p} \rangle = 0 \tag{1.148}$$

Since $p\dot{q} = 2T$ and $\dot{p} = -F$, we have

$$\langle 2T \rangle = -\langle qF \rangle \tag{1.149}$$

In Cartesian coordinates this leads to

$$\langle 2T \rangle = -\left\langle \sum_i x_i F_i \right\rangle \tag{1.150}$$

For a conservative system $F = -\nabla V$. Thus, if we have a centro-symmetric potential given by $V = Cr^n$, it is easy to show that

$$\langle 2T \rangle = n\langle V \rangle \tag{1.151}$$

For the case of the harmonic oscillator, $n = 2$ and $\langle T \rangle = \langle V \rangle$. So, for example, if we have a total energy equal to kT in this mode, then $\langle T \rangle + \langle V \rangle = kT$ and $\langle T \rangle = \langle V \rangle = kT/2$. Moreover, for the interaction between two opposite charges separated by r, $n = -1$ and

$$\langle 2T \rangle = -\langle V \rangle \tag{1.152}$$

1.4.5 ANGULAR MOMENTUM

We noted above that if we have a radial force, then the angular velocity and angular momentum are constants of the motion. In general, the angular momentum is defined as the cross product between a radial vector locating the particle and its linear momentum

$$\vec{M} = \vec{r} \times \vec{p} \tag{1.153}$$

Cross products are equivalent to taking the determinant of a matrix

$$\vec{M} = \begin{vmatrix} \hat{i} & \hat{j} & \hat{k} \\ x & y & z \\ p_x & p_y & p_z \end{vmatrix} \tag{1.154}$$

where $\hat{i}$, $\hat{j}$, and $\hat{k}$ are the unit vectors along the x, y, z axes. Evaluating the determinant gives

$$\vec{M} = \hat{i}(yp_z - zp_y) - \hat{j}(xp_z - zp_x) + \hat{k}(xp_y - yp_x) \tag{1.155}$$
$$= \hat{i}M_x + \hat{j}M_y + \hat{k}M_z \tag{1.156}$$

For motion in the x–y plane, the only term that remains is the M_z term, indicating that the angular momentum vector points perpendicular to the plane of rotation,

$$M_z = (xp_y - yp_x) = m(xv_y - yv_x) \tag{1.157}$$

Since we have noted that the angular momentum is a constant of the motion, we must have $dM_z/dt = 0$. Let us check:

$$\frac{dM_z}{dt} = m(v_x v_y - v_y v_x + xa_y - ya_x) \tag{1.158}$$

where $a_x = \dot{v}_x$ is the acceleration in x. Thus,

$$\frac{dM_z}{dt} = (xF_y - yF_x) \tag{1.159}$$

If the force is radial, $F_x = F_r \cos(\theta)$ and $F_y = F_r \sin(\theta)$. Likewise, $x = r\cos(\theta)$ and $y = r\sin(\theta)$. Putting this into the equations, we have

$$\frac{dM_z}{dt} = rF_r(\sin(\theta)\cos(\theta) - \sin(\theta)\cos(\theta)) = 0 \tag{1.160}$$

Taking $\theta = \omega t$ as above where ω is the angular frequency, and using $v_x = -r\omega\sin(\omega t)$ and $y_y = +r\omega\cos(\omega t)$, we can also write

$$M = m(v_x y - v_y x) = mvr(\sin^2(\omega t) + \cos^2(\omega t)) = mvr \tag{1.161}$$

1.4.6　Classical Motion of an Electron about a Positive Charge (Nucleus)

Now we are ready to describe the motion and energy of a charged particle about a nucleus. This will provide us with a classical description of the hydrogen atom and any hydrogenic (hydrogen-like) species. We shall need these results to begin our discussion of quantum theory.

For an electron bound to a nucleus of charge $+Ze$, the Coulomb force holding the electron to its orbit is counterbalanced by the centrifugal force as in the equations we derived above. The Coulomb force is

$$F = \frac{1}{4\pi\varepsilon_o} \frac{Ze^2}{r^2} \tag{1.162}$$

Note that we shall eventually use units such that $4\pi\varepsilon_o = 1$; for now, we keep the SI units. Since the Coulomb force counterbalances the centrifugal force,

$$\frac{1}{4\pi\varepsilon_o} \frac{Ze^2}{r^2} = mr\omega^2 = m\frac{v^2}{r} \tag{1.163}$$

As above, we can see that the virial relation between the kinetic and potential energies

$$T = \frac{1}{2}mv^2 = \frac{1}{4\pi\varepsilon_o} \frac{Ze^2}{2r} = -\frac{1}{2}V \tag{1.164}$$

where

$$V = -\frac{1}{4\pi\varepsilon_o} \frac{Ze^2}{r} \tag{1.165}$$

is the Coulomb potential. Since the classical energy is $E = T + V$, we can write

$$E = \frac{Z}{4\pi\varepsilon_o}\left(\frac{e^2}{2r} - \frac{e^2}{r}\right) = -\frac{1}{4\pi\varepsilon_o}\frac{Ze^2}{2r} \tag{1.166}$$

1.4.7　Birth of Quantum Theory

In spite of the elegance of classical mechanics, there is something clearly missing. According to electrodynamics, an accelerating charge radiates electromagnetic energy. The ramification of this is that if matter is composed of negatively charged electrons moving about positively charged nuclear cores, all matter would be unstable since every classically bound electron would spiral inwards towards the nucleus giving off a burst of x-ray and gamma ray radiation. This is clearly not observed! Moreover, there is no restriction about which radius we choose for the electron . . . we can have any energy we want. However, we also know that hydrogen emits and absorbs light only at specific wavelengths. This fact was demonstrated by Balmer back in 1885 when he showed that all of the absorption (and emission) lines of H could be fit to a single empirical equation

$$\lambda^{-1} = 109677 \text{ cm}^{-1}\left(\frac{1}{n^2} - \frac{1}{m^2}\right) \tag{1.167}$$

where n and m are integers $n = 1, 2, 3, 4, \ldots$ and $m = n + 1, n + 2, n + 3, \ldots$. The numerical constant $109,677 \text{ cm}^{-1}$ is the Rydberg constant (denoted R_H).

About the turn of the nineteenth century, results such as this and results demonstrating the photoelectric effect and Planck's theory of blackbody radiation indicated that there was something wrong with the way we viewed the physical world when it came to atoms and molecules. Niels Bohr attempted to explain the spectral observations and combine Planck's notion of quantized energies with a radation-less orbit of the electron about the nucleus. In doing so, he made a number of remarkable leaps of faith. Bohr postulated the following:

1. A discrete spectrum implies discrete energy levels and that the energy absorbed or emitted are the energy difference between these levels,

$$\Delta E = h\nu = W_f - W_i \tag{1.168}$$

2. Electrons undergo transition between these levels via absorbing or emitting light.
3. Electrons are bound to the nuclei via Coulombic forces and obey classical mechanics.
4. The Rydberg equation gives us the ΔE; hence, since $\lambda \nu = c$,

$$\Delta E = \frac{hc}{\lambda} = 109677 \text{ cm}^{-1} \left(\frac{1}{n^2} - \frac{1}{m^2} \right) \tag{1.169}$$

5. The energy levels are then

$$W_n = -hcR_H \frac{1}{n^2} \tag{1.170}$$

6. *Correspondence principle:* At large values of n, the classical emission frequency must be equal to the electron's orbital frequency, as required by classical electrodynamics.

Now Bohr makes a dramatic leap of faith. The W_n are the quantum energies. Bohr assumes that

$$\text{quantum} = \text{classical}$$

$$-hcR_H \frac{1}{n^2} = -\frac{1}{4\pi\varepsilon_o} \frac{Ze^2}{2r} \tag{1.171}$$

According to Bohr's correspondence principle, the angular frequency of an electron for large values of n must be equal to the classical radiation frequency (I now drop the $4\pi\varepsilon_o$):

$$\nu = \tau \sqrt{\frac{e^2}{4\pi m r^3}} \tag{1.172}$$

where $\tau = 1, 2, 3, \ldots$. For large values of n, the quantum frequency is

$$\nu = R_H c \left(-\frac{1}{n^2} + \frac{1}{(n - \tau)^2} \right) \tag{1.173}$$

$$= \frac{2R_H c\tau (1 - \tau/(2n))}{n^3 (1 - 2\tau/n + \tau^2/n^2)} \tag{1.174}$$

where $\tau = 1, 2, 3, \ldots$ is the change in quantum number as the electron goes from a higher energy orbit to a lower energy orbit. For large values of n,

$$\nu = \frac{2R_H\tau}{n^3} \tag{1.175}$$

Thus, the frequencies are integer multiples of some fundamental frequency ν_o. Again, we use

$$\text{quantum} = \text{classical}$$

$$\frac{2R_H\tau}{n^3} = \tau\sqrt{\frac{e^2}{4\pi mr^3}} \tag{1.176}$$

Now we start canceling terms. To eliminate n we use the expression for the energy levels

$$n = \sqrt{\frac{R_H hc}{|W_n|}} \tag{1.177}$$

and for r, the classical radius

$$W = -\frac{e^2}{2r} \tag{1.178}$$

which gives, $r = e^2/(2|W|)$. Turning the crank and eliminating variables where possible,

$$R_H = \frac{2\pi^2 me^4}{ch^3} \tag{1.179}$$

which gives the Rydberg constant entirely in terms of fundamental physical constants: m, mass of electron; c, speed of light; and h, Planck's constant. Thus, we can write the energy levels in terms of no adjustable parameters:

$$W_n = -\frac{R_H hc}{n^2} = -\frac{2\pi me^4}{h^2}\frac{1}{n^2} \tag{1.180}$$

Furthermore, we can go on to show that since

$$W = -\frac{e^2}{2r} = -\frac{R_H hc}{n^2} \tag{1.181}$$

we can solve for the orbital radius r,

$$r = \frac{e^2}{2}n^2 R_H hc \tag{1.182}$$

giving fixed circular orbitals for the electrons

$$r_n = \frac{h^2}{4\pi^2 me^2}n^2 = \frac{\hbar^2}{me^2}n^2 \tag{1.183}$$

where $n = 1, 2, 3, \ldots$. The innermost orbital, with $n = 1$, is the Bohr radius a_o. Hence, $r_n = n^2 a_o$.

Finally, since the electron is orbiting about the proton, it must have angular momentum. From classical mechanics $M = mr^2\omega = mvr$. Again, following our prescription

$$\text{quantum} = \text{classical}$$

$$-\frac{2\pi^2}{h}me^4\frac{1}{n^2} = -\frac{e^2}{2r} \tag{1.184}$$

from the energy expression. Now, take

$$m\omega^2 r = \frac{e^2}{r^2} \tag{1.185}$$

and solve for e^2

$$e^2 = mr^3\omega^2 \tag{1.186}$$

and plug this back into the energy equation above

$$\text{quantum} = \text{classical}$$

$$-\frac{2\pi^2}{h}me^4\frac{1}{n^2} = -\frac{mr^3\omega^2}{2r} = -\frac{M^2}{2mr^2} \tag{1.187}$$

where M is the angular momentum we derived above. Thus,

$$M^2 = \frac{2\pi^2 me^4}{h} \cdot \frac{2mr^2}{n^2} \tag{1.188}$$

Taking our expression for the quantized radii from above,

$$M^2 = \left(\frac{h}{2\pi}\right)n^2 = \hbar^2 n^2 \tag{1.189}$$

Thus, $M = \hbar n$ is the angular momentum. This, too, is quantized in units of Planck's constant over 2π.

1.4.8 Do the Electron's Orbitals Need to Be Circular?

Here we pose an interesting question. Is there any reason to believe that the electron's orbits need to be strictly circular (or rather elliptical if we account for motion about the center of mass of the electron-proton system)? It seems strange that only circular motion would be allowed. Is there a deeper underlying reason for the quantization? A bit of dimensional analysis indicates that the units of h are energy × time. That is also equivalent to momentum × length. If we imagine plotting the position versus the momentum of a particle on the $p - x$ plane, then momentum × length corresponds to

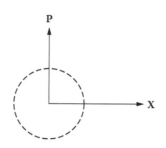

FIGURE 1.3 Closed-loop phase space trajectory corresponding to a bound particle.

some area encompassed by a closed path as shown in Figure 1.3. The closed dashed loop encompasses an area equal to

$$area = \oint p(x)dx \tag{1.190}$$

If we assume that energy is conserved, then the energy is a constant along the dashed loop, $H(p, x) = E = const.$

Planck required that $E = h\nu n$ for the quantized harmonic oscillator levels, but that equation is applicable only to harmonic systems. If the quantization condition is general, then we should see it appear in a more general context. Let us take a harmonic oscillator system as an example:

$$H(p, x) = \frac{p^2}{2m} + \frac{k}{2}x^2 = E \tag{1.191}$$

This is the equation for an ellipse:

$$1 = \frac{p^2}{2mE} + \frac{k}{2E}x^2 \tag{1.192}$$

with major and minor axes $a = \sqrt{2mE}$ and $b = \sqrt{2E/k}$. The area of an ellipse is

$$A = \pi ab \tag{1.193}$$

Hence, the arc on the px plane (called *phase space*) is

$$A = \pi \sqrt{2mE}\sqrt{2E/k} = 2\pi E \sqrt{\frac{m}{k}} = \frac{2\pi E}{\omega} = \frac{E}{\nu} \tag{1.194}$$

Since we assumed A to be quantized in multiples of $\hbar$,

$$A = \frac{E}{\nu} \tag{1.195}$$

From Planck: $E = h\nu n$; thus,

$$A = hn \tag{1.196}$$

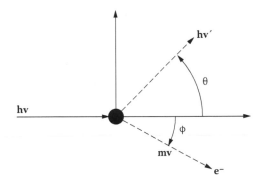

FIGURE 1.4 Compton scattering experiment.

Hence, the electrons need not move in strictly circular orbits; they only need to move in closed paths such that

$$nh = \oint p(x)dx \tag{1.197}$$

This is termed the Bohr–Sommerfield quantization condition. In Chapter 3, we shall return to the use of classical trajectories to analyze quantum mechanical states within a semiclassical theory.

1.4.9 WAVE – PARTICLE DUALITY

In 1922, Arthur Haley Compton at the University of Chicago performed a series of remarkable experiments by scattering x-rays from atoms. He observed that the scattering wavelength λ' was always greater than the incident wavelength. To explain this, he assumed that the energy lost was due to the particle-like collision between a photon and an electron and that some of the incident energy was transferred to the electron. Figure 1.4 gives the geometry of this experiment.

From energy conservation:

$$h\nu = h\nu' + \frac{mv^2}{2} \tag{1.198}$$

Furthermore, the x and y components of the momentum must also be conserved:

$$p_x = p' \cos(\theta) + mv \cos(\phi) = p \tag{1.199}$$

$$p_y = p' \sin(\theta) + mv \sin(\phi) = 0 \tag{1.200}$$

Here, p and p' are the incident and final momenta of the photon, v is the final velocity of the electron, and we take the incident momentum to be along the x axis,

$$mv \cos(\phi) = p - p' \cos(\theta) \tag{1.201}$$

$$mv \sin(\phi) = -p' \sin(\theta) \tag{1.202}$$

Squaring both and adding them together produces

$$m^2 v^2 = p'^2 + p^2 - 2pp' \cos \theta \tag{1.203}$$

Going back to the energy equation,

$$m^2 v^2 = 2m(h\nu - h\nu') \tag{1.204}$$

and equating the last two equations:

$$2m(h\nu - h\nu') = p'^2 + p^2 - 2pp' \cos \theta \tag{1.205}$$

Compton then postulated that the momentum of a photon is given by $h/\lambda = h\nu/c$. Thus, replacing the frequency with c/λ,

$$2mch \left(\frac{1}{\lambda} - \frac{1}{\lambda'} \right) = h^2 \left(\frac{1}{\lambda^2} + \frac{1}{\lambda'^2} \right) - \frac{2h^2}{\lambda\lambda'} \cos \theta \tag{1.206}$$

Taking common denominators,

$$\lambda' - \lambda = \frac{\lambda\lambda'h}{2mc} \left(\frac{1}{\lambda^2} + \frac{1}{\lambda'^2} \right) - \frac{h}{mc} \cos \theta \tag{1.207}$$

Finally, if the change in wavelength is small, $\lambda \approx \lambda'$ and $\lambda\lambda' \approx \lambda^2 \approx \lambda'^2$, we have

$$\lambda' = \lambda + \frac{h}{mc}(1 - \cos \theta) \tag{1.208}$$

which is the Compton scattering formula. The factor h/mc is the Compton wavelength ($\lambda_c = 0.024263$ Å). Since x-ray wavelengths are on the order of 1–3 Å, the assumption that $\lambda = \lambda'$ is pretty accurate. In fact, if we do a relativistic treatment, we do not need to even make that assumption.

If we rewrite this as an energy equation,

$$E' = \frac{Emc^2}{mc^2 + E(1 - \cos \theta)} \tag{1.209}$$

Since m is the mass of the electron, $mc^2 = 0.5$ MeV is the rest-energy of the electron. We can also write this as

$$\frac{1}{E'} = \frac{1}{E} + \frac{1}{mc^2}(1 - \cos \theta) \tag{1.210}$$

Plotting $1/E'$ vs $1 - \cos \theta$ should give a straight line with the slope being $1/m_e c^2$. In Table 1.1 and Figure 1.5 we show some data measured by the author as part of an Experimental Physics course. Using the data, we can calculate the rest-mass and incident energy of the x-ray photon emitted by the ^{137}Cs source.

TABLE 1.1
Compton Scattering Data

θ (degrees)	E'(MeV)
30	0.562
45	0.464
60	0.3955
75	0.339
90	0.29
105	0.244
120	0.223
135	0.205

Note: This data was measured by the author back in 1987 in an Experimental Physics course at Valparaiso University. Here we measured the energy of the scattered x-ray toward a target. From this, you can calculate the rest-mass of the electron and the incident energy of the x-ray emitted by the source (^{137}Cs).

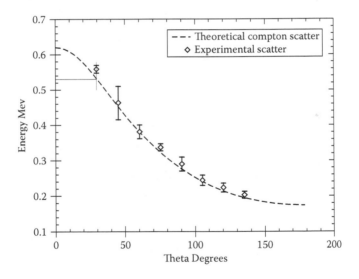

FIGURE 1.5 Experimental Compton scattering data taken by the author in an Experimental Physics course at Valparaiso University (1987).

1.4.10 DE BROGLIE'S MATTER WAVES

Louis de Broglie rationalized that if light could behave as both particle (as in the Compton experiment) and wavelike (as in diffraction experiments), then so should ordinary matter such as electrons, H atoms, neutrons, and so on. For light:

$$E = h\nu = pc \tag{1.211}$$

where $pc = E$ is from Einstein's relativity. Thus, $\lambda = c/v$, which is what Compton also used. Consequently, $\lambda = h/p$. For particles with mass, $p = \sqrt{2mE}$; thus, the wavelength of a particle with mass is

$$\lambda = \frac{h}{\sqrt{2mE}} \qquad (1.212)$$

This served as the basis of de Broglie's PhD thesis in 1923. When asked during his PhD examination just how one may observe such "matter waves," he replied that one should be able to diffract very light particles such as electrons or He atoms from a surface, just as one can do x-ray diffraction. This suggestion was tested by Davisson and Germer and eventually led to the development of a number of powerful analytical techniques for analyzing the structure of crystal surfaces (most commonly low energy electron diffraction, LEED). For this, both de Broglie (in 1929) and Davisson (with Thompson in 1937) were awarded Nobel Prizes.

To come full circle, we ask, "How many de Broglie wavelengths does an electron have in the H atom?"

$$\lambda = \frac{h}{\sqrt{2mE}} \qquad (1.213)$$

$$= h(2me^2/2r)^{-1/2} \qquad (1.214)$$

$$= h \left(2me^2 \left(\frac{me^2}{\hbar^2 n^2} \right) \right)^{-1/2} \qquad (1.215)$$

$$= \frac{h\hbar n}{me^2} \qquad (1.216)$$

$$= 2\pi n \frac{\hbar^2}{me^2} = 2\pi n a_o \qquad (1.217)$$

Thus, for $n = 1$, λ corresponds to the circumference of the first Bohr radius. Hence, we can imagine the electron as a standing wave on a ring of radius a_o

1.5   PROBLEMS AND EXERCISES

Problem 1.1   Find the force that each of the following potentials implies.

1. $V(x) = ax^2$
2. $V(x) = a \log \sin(x)$
3. $V(x, y) = a \cos(by) + c \sin(dx)$
4. $V(x, y, z) = e^{ax}(\tan z + b \sin(x/y))$
5. $V(x) = -a/x + b/x^3$
6. $V(r, \theta) = e^{-b\theta}/r$ in polar coordinates
7. $V(r, \theta, \phi) = r^2 \cos(\phi) \sin(2\theta)$ in spherical polar coordinates

Problem 1.2   The gravitational potential of the Earth is $V(r) = -GM/r$, where G is the gravitational constant, M is the mass of the Earth, and r is the distance from the center of the Earth to some point r. If we set $r = R + z$, where R, is the radius of the Earth and z is the altitude above the surface, show that for $z \ll R$ the resulting force is given by

$$F(z) = -\frac{GM}{R^3}\left(1 - 2\frac{z}{R}\right) \qquad (1.218)$$

Using values for G, M, and R, what is the gravitational force on an object 1 km above the surface of the Earth?

Problem 1.3   Calculate the work necessary to move a unit of mass $m = 1$ along the indicated paths in the xy plane from $(0, 1)$ to $(1, 0)$.
Path 1: Counterclockwise along a circle of radius 1.
Path 2: First from $(0, 1)$ to $(1, 1)$, then from $(1, 1)$ to $(1, 0)$. Use the following force fields normal to the xy plane for your calculations:

1. $F(x, y) = Axy$
2. $F(x, y) = B \log(y)$
3. $F(x, y) = A/\sqrt{x^2 + y^2}$
4. $F(x, y) = A(x^2 + y^2)^2$
5. $F(x, y) = A \exp(-\beta(x^2 + y^2)^2)$

Which of these are conservative fields?

Problem 1.4   Given the two vectors $\vec{\mu}_A = 3\hat{i} + 2\hat{j} - 6\hat{k}$ and $\vec{\mu}_B = -5\hat{i} + 7\hat{j} + 10\hat{k}$, find $\vec{\mu}_A \cdot \vec{\mu}_B$ and $\vec{\mu}_A \times \vec{\mu}_B$.

Problem 1.5   Given the vector fields $\vec{A} = 3x\hat{i} + 2xy\hat{j} + z^2\hat{k}$ and $\vec{B} = -4x\hat{i} + 3xz\hat{j} - .25\cos(z/y)\hat{k}$, find $\vec{\nabla} \cdot (\vec{A} + \vec{B})$, $\vec{\nabla} \cdot \vec{A}$, and $\vec{\nabla} \cdot \vec{B}$ at the point $(3, 1, 6)$.

Problem 1.6   Given the vector field $\vec{A} = (x + y)\hat{i} + x/z\hat{j} + e^{-x^2-y^2}\hat{k}$, find $\vec{\nabla} \cdot A$.

Problem 1.7   Compute the flux due to the vector $\vec{F} = 4xy\hat{i} + 3\hat{j} + z^3\hat{k}$ through the surface of a sphere of radius a centered at the origin.

Problem 1.8   A particle moves between two points on the x axis, from $A = (-a, 0)$ to $B = (+2a, 0)$ under the influence of a radial force $f = k/(x^2+y^2)$ directed toward the origin. Calculate by direct integration the work done along the following paths:

1. Over a rectangular path $(-a, 0) \to (-a, a) \to (+2a, a) \to (+2a, 0)$
2. Along the straight line from A to B
3. Along a circular arc from $(0, 2a)$ to $(2a, 0)$

Problem 1.9   Consider the paths in the previous problem. Assume that no force is present but that the particle moves in a viscous medium that slows the motion with a velocity-dependent force $\vec{F} = -\gamma\vec{v}$. Compute the work required to move the particle along each of the paths at constant speed.

Problem 1.10   Find the force field associated with the potential $V(x, y, z) = xy + 4z/x - 2z^2y^2/x^4$.

Problem 1.11   Given the force vector $\vec{F} = 3x^2 y\hat{i} - (4z^2 - 6y)\hat{j} + (\cos(x)/2 - ye^{-z})\hat{k}$, find the potential V, from which this force is derived.

Problem 1.12   Show that if the energy is independent of one or more coordinates, then the momentum associated with those coordinates is constant. Use this result to show that a classical electron moving in a Coulomb potential has constant angular velocity.

Problem 1.13   Derive the Euler-Lagrange equations of motion for a pendulum consisting of a mass suspended by a (massless) rigid rod attached to a ball and socket. Neglect any effects of the Earth's rotation.

Problem 1.14   Show that the Hamiltonian for an electron moving in a centro-symmetric potential can be written as

$$H = \frac{p_r^2}{2m} + \frac{p_\theta^2}{2r^2m} + V(r)$$

where p_r and p_θ are the respective radial and angular momenta and r is the radial coordinate.

Problem 1.15   Prove that if $\vec{F} = \vec{\nabla}\Phi$ where Φ is a potential function, then $\vec{F}$ is a conservative vector field. Show that if $\vec{F}$ is a conservative field, then $\vec{\nabla} \times \vec{F} = 0$.

Problem 1.16   In the previous problem, you showed that if a vector field is conservative, it is also irrotational. However, is the converse also true? Is an irrotational field also conservative? Consider the field

$$\vec{v} = \left(\frac{-y}{x^2 + y^2}, \frac{x}{x^2 + y^2}, 0 \right)$$

1. Show that $\vec{v}$ is irrotational at every point on the x, y plane.
2. Compute the integral

$$\oint_C \vec{v} \cdot d\vec{r}$$

   where C is the unit circle on the x, y plane.
3. Is $\vec{v}$ conservative?

2 Waves and Wave Functions

In the world of quantum physics, no phenomenon is a phenomenon until it is a recorded phenomenon.

John Archibald Wheeler

Bohr's model of the hydrogen atom was successful in that it gave us a radically new way to look at atoms. However, it has serious shortcomings. It could not be used to explain the spectra of He or any multielectron atom. It could not predict the intensities of the H absorption and emission lines. With de Broglie's hypothesis that matter was also wavelike,[1] there arose a question at the 1925 Solvey conference: What is the wave equation? De Broglie could not answer this; however, over the next year Erwin Schrödinger, working in Vienna, published a series of papers in which he deduced the general form of the equation that bears his name and applied it successfully to the hydrogen atom.[2,3] What emerged was a new set of postulates, much like Newton's, that laid the foundations of quantum theory.

The physical basis of quantum mechanics is

1. That matter, such as electrons, always arrives at a point as a discrete chunk, but that the probibility of finding a chunk at a specified position is like the intensity distribution of a wave.
2. The "quantum state" of a system is described by a mathematical object called a "wave function" or state vector and is denoted $|\psi\rangle$.
3. The state $|\psi\rangle$ can be expanded in terms of the basis states of a given vector space, $\{|\phi_i\rangle\}$, as

$$|\psi\rangle = \sum_i |\phi_i\rangle\langle\phi_i|\psi\rangle \qquad (2.1)$$

   where $\langle\phi_i|\psi\rangle$ denotes an inner product of the two vectors.
4. Observable quantities are associated with the expectation value of Hermitian operators and that the eigenvalues of such operators are always real.
5. If two operators commute, one can measure the two associated physical quantities simultaneously to arbitrary precision.
6. The result of a physical measurement projects $|\psi\rangle$ onto an eigenstate of the associated operator $|\phi_n\rangle$ yielding a measured value of a_n with probability $|\langle\phi_n|\psi\rangle|^2$.

2.1   POSITION AND MOMENTUM REPRESENTATION OF $|\psi\rangle$

Two common operators that we shall use extensively are the position and momentum operators.

The position operator acts on the state $|\psi\rangle$ to give the amplitude of the system to be at a given position:

$$\hat{x}|\psi\rangle = |x\rangle\langle x|\psi\rangle \tag{2.2}$$

$$= |x\rangle\psi(x) \tag{2.3}$$

We shall call $\psi(x)$ the wave *function* of the system since it is the amplitude of $|\psi\rangle$ at point x. Here we can see that $\psi(x)$ is an eigenstate of the position operator. We also define the momentum operator $\hat{p}$ as a derivative operator:

$$\hat{p} = \frac{\hbar}{i}\frac{\partial}{\partial x} \tag{2.4}$$

Thus,

$$\hat{p}\psi(x) = -i\hbar\psi'(x) \tag{2.5}$$

Note that $\psi'(x) \neq \psi(x)$; thus, an eigenstate of the position operator is not also an eigenstate of the momentum operator.

We can deduce this also from the fact that $\hat{x}$ and $\hat{p}$ do not commute. To see this, first consider

$$\frac{\partial}{\partial x}xf(x) = f(x) + xf'(x) \tag{2.6}$$

Thus (using the shorthand ∂_x as partial derivative with respect to x),

$$[\hat{x}, \hat{p}]f(x) = -i\hbar(x\partial_x f(x) - \partial_x(xf(x))) \tag{2.7}$$

$$= \frac{\hbar}{i}(xf'(x) - f(x) - xf'(x)) \tag{2.8}$$

$$= -\frac{\hbar}{i}f(x) \tag{2.9}$$

What are the eigenstates of the $\hat{p}$ operator? To find them, consider the following eigenvalue equation:

$$\hat{p}|\phi(k)\rangle = k|\phi(k)\rangle \tag{2.10}$$

Inserting a complete set of position states using the idempotent operator

$$I = \int |x\rangle\langle x|dx \tag{2.11}$$

and using the "coordinate" representation of the momentum operator, we get

$$-i\hbar\partial_x\phi(k, x) = k\phi(k, x) \tag{2.12}$$

Thus, the solution of this is (subject to normalization)

$$\phi(k, x) = C \exp(ikx/\hbar) = \langle x|\phi(k)\rangle \tag{2.13}$$

We can also use the $|\phi(k)\rangle = |k\rangle$ states as a basis for the state $|\psi\rangle$ by writing

$$|\psi\rangle = \int dk|k\rangle\langle k|\psi\rangle \tag{2.14}$$

$$= \int dk|k\rangle\overline{\psi}(k) \tag{2.15}$$

where $\overline{\psi}(k)$ is related to $\psi(x)$ via

$$\overline{\psi}(k) = \langle k|\psi\rangle = \int dx\langle k|x\rangle\langle x|\psi\rangle \tag{2.16}$$

$$= C\int dx \exp(ikx/\hbar)\psi(x) \tag{2.17}$$

This type of integral is called a "Fourier transform." There are a number of ways to define the normalization C when using this transform; for our purposes at the moment, we will set $C = 1/\sqrt{2\pi\hbar}$ so that

$$\psi(x) = \frac{1}{\sqrt{2\pi\hbar}}\int_{-\infty}^{+\infty} dk\overline{\psi}(k)\exp(-ikx/\hbar) \tag{2.18}$$

and

$$\overline{\psi}(x) = \frac{1}{\sqrt{2\pi\hbar}}\int_{-\infty}^{+\infty} dx\psi(x)\exp(ikx/\hbar) \tag{2.19}$$

Using this choice of normalization, the transform and the inverse transform have symmetric forms and we only need to remember the sign in the exponential.

2.2 THE SCHRÖDINGER EQUATION

Postulate 2.1
The quantum state of the system is a solution of the Schrödinger equation

$$i\hbar\partial_t|\psi(t)\rangle = H|\psi(t)\rangle \tag{2.20}$$

where H is the quantum mechanical analog of the classical Hamiltonian.

From classical mechanics, H is the sum of the kinetic and potential energy of a particle,

$$H = \frac{1}{2m}p^2 + V(x) \tag{2.21}$$

Thus, using the quantum analogs of the classical x and p, the quantum H is

$$H = \frac{1}{2m}\hat{p}^2 + V(\hat{x}) \tag{2.22}$$

To evaluate $V(\hat{x})$ we need a theorem that a function of an operator is the function evaluated at the eigenvalue of the operator. The proof is straightforward.
If

$$V(x) = \left(V(0) + xV'(0) + \frac{1}{2}V''(0)x^2 \cdots \right) \tag{2.23}$$

then

$$V(\hat{x}) = \left(V(0) + \hat{x}V'(0) + \frac{1}{2}V''(0)\hat{x}^2 \cdots \right) \tag{2.24}$$

Since for any operator

$$[\hat{f}, \hat{f}^p] = 0 \forall\, p \tag{2.25}$$

Thus, we have

$$\langle x|V(\hat{x})|\psi\rangle = V(x)\psi(x) \tag{2.26}$$

So, in coordinate form, the Schrödinger equation is written as

$$i\hbar\frac{\partial}{\partial t}\psi(x,t) = \left(-\frac{\hbar}{2m}\frac{\partial^2}{\partial x^2} + V(x) \right)\psi(x,t) \tag{2.27}$$

2.2.1 GAUSSIAN WAVE FUNCTIONS

Let us assume that our initial state is a Gaussian in x with some initial momentum k_o.

$$\psi(x,0) = \left(\frac{2}{\pi a^2} \right)^{1/4} \exp(ik_ox)\exp(-x^2/a^2) \tag{2.28}$$

The momentum representation of this is

$$\overline{\psi}(k,0) = \frac{1}{2\pi\hbar}\int dx\, e^{-ikx}\psi(x,0) \tag{2.29}$$

$$= (\pi a)^{1/2}e^{-(k-k_o)^2a^2/4} \tag{2.30}$$

In Figure 2.1, we see a Gaussian wave packet centered about $x = 0$ with $k_o = 10$ and $a = 1$. For now we will use dimensionless units. The gray components correspond to the real and imaginary components of ψ and the black curve is $|\psi(x)|^2$. Notice, that the wave function is pretty localized along the x axis.

In the next figure (Figure 2.2), we have the momentum distribution of the wave function, $\overline{\psi}(k,0)$. Again, we have chosen $k_o = 10$. Notice that the center of the distribution is shifted about k_o.

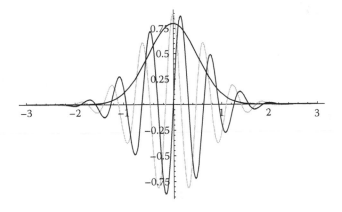

FIGURE 2.1 Real, imaginary, and absolute value of Gaussian wave packet $\psi(x)$.

So, for $f(x) = \exp(-x^2/b^2)$, $\Delta x = b/\sqrt{2}$. Thus, when x varies form 0 to $\pm\Delta x$, $f(x)$ is diminished by a factor of $1/\sqrt{e}$. [Δx is the root-mean-square deviation of $f(x)$.]

For the Gaussian wave packet,

$$\Delta x = a/2 \tag{2.31}$$

$$\Delta k = 1/a \tag{2.32}$$

or

$$\Delta p = \hbar/a \tag{2.33}$$

Thus, $\Delta x \Delta p = \hbar/2$ for the initial wave function.

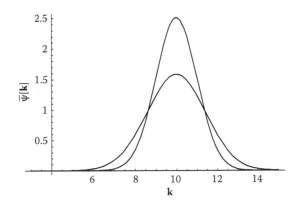

FIGURE 2.2 Momentum-space distribution of $\overline{\psi}(k)$.

2.2.2 EVOLUTION OF $\psi(x)$

Now, let us consider the evolution of a free particle. By a "free" particle, we mean a particle whose potential energy does not change; that is, we set $V(x) = 0$ for all x and solve

$$i\hbar \frac{\partial}{\partial t}\psi(x, t) = \left(-\frac{\hbar}{2m}\frac{\partial^2}{\partial x^2}\right)\psi(x, t) \tag{2.34}$$

This equation is actually easier to solve in k-space. Taking the Fourier transform (FT),

$$i\hbar \partial_t \overline{\psi}(k, t) = \frac{k^2}{2m}\overline{\psi}(k, t) \tag{2.35}$$

Thus, the temporal solution of the equation is

$$\overline{\psi}(k, t) = \exp(-ik^2/(2m)t/\hbar)\overline{\psi}(k, 0) \tag{2.36}$$

This is subject to some initial function $\overline{\psi}(k, 0)$. To get the coordinate x-representation of the solution, we can use the FT relations above:

$$\psi(x, t) = \frac{1}{\sqrt{2\pi\hbar}}\int dk\overline{\psi}(k, t)\exp(-ikx) \tag{2.37}$$

$$= \int dx' \langle x|\exp(-i\hat{p}^2/(2m)t/\hbar)|x'\rangle\psi(x', 0) \tag{2.38}$$

$$= \sqrt{\frac{m}{2\pi i\hbar t}}\int dx'\exp\left(\frac{im(x - x')^2}{2\hbar t}\right)\psi(x', 0) \tag{2.39}$$

$$= \int dx' G_o(x, x')\psi(x', 0) \tag{2.40}$$

The function $G_o(x, x')$ is called the free particle propagator or Green's function. This gives the amplitude for a particle at x to be found at x' some time, t, later. A plot of $G_o(x, x')$ is shown in Figure 2.3 for a particle starting at the origin. Notice, that as $|x|$ increases, the ascillation period decreases rapidly. Since momentum is inversely proportional to wavelength $p = h/\lambda$ through the de Broglie relationship, in order for a particle starting at the origin to move distance (x) away in time t, it must have sufficient momentum.

The sketch tells us that in order to get far away from the initial point in time t, we need to have a lot of energy (wiggles get closer together implying higher Fourier component).

Consequently, to find a particle at the initial point decreases with time. Since the period of oscillation (T) is the time required to increase the phase by 2π,

$$2\pi = \frac{mx^2}{2\hbar t} - \frac{mx^2}{2\hbar(t + T)} \tag{2.41}$$

$$= \frac{mx^2}{2\hbar t^2}\left(\frac{T^2}{1 + T/t}\right) \tag{2.42}$$

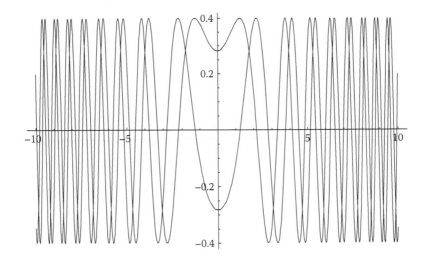

FIGURE 2.3 G_o for fixed t as a function of x.

Let $\omega = 2\pi/T$ and take the long time limit $t \gg T$; we can estimate

$$\omega \approx \frac{m}{2\hbar}\left(\frac{x}{t}\right)^2 \tag{2.43}$$

Since the classical kinetic energy is given by $E = m/2v^2$, we obtain

$$E = \hbar\omega \tag{2.44}$$

Thus, the energy of the wave is proportional to the period of oscillation.

We can evaluate the evolution in x using either the G_o we derived above, or by taking the FT of the wave function evolving in k-space. Recall that the solution in k-space was

$$\overline{\psi}(k, t) = \exp(-ik^2/(2m)t/\hbar)\overline{\psi}(k, 0) \tag{2.45}$$

Assuming a Gaussian form for $\overline{\psi}(k)$ as above,

$$\psi(x, t) = \frac{\sqrt{a}}{(2\pi)^{3/4}} \int dk e^{-a^2/4(k-k_o)^2} e^{i(kx-\omega(k)t)} \tag{2.46}$$

where $\omega(k)$ is the dispersion relation for a free particle:

$$\omega(k) = \frac{\hbar k^2}{2m} \tag{2.47}$$

Cranking through the integral,

$$\psi(x, t) = \left(\frac{2a^2}{\pi}\right)^{1/4} \frac{e^{i\phi}}{\left(a^4 + \frac{4\hbar^2 t^2}{m^2}\right)^{1/4}} e^{ik_o x} \exp\left[\frac{(x - \hbar k_o/mt)^2}{a^2 + 2i\hbar t/m}\right] \tag{2.48}$$

where $\phi = -\theta - \hbar k_o^2/(2m)t$ and $\tan 2\theta = 2\hbar t/(ma^2)$.

Likewise, for the amplitude

$$|\psi(x,t)|^2 = \sqrt{\frac{1}{2\pi \Delta x(t)^2}} \exp\left[-\frac{(x-v_o t)^2}{2\Delta x(t)^2}\right] \qquad (2.49)$$

where we define

$$\Delta x(t) = \frac{a}{2}\sqrt{1 + \frac{4\hbar^2 t^2}{m^2 a^4}} \qquad (2.50)$$

and the time-dependent root-mean-square (rms) width of the wave and the group velocity as

$$v_o = \frac{\hbar k_o}{m} \qquad (2.51)$$

Now, since $\Delta p = \hbar \Delta k = \hbar/a$ is a constant for all time, the uncertainty relation becomes

$$\Delta x(t)\Delta p \geq \hbar/2 \qquad (2.52)$$

corresponding to the particle's wave function becoming more and more diffuse as it evolves in time.

2.3 PARTICLE IN A BOX

2.3.1 INFINITE BOX

The potential we will work with for this example consists of two infinitely steep walls placed at $x = \ell$ and $x = 0$ such that between the two walls, $V(x) = 0$. Within this region, we seek solutions to the differential equation

$$\partial_x^2 \psi(x) = -2mE/\hbar^2 \psi(x) \qquad (2.53)$$

The solutions of this are plane waves traveling to the left and to the right:

$$\psi(x) = A\exp(-ikx) + B\exp(+ikx) \qquad (2.54)$$

The coefficients A and B we will have to determine; k is determined by substitution back into the differential equation

$$\psi''(x) = -k^2\psi(x) \qquad (2.55)$$

Thus, $k^2 = 2mE/\hbar^2$, or $\hbar k = \sqrt{2mE}$. Let us work in units in which $\hbar = 1$ and $m_e = 1$. Energy in these units is the hartree (≈ 27. eV).

Since $\psi(x)$ must vanish at $x = 0$ and $x = \ell$,

$$A + B = 0 \qquad (2.56)$$

$$A\exp(ik\ell) + B\exp(-ik\ell) = 0 \qquad (2.57)$$

We can see immediately that $A = -B$ and that the solutions must correspond to a family of sine functions:

$$\psi(x) = A \sin(n\pi/\ell x) \tag{2.58}$$

Just a check,

$$\psi(\ell) = A \sin(n\pi/\ell\ell) = A \sin(n\pi) = 0 \tag{2.59}$$

To obtain the coefficient, we simply require that the wave functions be normalized over the range $x = [0, \ell]$:

$$\int_0^\ell \sin(n\pi x/\ell)^2 dx = \frac{\ell}{2} \tag{2.60}$$

Thus, the normalized solutions are

$$\psi_n(x) = \sqrt{\frac{2}{\ell}} \sin(n\pi/\ell x) \tag{2.61}$$

The eigenenergies are obtained by applying the Hamiltonian to the wave-function solution

$$E_n \psi_n(x) = -\frac{\hbar^2}{2m} \partial_x^2 \psi_n(x) \tag{2.62}$$

$$= \frac{\hbar^2 n^2 \pi^2}{2a^2 m} \psi_n(x) \tag{2.63}$$

Thus we can write E_n as a function of n:

$$E_n = \frac{\hbar^2 \pi^2}{2a^2 m} n^2 \tag{2.64}$$

for $n = 0, 1, 2, \ldots$. What about the case where $n = 0$? Clearly it is an allowed solution of the Schrödinger equation. However, we also required that the probability to find the particle *anywhere* must be 1. Thus, the $n = 0$ solution cannot be permitted.

Note that the cosine functions are also allowed solutions. However, the restriction of $\psi(0) = 0$ and $\psi(\ell) = 0$ discounts these solutions.

In Figure 2.4 we show the first few eigenstates for an electron trapped in a well of length $a = \pi$. Notice that the number of nodes increases as the energy increases. In fact, we can determine the state of the system by simply counting nodes.

What about orthonormality? We stated that the solution of the eigenvalue problem forms an orthonormal basis. In Dirac notation we can write

$$\langle \psi_n | \psi_m \rangle = \int dx \langle \psi_n | x \rangle \langle x | \psi_m \rangle \tag{2.65}$$

$$= \int_0^\ell dx \psi_n^*(x) \psi_m(x) \tag{2.66}$$

$$= \frac{2}{\ell} \int_0^\ell dx \sin(n\pi x/\ell) \sin(m\pi x/\ell) \tag{2.67}$$

$$= \delta_{nm} \tag{2.68}$$

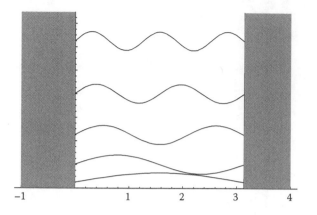

FIGURE 2.4 Particle in a box states.

Thus, we can see in fact that these solutions do form a complete set of orthogonal states on the range $x = [0, \ell]$. Note that it is important to specify "on the range ..." since clearly the sine functions are not a set of orthogonal functions over the entire x axis.

2.3.2 PARTICLE IN A FINITE BOX

Now, suppose our box is finite. That is,

$$V(x) = \begin{cases} -V_o & \text{if } -a < x < a \\ 0 & \text{otherwise} \end{cases} \qquad (2.69)$$

Let us consider the case for $E < 0$. The case $E > 0$ will correspond to scattering solutions. Inside the well, the wave function oscillates, much as in the previous case,

$$\psi_W(x) = A \sin(k_i x) + B \cos(k_i x) \qquad (2.70)$$

where k_i comes from the equation for the momentum inside the well

$$\hbar k_i = \sqrt{2m(E_n + V_o)} \qquad (2.71)$$

We actually have two classes of solution: a symmetric solution when $A = 0$ and an antisymmetric solution when $B = 0$. Outside the well the potential is 0 and we have the solutions

$$\psi_O(x) = c_1 e^{\rho x} \quad \text{and} \quad c_2 e^{-\rho x} \qquad (2.72)$$

We will choose the coefficients c_1 and c_2 to create two cases, ψ_L and ψ_R on the left- and right-hand sides of the well. Also,

$$\hbar \rho = \sqrt{-2mE} \qquad (2.73)$$

Thus, we have three pieces of the full solution that we must connect together:

$$\psi_L(x) = Ce^{\rho x} \quad \text{for } x < -a \tag{2.74}$$

$$\psi_R(x) = De^{-\rho x} \quad \text{for } x > +a \tag{2.75}$$

$$\psi_W(x) = A\sin(k_i x) + B\cos(k_i x) \quad \text{for inside the well} \tag{2.76}$$

To find the coefficients, we need to set up a series of simultaneous equations by applying the conditions that (a) the wave function is a continuous function of x and that (b) it has continuous first derivatives with respect to x. Thus, applying the two conditions at the boundaries, we have

$$\psi_L(-a) = \psi_W(-a) \tag{2.77}$$

$$\psi_R(a) = \psi_W(a) \tag{2.78}$$

$$\psi'_L(-a) = \psi'_W(-a) \tag{2.79}$$

$$\psi'_R(a) = \psi'_W(a) \tag{2.80}$$

Since the well is symmetric about $x = 0$, we have either symmetric or antisymmetric solutions. For the symmetric case: $A = 0$ and $C = D$. Thus we have

$$B = C\sec(ak_i)e^{-a\rho} \tag{2.81}$$

$$-k_i B = -\rho Ce^{-a\rho} \tag{2.82}$$

Eliminating B/C, we have the condition

$$\frac{\rho}{k_i} = \tan(ak_i) \tag{2.83}$$

Both k_i and ρ are functions of the energy

$$\sqrt{\frac{-E}{E+V_o}} = \tan\left(a\sqrt{2m(E+V_o)}/\hbar\right) \tag{2.84}$$

Similarly, for the antisymmetric case, we have

$$\frac{\rho}{k_i} = \cot(ak_i) \tag{2.85}$$

Again, k_i and ρ are functions of energy E:

$$\sqrt{\frac{-E}{V_o+E}} = -\cot\left(a\sqrt{2m(V_o+E)}/\hbar\right) \tag{2.86}$$

These equalities can only be satisfied by stationary solutions of the Schrödinger equation. However, try as we may, we cannot obtain a closed-form equation for the stationary energies. Equations such as these are called "transcendental" equations and closed-form solutions are generally impossible to obtain. Consequently,

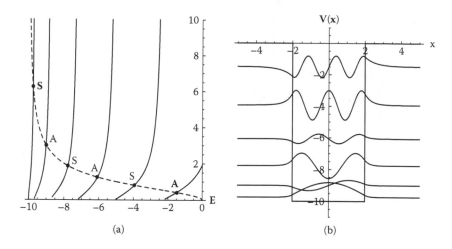

FIGURE 2.5 (a) Graphical solution to transendental equations for an electron in a truncated hard well of depth $V_o = 10$ and width $a = 2$. (b) Wave functions corresponding to stationary states for the finite well.

we need to perform a numerical root search or use graphical techniques. One of the tricks to performing an efficient root search is knowing where to start. For transcendental functions such as $\cot(x)$ and $\tan(x)$, we need to start a root search on a given branch. In Figure 2.5 we show the graphical solution to the transendental equations for an electron in a $V_o = -10$ well of width $a = 2$. The intersections indicate the presence of six bound states. The symmetric states are located at $E_n = -9.75, -7.77, -3.94$ and the asymmetric states at $E_n = -9.01, -6.07$ and -1.49.

2.3.3 SCATTERING STATES AND RESONANCES

Now let us take the same example as above, except look at states for which $E > 0$. In this case, we have to consider where the particles are coming from and where they are going. We will assume that the particles are emitted with precise energy E toward the well from $-\infty$ and travel from left to right. As in the case above, we have three distinct regions:

    1. $x > -a$, where $\psi(x) = e^{ik_1x} + Re^{-ik_1x} = \psi_L(x)$
    2. $-a \leq x \leq +a$, where $\psi(x) = Ae^{-ik_2x} + Be^{+ik_2x} = \psi_W(x)$
    3. $x > +a$, where $\psi(x) = Te^{+ik_1x} = \psi_R(x)$

where $k_1 = \sqrt{2mE}/\hbar$ is the momentum outside the well, $k_2 = \sqrt{2m(E-V)}/\hbar$ is the momentum inside the well, and A, B, T, and R are coefficients we need to determine.

We also have the matching conditions:

$$\psi_L(-a) - \psi_W(-a) = 0$$

$$\psi_L'(-a) - \psi_W'(-a) = 0$$

$$\psi_R(a) - \psi_W(a) = 0$$

$$\psi_R'(a) - \psi_W'(a) = 0$$

This can be solved by hand; however, Mathematica keeps the bookkeeping easy. The result is a series of rules that we can use to determine the transmission and reflection coefficients:

$$T \rightarrow \frac{-4e^{-2iak_1 + 2iak_2} k_1 k_2}{-k_1{}^2 + e^{4iak_2} k_1{}^2 - 2k_1 k_2 - 2e^{4iak_2} k_1 k_2 - k_2{}^2 + e^{4iak_2} k_2{}^2}$$

$$A \rightarrow \frac{2e^{-iak_1 + 3iak_2} k_1 (k_1 - k_2)}{-k_1{}^2 + e^{4iak_2} k_1{}^2 - 2k_1 k_2 - 2e^{4iak_2} k_1 k_2 - k_2{}^2 + e^{4iak_2} k_2{}^2}$$

$$B \rightarrow \frac{-2e^{-iak_1 + iak_2} k_1 (k_1 + k_2)}{-k_1{}^2 + e^{4iak_2} k_1{}^2 - 2k_1 k_2 - 2e^{4iak_2} k_1 k_2 - k_2{}^2 + e^{4iak_2} k_2{}^2}$$

$$R \rightarrow \frac{\left(-1 + e^{4iak_2}\right) \left(k_1{}^2 - k_2{}^2\right)}{e^{2iak_1} \left(-k_1{}^2 + e^{4iak_2} k_1{}^2 - 2k_1 k_2 - 2e^{4iak_2} k_1 k_2 - k_2{}^2 + e^{4iak_2} k_2{}^2\right)}$$

The R and T coefficients are related to the ratios of the reflected and transimitted flux to the incoming flux. The current operator is given by

$$j(x) = \frac{\hbar}{2mi} \left(\psi^* \nabla \psi - \psi \nabla \psi^*\right) \qquad (2.87)$$

Inserting the wave functions above yields

$$j_{in} = \frac{\hbar k_1}{m}$$

$$j_{ref} = -\frac{\hbar k_1 R^2}{m}$$

$$j_{trans} = \frac{\hbar k_1 T^2}{m}$$

Thus, $R^2 = -j_{ref}/j_{in}$ and $T^2 = j_{trans}/j_{in}$. In Figure 2.6 we show the transmitted and reflection coefficients for an electron passing over a well of depth $V = -40$ and $a = 1$ as a function of incident energy E.

Notice that the transmission and reflection coefficients undergo a series of oscillations as the incident energy is increased. These are due to resonance states that lie in the continuum. The condition for these states is such that an integer number of the de Broglie wavelength of the wave in the well matches the total length of the well:

$$\lambda/2 = na$$

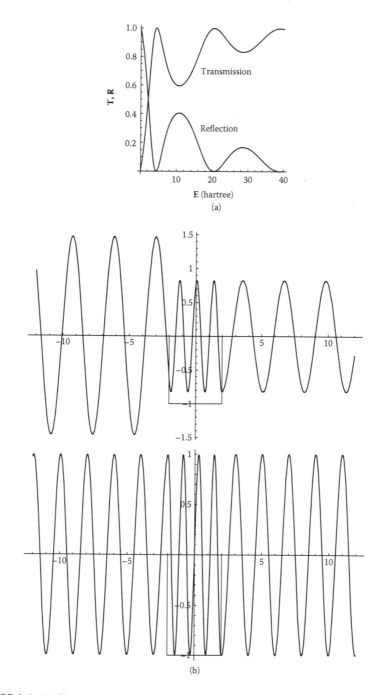

FIGURE 2.6 (a) Transmission and reflection coefficients for an electron scattering over a square well ($V = -40$ and $a = 1$). (b) Scattering waves for particle passing over a well. In the top graphic, the particle is partially reflected from the well ($V < 0$); in the bottom graphic, the particle passes over the well with a slightly different energy than above, this time with little reflection.

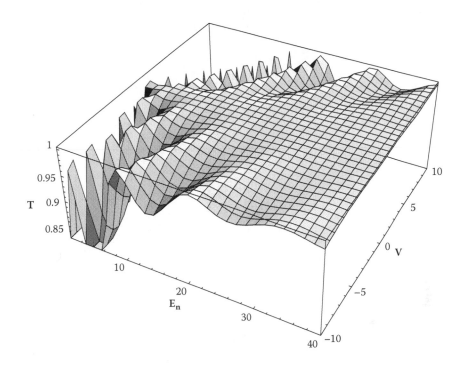

FIGURE 2.7 Transmission coefficient for particle passing over a bump. Here we have plotted T as a function of V and incident energy E_n. The oscillations correspond to resonance states that occur as the particle passes over the well (for $V < 0$) or bump $V > 0$.

Figure 2.7 shows the transmission coefficient as a function of both incident energy and the well depth and (or height) over a wide range, indicating that resonances can occur for both wells and bumps.

2.3.4 APPLICATION: QUANTUM DOTS

One of the most active areas of research in soft condensed matter is that of designing physical systems that can confine a quantum state in some controllable way. The idea of engineering a quantum state is extremely appealing and has numerous technological applications from small logic gates in computers to optically active materials for biomedical applications. The basic physics of these materials is relatively simple, and we can use the basic ideas presented in this chapter. The basic idea is to layer a series of materials such that electrons can be trapped in a geometrically confined region. This can be accomplished by insulator–metal–insulator layers and etching, creating disclinations in semiconductors, growing semiconductor or metal clusters, and so on. A quantum dot can even be a defect site.

We will assume throughout that our quantum well contains a single electron so that we can treat the system as simply as possible. For a square or cubic quantum well, energy levels are simply those of an n-dimensional particle in a box. For example, for a three-dimensional (3D) system,

$$E_{n_x,n_y,n_z} = \frac{\hbar^2\pi^2}{2m}\left(\left(\frac{n_x}{L_x}\right)^2 + \left(\frac{n_y}{L_y}\right)^2 + \left(\frac{n_z}{L_z}\right)^2\right) \qquad (2.88)$$

where L_x, L_y, and L_z are the lengths of the sides of the box and m is the mass of an electron.

The density of states is the number of energy levels per unit energy. If we take the box to be a cube $L_x = L_y = L_z$, we can relate n to a radius of a sphere and write the density of states as

$$\rho(n) = 4\pi^2 n^2 \frac{dn}{dE} = 4\pi^2 n^2 \left(\frac{dE}{dn}\right)^{-1}$$

Thus, for a 3D cube, the density of states is

$$\rho(n) = \left(\frac{4mL^2}{\pi\hbar^2}\right) n$$

that is, for a three-dimensional cube, the density of states increases as n and hence as $E^{1/2}$ (Figure 2.8).

Note that the scaling of the density of states with energy depends strongly upon the dimensionality of the system. For example, in one dimension,

$$\rho(n) = \frac{2mL^2}{\hbar^2\pi^2}\frac{1}{n}$$

and in two dimensions

$$\rho(n) = const$$

The reason for this lies in the way the volume element for linear, circular, and spherical integration scales with radius n. Thus, measuring the density of states tells us not only the size of the system but also its dimensionality.

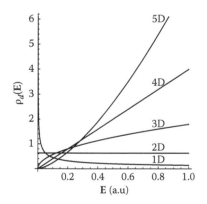

FIGURE 2.8 Density of states versus dimensionality of the system. For $D > 3$, these are effective dimensions reflecting the number of free-particle degrees of freedom carried by a given particle.

We can generalize the results here by realizing that the volume of a d-dimensional sphere in k space is given by

$$V_d = \frac{k^d \pi^{d/2}}{\Gamma(1 + d/2)}$$

where $\Gamma(x)$ is the gamma function. The total number of states per unit volume in a d-dimensional space is then

$$n_k = 2 \frac{1}{2\pi^2} V_d$$

and the density is then the number of states per unit of energy. The relation between energy and k is

$$E_k = \frac{\hbar^2}{2m} k^2$$

that is,

$$k = \frac{\sqrt{2E_k m}}{\hbar}$$

which gives

$$\rho_d(E) = \frac{2^{\frac{d}{2}-1} \, d \, \pi^{\frac{d}{2}-2} \left(\frac{\sqrt{mE}}{\hbar}\right)^d}{E \, \Gamma\left(1 + \frac{d}{2}\right)}$$

A quantum well is typically constructed so that the system is confined in one dimension and unconfined in the other two. Thus, a quantum well will typically have a discrete state only in the confined direction. The density of states for this system will be identical to that of the three-dimensional system at energies where the k vectors coincide. If we take the thickness to be s, then the density of states for the quantum well is

$$\rho = \frac{L}{s} \rho_2(E) \left\lfloor L \frac{\rho_3(E)}{L\rho_2(E)/s} \right\rfloor$$

where $\lfloor x \rfloor$ is the "floor" function, which means it takes the largest integer less than x. This is plotted in Figure 2.9a and the stair-step density of states (DOS) is indicative of the embedded confined structure.

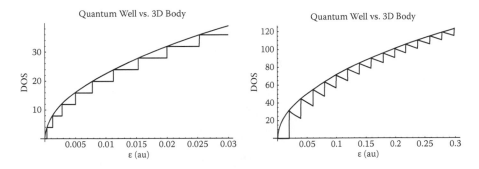

FIGURE 2.9 Density of states for a quantum well and quantum wire compared to a 3D space. Here $L = 5$ and $s = 2$ for comparison.

Next, we consider a quantum wire of thickness s along each of its two confined directions (Figure 2.9b). The DOS along the unconfined direction is one dimensional. As above, the total DOS will be identical to the 3D case when the wave vectors coincide. Increasing the radius of the wire eventually leads to the case where the steps decrease and merge into the 3D curve,

$$\rho = \left(\frac{L}{s}\right)^2 \rho_1(E) \left\lfloor \frac{L^2 \rho_2(E)}{L^2 \rho_2(E)/s} \right\rfloor$$

For a spherical dot, we consider the case in which the radius of the quantum dot is small enough to support discrete rather than continuous energy levels. In a later chapter, we will derive this result in more detail, for now, we consider just the results. First, an electron in a spherical dot obeys the Schrödinger equation:

$$-\frac{\hbar^2}{2m}\nabla^2\psi = E\psi \tag{2.89}$$

where ∇^2 is the Laplacian operator in spherical coordinates

$$\nabla^2 = \frac{1}{r}\frac{\partial^2}{\partial r^2}r + \frac{1}{r^2\sin\theta}\frac{\partial}{\partial\theta}\sin\theta\frac{\partial}{\partial\theta} + \frac{1}{r^2\sin^2\theta}\frac{\partial^2}{\partial\phi^2}$$

The solution of the Schrödinger equation is subject to the boundary condition that for $r \geq R$, $\psi(r) = 0$, where R is the radius of the sphere and is given in terms of the spherical Bessel function, $j_l(r)$, and spherical harmonic functions, Y_{lm},

$$\psi_{nlm} = \frac{2^{1/2}}{R^{3/2}}\frac{j_l(\alpha r/R)}{j_{l+1}(\alpha)}Y_{lm}(\Omega) \tag{2.90}$$

with energy

$$E = \frac{\hbar^2}{2m}\frac{\alpha^2}{R^2} \tag{2.91}$$

Note that the spherical Bessel functions (of the first kind) are related to the Bessel functions via

$$j_l(x) = \sqrt{\frac{\pi}{2x}}J_{l+1/2}(x) \tag{2.92}$$

The first few of these are shown in Figure 2.10,

$$j_0(x) = \frac{\sin x}{x} \tag{2.93}$$

$$j_1(x) = \frac{\sin x}{x^2} - \frac{\cos x}{x} \tag{2.94}$$

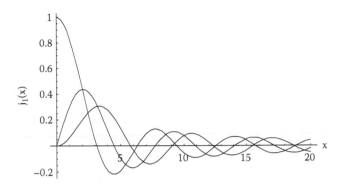

FIGURE 2.10 Spherical Bessel functions, j_0, j_1, and j_2.

$$j_2(x) = \left(\frac{3}{x^3} - \frac{1}{x}\right)\sin x - \frac{3}{x^2}\cos x \qquad (2.95)$$

$$j_n(x) = (-1)^n x^n \left(\frac{1}{x}\frac{d}{dx}\right)^n j_0(x) \qquad (2.96)$$

where Equation 2.97 provides a way to generate j_n from j_0.

The α's appearing in the wave function and in the energy expression are determined by the boundary condition that $\psi(R) = 0$. Thus, for the lowest energy state we require

$$j_0(\alpha) = 0, \qquad (2.97)$$

that is, $\alpha = \pi$. For the next state ($l = 1$),

$$j_1(\alpha) = \frac{\sin\alpha}{\alpha^2} - \frac{\cos\alpha}{\alpha} = 0. \qquad (2.98)$$

This can be solved to give $\alpha = 4.4934$. These correspond to where the spherical Bessel functions pass through zero. The first six of these are 3.14159, 4.49341, 5.76346, 6.98793, 8.18256 and 9.35581. These correspond to where the first zeros occur and give the condition for the radial quantization, $n = 1$ with angular momentum $l = 0, 1, 2, 3, 4, 5$. There are more zeros, and these correspond to the case where $n > 1$.

In the next set of figures (Figure 2.11), we look at the radial wave functions for an electron in a 0.5 Å quantum dot. First, the case where $n = 1, l = 0$ and $n = 0, l = 1$. In both cases, the wave functions vanish at the radius of the dot. The radial probability distribution function (PDF) is given by $P = r^2|\psi_{nl}(r)|^2$. Note that increasing the angular momentum l from 0 to 1 causes the electron's most probable position to shift outwards. This is due to the centrifugal force due to the angular motion of the electron. For the $n, l = (2, 0)$ and $(2, 1)$ states, we have one node in the system and two peaks in the PDF functions.

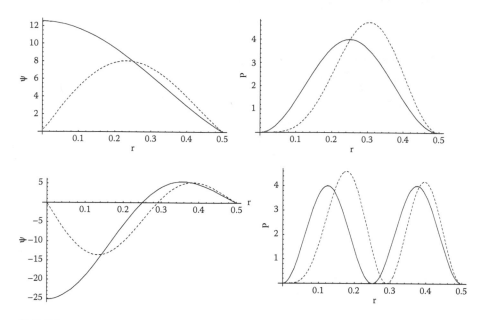

FIGURE 2.11 Radial wave functions (left column) and corresponding PDFs (right column) for an electron in an $R = 0.5$ Å quantum dot. The upper two correspond to $(n, l) = (1, 0)$ (solid) and $(n, l) = (1, 1)$ (dashed) while the lower correspond to $(n, l) = (2, 0)$ (solid) and $(n, l) = (2, 1)$ (dashed).

2.4 PROBLEMS AND EXERCISES

Problem 2.1 Derive the expression for

$$G_o(x, x') = \langle x | \exp(-i h_o t / \hbar) | x' \rangle \qquad (2.99)$$

where h_o is the free particle Hamiltonian,

$$h_o = -\frac{\hbar^2}{2m} \frac{\partial^2}{\partial x^2} \qquad (2.100)$$

Problem 2.2 Show that G_o is a solution of the free particle Schrödinger equation

$$i\hbar \partial_t G_o(t) = h_o G_o(t) \qquad (2.101)$$

Problem 2.3 Show that the normalization of a wave function is independent of time.

Problem 2.4 Compute the bound-state solutions ($E < 0$) for a square well of depth V_o where

$$V(x) = \begin{cases} -V_o & -a/2 \le x \le a/2 \\ 0 & \text{otherwise} \end{cases} \qquad (2.102)$$

1. How many energy levels are supported by a well of width a?
2. Show that a very narrow well can support only one bound state, and that this state is an even function of x.
3. Show that the energy of the lowest bound state is

$$E \approx \frac{mV_o^2a^2}{2\hbar^2} \tag{2.103}$$

4. Show that as

$$\rho = \sqrt{-\frac{2mE}{\hbar^2}} \rightarrow 0 \tag{2.104}$$

the probability of finding the particle inside the well vanishes.

Problem 2.5 Consider a particle with the potential

$$V(x) = \begin{cases} 0 & \text{for } x > a \\ -V_o & \text{for } 0 \le x \le a \\ \infty & \text{for } x < 0 \end{cases} \tag{2.105}$$

1. Let $\phi(x)$ be a stationary state. Show that $\phi(x)$ can be extended to give an odd wave function corresponding to a stationary state of the symmetric well of width $2a$ (that is, the one studied above) and depth V_o.
2. Discuss with respect to a and V_o the number of bound states and argue that there is always at least one such state.
3. Now turn your attention toward the $E > 0$ states of the well. Show that the transmission of the particle into the well region vanishes as $E \rightarrow 0$ and that the wave function is perfectly reflected off the sudden change in potential at $x = a$.

Problem 2.6 Which of the following are eigenfunctions of the kinetic energy operator:

$$\hat{T} = -\frac{\hbar^2}{2m}\frac{\partial^2}{\partial x^2} \tag{2.106}$$

$$e^x, x^2, x^n, 3\cos(2x), \sin(x) + \cos(x), e^{-ikx}$$

$$f(x - x') = \int_{-\infty}^{\infty} dk e^{-ik(x-x')} e^{-ik^2/(2m)} \tag{2.107}$$

Problem 2.7 Which of the following would be acceptable one-dimensional wave functions for a bound particle (upon normalization): $f(x) = e^{-x}$, $f(x) = e^{-x^2}$, $f(x) = xe^{-x^2}$, or

$$f(x) = \begin{cases} e^{-x^2} & x \ge 0 \\ 2e^{-x^2} & x < 0 \end{cases} \tag{2.108}$$

Problem 2.8   For a one-dimensional problem, consider a particle with wave function

$$\psi(x) = N \frac{\exp(ip_o x / \hbar)}{\sqrt{x^2 + a^2}} \tag{2.109}$$

where a and p_o are real constants and N the normalization.

1. Determine N so that $\psi(x)$ is normalized

$$\int_{-\infty}^{\infty} dx |\psi(x)|^2 = N^2 \int_{-\infty}^{\infty} dx \frac{1}{x^2 + a^2} \tag{2.110}$$

$$= N^2 \frac{\pi}{a} \tag{2.111}$$

Thus $\psi(x)$ is normalized when

$$N = \sqrt{\frac{a}{\pi}} \tag{2.112}$$

2. The position of the particle is measured. What is the probability of finding a result between $-a/\sqrt{3}$ and $+a/\sqrt{3}$?

$$\frac{a}{\pi} \int_{-a/\sqrt{3}}^{+a/\sqrt{3}} dx |\psi(x)|^2 = \int_{-a/\sqrt{3}}^{+a/\sqrt{3}} dx \frac{1}{x^2 + a^2} \tag{2.113}$$

$$= \frac{1}{\pi} \tan^{-1}(x/a) \Big|_{-a/\sqrt{3}}^{+a/\sqrt{3}} \tag{2.114}$$

$$= \frac{1}{3} \tag{2.115}$$

3. Compute the mean value of a particle that has $\psi(x)$ as its wave function.

$$\langle x \rangle = \frac{a}{\pi} \int_{-\infty}^{\infty} dx \frac{x}{x^2 + a^2} \tag{2.116}$$

$$= 0 \tag{2.117}$$

Problem 2.9   Consider the Hamiltonian of a particle in a one-dimensional well given by

$$H = \frac{1}{2m} \hat{p}^2 + \hat{x}^2 \tag{2.118}$$

where $\hat{x}$ and $\hat{p}$ are position and momentum operators. Let $|\phi_n\rangle$ be a solution of

$$H |\phi_n\rangle = E_n |\phi_n\rangle \tag{2.119}$$

for $n = 0, 1, 2, \ldots$. Show that

$$\langle \phi_n | \hat{p} | \phi_m \rangle = \alpha_{nm} \langle \phi_n | \hat{x} | \phi_m \rangle \tag{2.120}$$

where α_{nm} is a coefficient depending upon $E_n - E_m$. Compute α_{nm}. (Hint: You will need to use the commutation relations of $[\hat{x}, H]$ and $[\hat{p}, H]$ to get this.) Finally, from all this, deduce that

$$\sum_m (E_n - E_m)^2 |\langle \phi_n | \hat{x} | \phi_m \rangle|^2 = \frac{\hbar^2}{2m} \langle \phi_n | \hat{p}^2 | \phi_n \rangle \qquad (2.121)$$

Problem 2.10   The state space of a certain physical system is three dimensional. Let $|u_1\rangle$, $|u_2\rangle$, and $|u_3\rangle$ be an orthonormal basis of the space in which kets $|\psi_1\rangle$ and $|\psi_2\rangle$ are defined by

$$|\psi_1\rangle = \frac{1}{\sqrt{2}}|u_1\rangle + \frac{i}{2}|u_2\rangle + \frac{1}{2}|u_3\rangle \qquad (2.122)$$

$$|\psi_2\rangle = \frac{1}{\sqrt{3}}|u_1\rangle + \frac{i}{\sqrt{3}}|u_3\rangle \qquad (2.123)$$

1. Are the states normalized?
2. Determine the matrices ρ_l and ρ_z as represented in the $\{|u_i\rangle\}$ basis, the projection operators onto $|\psi_1\rangle$ and $|\psi_2\rangle$. Verify that these matrices are Hermitian.

Problem 2.11   Let $\psi(r) = \psi(x, y, z)$ be the normalized wave function of a particle. Express in terms of $\psi(r)$:

1. A measurement along the x axis to yield a result between x_1 and x_2.
2. A measurement of momentum component p_x to yield a result between p_1 and p_2.
3. Simultaneous measurements of x and p_z to yield $x_1 \leq x \leq x_2$ and $p_z > 0$.
4. Simultaneous measurements of p_x, p_y, and p_z, to yield

$$p_1 \leq p_x \leq p_2 \qquad (2.124)$$
$$p_3 \leq p_y \leq p_4 \qquad (2.125)$$
$$p_5 \leq p_z \leq p_6 \qquad (2.126)$$

Show that this result is equal to the result of part 2 when $p_3, p_5 \to -\infty$ and $p_4, p_6 \to +\infty$.

Problem 2.12   Consider a particle of mass m whose potential energy is

$$V(x) = -\alpha(\delta(x + l/2) + \delta(x - l/2)) \qquad (2.127)$$

1. Calculate the bound states of the particle, setting

$$E = -\frac{\hbar^2 \rho^2}{2m} \qquad (2.128)$$

Show that the possible energies are given by

$$e^{-\rho l} = \pm \left(1 - \frac{2\rho}{\mu}\right) \qquad (2.129)$$

where $\mu = 2m\alpha/\hbar^2$. Give a graphic solution of this equation.

(a) *Ground State.* Show that the ground state is even about the origin and that its energy E_s is less than the bound state of a particle in a single δ-function potential $-E_L$. Interpret this physically. Plot the corresponding wave function.

(b) *Excited State.* Show that when l is greater than some value (which you need to determine), there exists an odd excited state of energy E_A with energy greater than $-E_L$. Determine and plot the corresponding wave function.

(c) Explain how the preceeding calculations enable us to construct a model for an ionized diatomic molecule, for example, H_2^+, whose nuclei are separated by l. Plot the energies of the two states as functions of l. What happens as $l \to \infty$ and $l \to 0$?

(d) If we take Coulombic repulsion of the nuclei into account, what is the total energy of the system? Show that a curve that gives the variation with respect to l of the energies thus obtained enables us to predict in certain cases the existence of bound states of H_2^+ and to determine the equilibrium bond length.

2. Calculate the reflection and transmission coefficients for this system. Plot R and T as functions of l. Show that resonances occur when l is an integer multiple of the de Broglie wavelength of the particle. Why?

Problem 2.13   Write down the Schrödinger equation for an oscillator in the momentum representation and determine the momentum wave functions.

3 Semiclassical Quantum Mechanics

Good actions ennoble us, and we are the sons of our own deeds.

Miguel de Cervantes

The use of classical mechanical analogs for quantum behavior holds a long and proud tradition in the development and application of quantum theory. In Bohr's original formulation of quantum mechanics to explain the spectra of the hydrogen atom, Bohr used purely classical mechanical notions of angular momentum and rotation for the basic theory and *imposed* a quantization condition that the angular momentum should come in integer multiples of $\hbar$. Bohr worked under the assumption that at some point the laws of quantum mechanics that govern atoms and molecules should correspond to the classical mechanical laws of ordinary objects like rocks and stones. Bohr's *Principle of Correspondence* states that quantum mechanics is not completely separate from classical mechanics; rather, it incorporates classical theory.

From a computational viewpoint, this is an extremely powerful notion since performing a classical trajectory calculation (even running thousands of them) is simpler than a single quantum calculation of a similar dimension. Consequently, the development of semiclassical methods has been and remains an important part of the development and utilization of quantum theory. In fact even in the most recent issues of leading physics and chemical physics journals, one finds new developments and applications of this very old idea.

In this chapter we will explore this idea in some detail. The field of semiclassical mechanics is vast and I would recommend the following for more information:

1. *Chaos in Classical and Quantum Mechanics*, Martin Gutzwiller (New York: Springer-Verlag, 1990). Chaos in quantum mechanics is a touchy subject and really has no clear-cut definition that anyone seems to agree upon. Gutzwiller is one of the key figures in sorting all this out. This is very nice and a not-too-technical monograph on quantum and classical correspondence.
2. *Semiclassical Physics*, M. Brack and R. Bhaduri (Reading, MA: Addison-Wesley, 1997). Very interesting book, mostly focusing upon many-body applications and Thomas-Fermi approximations.
3. *Computer Simulations of Liquids*, M. P. Allen and D. J. Tildesley (New York: Oxford, 1994). This book mostly focuses upon classical molecular dynamics (MD) methods, but has a nice chapter on the quantum methods that were state of the art in 1994. Methods come and methods go.

There are *many* others, of course. These are just the ones on my bookshelf.

3.1 BOHR–SOMMERFIELD QUANTIZATION

Let us first review Bohr's original derivation of the hydrogen atom. We will go through this a bit differently than Bohr since we already know part of the answer. In the chapter on the hydrogen atom we derived the energy levels in terms of the principle quantum number n:

$$E_n = -\frac{me^4}{2\hbar^2}\frac{1}{n^2} \qquad (3.1)$$

In Bohr's correspondence principle, the quantum energy must equal the classical energy. So for an electron moving about a proton, that energy is inversely proportional to the distance of separation. So, we can write

$$-\frac{me^4}{2\hbar^2}\frac{1}{n^2} = -\frac{e^2}{2r} \qquad (3.2)$$

Now we need to figure out how angular momentum gets pulled into this. For an orbiting body the centrifugal force, which pulls the body outward, is counterbalanced by the inward tugs of the centripetal force coming from the attractive Coulomb potential. Thus,

$$mr\omega^2 = \frac{e^2}{r^2} \qquad (3.3)$$

where ω is the angular frequency of the rotation. Rearranging this a bit, we can plug this into the right-hand side (rhs) of Equation 3.2 and write

$$-\frac{me^4}{2\hbar^2}\frac{1}{n^2} = -\frac{mr^3\omega^2}{2r} \qquad (3.4)$$

The numerator now looks amost like the classical definition of angular momentum: $L = mr^2\omega$. So we can write the last equation as

$$-\frac{me^4}{2\hbar^2}\frac{1}{n^2} = -\frac{L^2}{2mr^2} \qquad (3.5)$$

Solving for L^2:

$$L^2 = \frac{me^4}{2\hbar^2}\frac{2mr^2}{n^2} \qquad (3.6)$$

Now, we need to pull in another one of Bohr's results for the orbital radius of the H atom:

$$r = \frac{\hbar^2}{me^2}n^2 \qquad (3.7)$$

Plug this into Equation 3.6 and after the dust settles, we find

$$L = \hbar n \qquad (3.8)$$

But, why should electrons be confined to circular orbits? Equation 3.8 should be applicable to *any closed path* the electron should choose to take. If the quantization condition only holds for circular orbits, then the theory itself is in deep trouble. At least that is what Sommerfield thought.

The numerical units of $\hbar$ are energy times time. That is the unit of action in classical mechanics. In classical mechanics, the action of a mechanical system is given by the integral of the classical momentum along a classical path:

$$S = \int_{x_1}^{x_2} p\,dx \qquad (3.9)$$

For an orbit, the initial point and the final point must coincide, $x_1 = x_2$, so the action integral must describe some of the area circumscribed by a closed loop on the $p - x$ plane called *phase space*

$$S = \oint p\,dx \qquad (3.10)$$

So, Bohr and Sommerfield's idea was that the circumscribed area in phase space was quantized as well.

As a check, let us consider the harmonic oscillator. The classical energy is given by

$$E(p, q) = \frac{p^2}{2m} + \frac{k}{2}q^2$$

This is the equation for an ellipse in phase space since we can rearrange this to read

$$1 = \frac{p^2}{2mE} + \frac{k}{2E}q^2$$

$$= \frac{p^2}{a^2} + \frac{q^2}{b^2} \qquad (3.11)$$

where $a = \sqrt{2mE}$ and $b = \sqrt{2E/k}$ describe the major and minor axes of the ellipse. The area of an ellipse is $A = \pi ab$, so the area circumscribed by a classical trajectory with energy E is

$$S(E) = 2E\pi\sqrt{m/k} \qquad (3.12)$$

Since $\sqrt{k/m} = \omega$, $S = 2\pi E/\omega = E/\nu$. Finally, since E/ν must be an integer multiple of h, the Bohr–Sommerfield condition for quantization becomes

$$\oint p\,dx = nh \qquad (3.13)$$

where p is the classical momentum for a path of energy E, $p = \sqrt{2m(E - V(x))}$. Taking this a bit further, the de Broglie wavelength is p/h, so the Bohr–Sommerfield rule basically states that stationary energies correspond to classical paths for which there are an integer number of de Broglie wavelengths.

Now, perhaps you can anticipate a problem with the quantum description of a classically chaotic system. In classical chaos, chaotic trajectories *never* return to their exact staring point in phase space. They may come close, but there are no closed orbits. For 1D systems, this does not occur since the trajectories are the contours of the energy function. For higher dimensions, the dimensionality of the system makes it possible to have extremely complex trajectories that never return to their starting point.

Problem 3.1   Apply the Bohr–Sommerfield procedure to determine the stationary energies for a particle in a box of length l.

3.2   THE WENTZEL, KRAMERS, AND BRILLOUIN APPROXIMATION

The original Bohr-Sommerfield idea can be improved upon considerably to produce an asymptotic ($\hbar \to 0$) approximation to the Schrödinger wave function. The idea was put forward at about the same time by three different theoreticians: Brillouin (in Belgium), Kramers (in the Netherlands), and Wentzel (in Germany). Depending upon your point of origin, this method is the WKB (US & Germany), BWK (France, Belgium), JWKB (UK)—you get the idea. The original references are

1. "La mécanique odularatoire de Schrödinger; une méthode générale de résolution par approximations successives," L. Brillouin, *Comptes rendus* (Paris) 183, 24 (1926).
2. "Wellenmechanik und halbzahlige Quantisierung," H. A. Kramers, *Zeitschrift für Physik* 39, 828 (1926).
3. "Eine Verallgemeinerung der Quantenbedingungen für die Zwecke der Wellenmechanik," G. Wentzel *Zeitschrift für Physik* 38, 518 (1926).

We will first go through how one can use the approach to determine the eigenvalues of the Schrödinger equation via semiclassical methods, and then show how one can approximate the actual wave functions themselves.

3.2.1   ASYMPTOTIC EXPANSION FOR EIGENVALUE SPECTRUM

The WKB proceedure is initiated by writing the solution to the Schrödinger equation

$$\psi'' + \frac{2m}{\hbar^2}(E - V(x))\psi = 0$$

as

$$\psi(x) = \exp\left(\frac{i}{\hbar}\int \chi\,dx\right) \tag{3.14}$$

We will soon discover that χ is the classical momentum of the system, but for now, let us consider it to be a function of the energy of the system. Substituting this into the Schrödinger equation produces a new differential equation for χ:

$$\frac{\hbar}{i}\frac{d\chi}{dx} = 2m(E-V) - \chi^2 \tag{3.15}$$

If we take $\hbar \to 0$, it follows then that

$$\chi = \chi_o = \sqrt{2m(E-V)} = |p| \tag{3.16}$$

which is the magnitude of the classical momentum of a particle. So, if we assume
that this is simply the leading order term in a series expansion in $\hbar$, we would have

$$\chi = \chi_o + \frac{\hbar}{i}\chi_1 + \left(\frac{\hbar}{i}\right)^2 \chi_2 \cdots \tag{3.17}$$

Substituting Equation 3.17 into

$$\chi = \frac{\hbar}{i}\frac{1}{\psi}\frac{\partial\psi}{x} \tag{3.18}$$

and equating to zero coefficients with different powers of $\hbar$, we obtain equations that
determine the χ_n corrections in succession:

$$\frac{d}{dx}\chi_{n-1} = -\sum_{m=0}^{n}\chi_{n-m}\chi_m \tag{3.19}$$

for $n = 1, 2, 3, \ldots$. For example,

$$\chi_1 = -\frac{1}{2}\frac{\chi_o'}{\chi_o} = \frac{1}{4}\frac{V'}{E-V} \tag{3.20}$$

$$\chi_2 = -\frac{\chi_1^2 + \chi_1'}{2\chi_o}$$

$$= -\frac{1}{2\chi_o}\left\{\frac{V'^2}{16(E-V)^2} + \frac{V'^2}{4(E-V)^2} + \frac{V''}{4(E-V)}\right\}$$

$$= -\frac{5V'^2}{32(2m)^{1/2}(E-V)^{5/2}} - \frac{V''}{8(2m)^{1/2}(E-V)^{3/2}} \tag{3.21}$$

and so forth.

Problem 3.2   Verify Equation 3.19 and derive the first-order correction in Equation 3.20.

Now, to use these equations to determine the spectrum, we replace x everywhere
by a complex coordinate z and suppose that $V(z)$ is a regular and analytic function
of z in any physically relevant region.* Consequently, we can then say that $\psi(z)$ is an

---

* Note: An analytic function is such that it can be expanded in a polynomial series about some local point.

analytic function of z. So, we can write the phase integral as

$$n = \frac{1}{h} \int_C \chi(z) dz$$

$$= \frac{1}{2\pi i} \int_C \frac{\psi'_n(z)}{\psi_n(z)} dz \qquad (3.22)$$

where ψ_n is the nth discrete stationary solution to the Schrödinger equation and C is some contour of integration on the z plane. If there is a discrete spectrum, we know that the number of zeros, n, in the wave function is related to the quantum number corresponding to a given energy level. So if ψ has no real zeros, this is the ground-state wave function with energy E_o; one real zero corresponds to energy level E_1 and so forth.

Suppose the contour of integration, C, is taken such that it includes only these zeros and no others, then we can write

$$n = \frac{1}{\hbar} \int_C \chi_o dz + \frac{1}{2\pi i} \int_c \chi_1 - \hbar \int_C \chi_2 dz + \dots \qquad (3.23)$$

Each of these terms involves $E - V$ in the denominator. At the classical turning points where $V(z) = E$, we have poles and we can use the Cauchy integral theorem to evaluate the integrals. These poles are located at the classical turning points.

For example, from above, the integral

$$\frac{1}{2\pi i} \int_c \chi_1 = -\frac{1}{2\pi i} \frac{1}{4} \oint \frac{V'}{E - V} dz \qquad (3.24)$$

We can make a change of variables $Z = V(z)$ and $dZ = V' dz$ and write the integral as

$$\frac{1}{2\pi i} \int_c \chi_1 = -\frac{1}{2\pi i} \frac{1}{4} \oint \frac{dZ}{E - Z} = -\frac{1}{2} \qquad (3.25)$$

since each classical turning point contributes $-1/4$.

The next term we evaluate by integration by parts

$$\int_C \frac{V''}{(E - V(z))^{3/2}} dz = -\frac{3}{2} \int_C \frac{V'^2}{(E - V(z))^{5/2}} dz \qquad (3.26)$$

Hence, we can write

$$\int_C \chi_2(z) dz = \frac{1}{32(2m)^{1/2}} \int_C \frac{V'^2}{(E - V(z))^{5/2}} dz \qquad (3.27)$$

Putting it all together

$$n + 1/2 = \frac{1}{h} \int_c \sqrt{2m(E - V(z))}\,dz$$

$$- \frac{h}{128\pi^2(2m)^{1/2}} \int_c \frac{V'^2}{(E - V(z))^{5/2}}\,dz + \ldots \qquad (3.28)$$

The above analysis is pretty formal. But what we have is something new. Notice that we have an extra $1/2$ added here that we did not have in the original Bohr–Sommerfield (BS) theory. What we have is something even more general. The original BS idea came from the notion that energies and frequencies were related by integer multiples of h. But this is really only valid for transitions between states. If we go back and ask what happens at $n = 0$ in the Bohr–Sommerfield theory, this corresponds to a phase-space ellipse with major and minor axes both of length 0—which violates the Heisenberg uncertainly rule. This new quantization condition forces the system to have some lowest energy state with a phase-space area $h/2$.

Where did this extra $1/2$ come from? It originates from the classical turning points where $V(x) = E$. Recall that for a 1D system bound by a potential, there are at least two such points. Each contributes a $\pi/4$ to the phase for a total contribution of $\pi/2$. We will see this more explicitly in the next section when evaluating the matching conditions at the turning points.

3.2.2 Example: Semiclassical Estimate of Spectrum for Harmonic Oscillator

As an example in using this approach, let us consider the simple case of a harmonic oscillator. Recall in the discussion of the Bohr–Sommerfield approach, we noted that the momentum integral over a closed trajectory was equal to the area in phase space enclosed by an ellipse

$$nh = \oint p(x)\,dx = \pi ab = 2\pi E/\omega$$

where a and b are the major and minor axes along the x and p directions and E is the energy. The semiclassical treatment adds an additional factor of $1/2$ to the Bohr–Sommerfield expression so that

$$n + 1/2 = E/\hbar\omega$$

This agrees with the exact result for the harmonic oscillator energies: $E_n = \hbar\omega(n + 1/2)$ with $n = 0, 1, 2, \ldots$.

3.2.3 The Wentzel, Kramers, and Brillouin Wave Function

Going back to our original wave function in Equation 3.14 and writing

$$\psi = e^{iS/\hbar}$$

where S is the integral of χ, we can derive equations for S:

$$\frac{1}{2m}\left(\frac{\partial S}{\partial x}\right) - \frac{i\hbar}{2m}\frac{\partial^2 S}{\partial x^2} + V(x) = E \tag{3.29}$$

If we neglect the term involving $\hbar$, we recover the classical Hamilton–Jacobi equation for the action S,

$$\frac{1}{2m}\left(\frac{\partial S}{\partial x}\right) + V(x) = E \tag{3.30}$$

and can identify $\partial S/\partial x = \chi_o = p$ with the classical momentum. Again, as above, we can seek a series expansion of S in powers of $\hbar$. The result is simply the integral of Equation 3.17:

$$S = S_o + \frac{\hbar}{i}S_1 + \cdots \tag{3.31}$$

Looking at Equation 3.29: it is clear that the classical approximation is valid when the second term is very small compared to the first. That is,

$$\hbar\frac{S''}{S'^2} \ll 1$$

$$\hbar\frac{d}{dx}\left(\frac{dS}{dx}\right)\left(\frac{dx}{dS}\right)^2 \ll 1$$

$$\hbar\frac{d}{dx}\frac{1}{p} \ll 1 \tag{3.32}$$

where we equate $dS/dx = p$. Since p is related to the de Broglie wavelength of the particle $\lambda = h/p$, the same condition implies that

$$\left|\frac{1}{2\pi}\frac{d\lambda}{dx}\right| \ll 1 \tag{3.33}$$

Thus the semiclassical approximation is only valid when the wavelength of the particle as determined by $\lambda(x) = h/p(x)$ varies slightly over distances on the order of the wavelength itself. Noting the gradient of the momentum, this can be written another way:

$$\frac{dp}{dx} = \frac{d}{dx}\sqrt{2m(E - V(x))} = -\frac{m}{p}\frac{dV}{dx}$$

Thus, we can write the classical condition as

$$m\hbar|F|/p^3 \ll 1 \tag{3.34}$$

Alternatively, in terms of the de Broglie wavelength $\lambda = h/p$,

$$\frac{d}{dx}\sqrt{2m(E - V(x))} = -m\lambda\frac{dV}{dx}$$

Thus, the semiclassical condition is met when the potential changes slowly over a length-scale comparable to the local de Broglie wavelength $\lambda(x) = h/p(x)$.

Going back to the expansion for χ

$$\chi_1 = -\frac{1}{2}\frac{\chi_o'}{\chi_o} = \frac{1}{4}\frac{V'}{E-V} \tag{3.35}$$

or equivalently for S_1

$$S_1' = -\frac{S_o''}{2S'} = -\frac{p'}{2p} \tag{3.36}$$

So,

$$S_1(x) = -\frac{1}{2}\log p(x)$$

If we stick to regions where the semi-classical condition is met, then the wave function becomes

$$\psi(x) \approx \frac{C_1}{\sqrt{p(x)}}e^{\frac{i}{\hbar}\int p(x)dx} + \frac{C_2}{\sqrt{p(x)}}e^{-\frac{i}{\hbar}\int p(x)dx} \tag{3.37}$$

The $1/\sqrt{p}$ prefactor has a remarkably simple interpretation. The probability of finding the particle in some region between x and $x+dx$ is given by $|\psi|^2$ so that the classical probability is essentially proportional to $1/p$. So, the faster the particle is moving, the less likely it is to be found in some small region of space. Conversely, the slower a particle moves, the more likely it is to be found in that region. So the time spent in a small dx is inversely proportional to the momentum of the particle. We will return to this concept in a bit when we consider the idea of time in quantum mechanics.

The C_1 and C_2 coefficients are yet to be determined. If we take $x = a$ to be one classical turning point so that $x > a$ corresponds to the classically inaccessible region where $E < V(x)$, then the wave function in that region must be exponentially damped:

$$\psi(x) \approx \frac{C}{\sqrt{|p|}}\exp\left(-\frac{1}{\hbar}\int_a^x |p(x)|dx\right) \tag{3.38}$$

To the left of $x = a$, we have a combination of incoming and reflected components:

$$\psi(x) = \frac{C_1}{\sqrt{p}}\exp\left(\frac{i}{\hbar}\int_x^a pdx\right) + \frac{C_2}{\sqrt{p}}\exp\left(-\frac{i}{\hbar}\int_x^a pdx\right) \tag{3.39}$$

3.2.4 SEMICLASSICAL TUNNELING AND BARRIER PENETRATION

Before solving the general problem of how to use this in an arbitrary well, let us consider the case for tunneling through a potential barrier that has some bumpy top or corresponds to some simple potential. So, to the left of the barrier the wave function has incoming and reflected components:

$$\psi_L(x) = Ae^{ikx} + Be^{-ikx} \tag{3.40}$$

Inside we have

$$\psi_B(x) = \frac{C}{\sqrt{|p(x)|}} e^{+\frac{i}{\hbar} \int |p| dx} + \frac{D}{\sqrt{|p(x)|}} e^{-\frac{i}{\hbar} \int |p| dx} \tag{3.41}$$

and to the right of the barrier:

$$\psi_R(x) = F e^{+ikx} \tag{3.42}$$

If F is the transmitted amplitude, then the tunneling probability is the ratio of the transmitted probability to the incident probability: $T = |F|^2/|A|^2$. If we assume that the barrier is high or broad, then $C = 0$ and we obtain the semiclassical estimate for the tunneling probability:

$$T \approx \exp\left(-\frac{2}{\hbar} \int_a^b |p(x)| dx\right) \tag{3.43}$$

where a and b are the turning points on either side of the barrier.

Mathematically, we can "flip the potential upside down" and work in imaginary time. In this case the action integral becomes

$$S = \int_a^b \sqrt{2m(V(x) - E)} dx \tag{3.44}$$

So we can think of tunneling as motion under the barrier in imaginary time.

There are a number of useful applications of this formula. Gamow's theory of alpha decay is a common example. Another useful application is in the theory of reaction rates where we want to determine tunneling corrections to the rate constant for a particular reaction. Close to the top of the barrier, where tunneling may be important, we can expand the potential and approximate the peak as an upside-down parabola:

$$V(x) \approx V_o - \frac{k}{2} x^2$$

where $+x$ represents the product side and $-x$ represents the reactant side. The Eckart potential (Figure 3.1) is usually used to approximate the potential energy along a one-dimensional reaction path:

$$V_{eck}(x) = V_o \mathrm{sech}^2(x/a) \approx V_o(1 - (x/a)^2 + \cdots)$$

For convenience, set the zero in energy to be the barrier height V_o so that any transmission for $E < 0$ corresponds to tunneling.[*]

At sufficiently large distances from the turning point, the motion is purely quasi classical and we can write the momentum as

$$p = \sqrt{2m(E + kx^2/2)} \approx x\sqrt{mk} + E\sqrt{m/k}/x \tag{3.45}$$

---

[*] The analysis is from Kembel (1935) as discussed in Landau and Lifshitz, *Quantum Mechanics* (nonrelativistic theory), third edition. (New York: Oxford, Pergamon Press, 1977.)

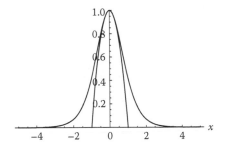

FIGURE 3.1 Eckart barrier and parabolic approximation of the transition state.

and the asymptotic of the Schrödinger wave function is

$$\psi = Ae^{+i\xi^2/2}\xi^{+i\epsilon-1/2} + Be^{-i\xi^2/2}\xi^{-i\epsilon-1/2} \qquad (3.46)$$

where A and B are the coefficients we need to determine by the matching condition and ξ and ϵ are dimensionless lengths and energies given by $\xi = x(mk/\hbar)^{1/4}$ and $\epsilon = (E/\hbar)\sqrt{m/k}$.

The particular case we are interested in is for a particle coming from the left and passing to the right with the barrier in between. So, the wave functions in each of these regions must be

$$\psi_R = Be^{+i\xi^2/2}\xi^{i\epsilon-1/2} \qquad (3.47)$$

and

$$\psi_L = e^{-i\xi^2/2}(-\xi)^{-i\epsilon-1/2} + Ae^{+i\xi^2/2}(-\xi)^{i\epsilon-1/2} \qquad (3.48)$$

where the first term is the incident wave and the second term is the reflected component. So, $|A|^2$ is the reflection coefficient and $|B|^2$ is the transmission coefficient normalized so that

$$|A|^2 + |B|^2 = 1$$

Let us move to the complex plane, write a new coordinate, $\xi = \rho e^{i\phi}$, and consider what happens as we rotate around in ϕ and take ρ to be large. Since $i\xi^2 = \rho^2(i\cos 2\phi - \sin 2\phi)$, we have

$$\psi_R(\phi = 0) = Be^{i\rho^2}\rho^{+i\epsilon-1/2}$$

$$\psi_L(\phi = 0) = Ae^{i\rho^2}(-\rho)^{+i\epsilon-1/2} \qquad (3.49)$$

and at $\phi = \pi$

$$\psi_R(\phi = \pi) = Be^{i\rho^2}(-\rho)^{+i\epsilon-1/2}$$

$$\psi_L(\phi = \pi) = Ae^{i\rho^2}\rho^{+i\epsilon-1/2} \qquad (3.50)$$

So, in other words, $\psi_R(\phi = \pi)$ looks like $\psi_L(\phi = 0)$ when

$$A = B(e^{i\pi})^{i\epsilon-1/2}$$

So, we have the relation $A = -iBe^{-\pi\epsilon}$. Finally, after we normalize this we get the transmission coefficient:

$$T = |B|^2 = \frac{1}{1 + e^{-2\pi\epsilon}}$$

which must hold for any energy. If the energy is large and negative, then

$$T \approx e^{-2\pi\epsilon}$$

Also, we can compute the reflection coefficient for $E > 0$ as $1 - D$,

$$R = \frac{1}{1 + e^{+2\pi\epsilon}}$$

This gives us the transmission probabilty as a function of incident energy. But normal chemical reactions are not done at constant energy, they are done at constant temperature. To get the thermal transmission coefficient, we need to take a Boltzmann weighted average of transmission coefficients

$$T_{th}(\beta) = \frac{1}{Z} \int dE e^{-E\beta} T(E) \tag{3.51}$$

where $\beta = 1/kT$ and Z is the partition function. If E represents a continuum of energy states, then

$$T_{th}(\beta) = -\frac{\beta\omega\hbar\left(\psi^{(0)}\left(\frac{\beta\omega\hbar}{4\pi}\right) - \psi^{(0)}\left(\frac{1}{4}\left(\frac{\beta\omega\hbar}{\pi} + 2\right)\right)\right)}{4\pi} \tag{3.52}$$

where $\psi^{(n)}(z)$ is the polygamma function, which is the nth derivative of the digamma function $\psi^{(0)}(z)$, which is the logarithmic derivative of Euler's gamma function $\psi^{(0)}(z) = \Gamma'(z)/\Gamma(z)$.

3.3 CONNECTION FORMULAS

In what we have considered thus far, we have assumed that up until the turning point the wave function was well behaved and smooth. We can think of the problem as having two domains: an exterior and an interior. The exterior part we assumed to be simple and the boundary conditions trivial to impose. The next task is to figure out the matching condition at the turning point for an arbitrary system. So far what we have are two pieces, ψ_L and ψ_R, in the notation above. What we need is a patch. To do so, we make a linearizing assumption for the force at the classical turning point:

$$E - V(x) \approx F_o(x - a) \tag{3.53}$$

where $F_o = -dV/dx$ evaluated at $x = a$. Thus, the phase integral is easy:

$$\frac{1}{\hbar} \int_a^x p \, dx = \frac{2}{3\hbar} \sqrt{2mF_o}(x - a)^{3/2} \tag{3.54}$$

Problem 3.3 Verify the relations for the transmission and reflection coefficients for the Eckart barrier problem.

But, we can do better than that. We can actually solve the Schrödinger equation for the linear potential and use the linearized solutions as our patch. The *Mathematica* Notebook for this chapter (*Chapter3.nb*) determines the solution of the linearized Schrödinger equation

$$-\frac{\hbar^2}{2m}\frac{d\psi}{dx^2} + (E + V')\psi = 0 \tag{3.55}$$

which can be rewritten as

$$\psi'' = \alpha^3 x \psi \tag{3.56}$$

with

$$\alpha = \left(\frac{2m}{\hbar^2}V'(0)\right)^{1/3}$$

Absorbing the coefficient into a new variable y, we get Airy's equation

$$\psi''(y) = y\psi$$

The solutions of Airy's equation are Airy functions, $Ai(y)$ and $Bi(y)$ for the regular and irregular cases. The integral representation of the Ai and Bi are

$$Ai(y) = \frac{1}{\pi}\int_0^\infty \cos\left(\frac{s^3}{3} + sy\right)ds \tag{3.57}$$

and

$$Bi(y) = \frac{1}{\pi}\int_0^\infty \left[e^{-s^3/3+sy} + \sin\left(\frac{s^3}{3} + sy\right)\right]ds \tag{3.58}$$

Plots of these functions are shown in Figure 3.2.

Since both Ai and Bi are acceptable solutions, we will take a linear combination of the two as our patching function and figure out the coefficients later:

$$\psi_P = a\,Ai(\alpha x) + b\,Bi(\alpha x) \tag{3.59}$$

We now have to determine those coefficients. We need to make two assumptions: (1) that the overlap zones are sufficiently close to the turning point that a linearized

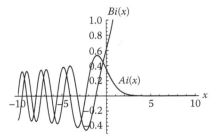

FIGURE 3.2 Airy functions, Ai(x) and Bi(x).

potential is reasonable and (2) that the overlap zone is far enough from the turning point (at the origin) that the WKB approximation is accurate and reliable. You can certainly cook up some potential for which this will not work, but we will assume it is reasonable. In the linearized region, the momentum is

$$p(x) = \hbar \alpha^{3/2}(-x)^{3/2} \tag{3.60}$$

So for $+x$,

$$\int_0^x |p(x)| dx = 2\hbar(\alpha x)^{3/2}/3 \tag{3.61}$$

and the WKB wave function becomes

$$\psi_R(x) = \frac{D}{\sqrt{\hbar}\alpha^{3/4}x^{1/4}} e^{-2(\alpha x)^{3/2}/3} \tag{3.62}$$

In order to extend into this region, we will use the asymptotic form of the Ai and Bi functions for $y \gg 0$:

$$Ai(y) \approx \frac{e^{-2y^{3/2}/3}}{2\sqrt{\pi}y^{1/4}} \tag{3.63}$$

$$Bi(y) \approx \frac{e^{+2y^{3/2}/3}}{\sqrt{\pi}y^{1/4}} \tag{3.64}$$

Clearly, the $Bi(y)$ term will not contribute, so $b = 0$ and

$$a = \sqrt{\frac{4\pi}{\alpha\hbar}} D$$

Now, for the other side, we do the same procedure, except this time $x < 0$ so the phase integral is

$$\int_x^0 p dx = 2\hbar(-\alpha x)^{3/2}/3 \tag{3.65}$$

Thus the WKB wave function on the left-hand side is

$$\psi_L(x) = \frac{1}{\sqrt{p}} \left(Be^{2i(-\alpha x)^{3/2}/3} + Ce^{-2i(-\alpha x)^{3/2}/3} \right) \tag{3.66}$$

$$= \frac{1}{\sqrt{\hbar}\alpha^{3/4}(-x)^{1/4}} \left(Be^{2i(-\alpha x)^{3/2}/3} + Ce^{-2i(-\alpha x)^{3/2}/3} \right) \tag{3.67}$$

That is the WKB part, to connect with the patching part, so we again use the asymptotic forms for $y \ll 0$ and take only the regular solution,

$$Ai(y) \approx \frac{1}{\sqrt{\pi}(-y)^{1/4}} \sin\left(2(-y)^{3/2}/3 + \pi/4 \right)$$

$$\approx \frac{1}{2i\sqrt{\pi}(-y)^{1/4}} \left(e^{i\pi/4} e^{i2(-y)^{3/2}/3} - e^{-i\pi/4} e^{-i2(-y)^{3/2}/3} \right) \tag{3.68}$$

Comparing the WKB wave and the patching wave, we can match term-by-term

$$\frac{a}{2i\sqrt{\pi}}e^{i\pi/4} = \frac{B}{\sqrt{\hbar\alpha}} \tag{3.69}$$

$$\frac{-a}{2i\sqrt{\pi}}e^{-i\pi/4} = \frac{C}{\sqrt{\hbar\alpha}} \tag{3.70}$$

Since we know a in terms of the normalization constant D, $B = ie^{i\pi/4}D$ and $C = ie^{-i\pi/4}$. This is the connection! We can write the WKB function across the turning point as

$$\psi_{WKB}(x) = \begin{cases} \dfrac{2D}{\sqrt{p(x)}} \sin\left[\dfrac{1}{\hbar} \displaystyle\int_x^0 pdx + \pi/4\right] & x < 0 \\[12pt] \dfrac{2D}{\sqrt{|p(x)|}} e^{-\frac{1}{\hbar}\int_x^0 pdx} & x > 0 \end{cases} \tag{3.71}$$

Example: Bound States in the Linear Potential

Since we worked so hard, we have to use the results. So, consider a model problem for a particle in a gravitational field. Actually, this problem is not so far-fetched since one can prepare trapped atoms above a parabolic reflector and make a quantum bouncing ball. Here the potential is $V(x) = mgx$ where m is the particle mass and g is the gravitational constant ($g = 9.80$ m/s). We will take the case where the reflector is infinite so that the particle cannot penetrate into it. The Schrödinger equation for this potential is

$$-\frac{\hbar^2}{2m}\psi'' + (E - mgx)\psi = 0 \tag{3.72}$$

The solutions are the Airy $Ai(x)$ functions. Setting, $\beta = mg$ and $c = \hbar^2/2m$, the solutions are

$$\psi = C\,Ai\left(\left(-\frac{\beta}{c}\right)^{1/3}(x - E/\beta)\right) \tag{3.73}$$

However, there is one caveat: $\psi(0) = 0$, thus the Airy functions must have their nodes at $x = 0$. So we have to systematically shift the $Ai(x)$ function in x until a node lines up at $x = 0$. The nodes of the $Ai(x)$ function can be determined and the first seven of them are listed in Table 3.1. To find the energy levels, we systematically solve the equation

$$\left(-\frac{\beta}{c}\right)^{1/3}\frac{E_n}{\beta} = x_n$$

So the ground state is where the first node lands at $x = 0$,

$$E_1 = \frac{2.33811\beta}{(\beta/c)^{1/3}}$$

$$= \frac{2.33811mg}{(2m^2g/\hbar^2)^{1/3}} \tag{3.74}$$

TABLE 3.1
Location of Nodes for Airy, $Ai(x)$ Function

node	x_n
1	-2.33811
2	-4.08795
3	-5.52056
4	-6.78671
5	-7.94413
6	-9.02265
7	-10.0402

and so on. Of course, we still have to normalize the wave function to get the correct energy.

We can make life a bit easier by using the quantization condition derived from the WKB approximation. Since we require the wave function to vanish exactly at $x = 0$, we have

$$\frac{1}{\hbar} \int_0^{x_t} p(x)dx + \frac{\pi}{4} = n\pi \tag{3.75}$$

This assures us that the wave vanishes at $x = 0$. In this case x_t is the turning point $E = mgx_t$ (see Figure 3.3). As a consequence,

$$\int_0^{x_t} p(x)dx = (n - 1/4)\pi$$

Since $p(x) = \sqrt{2m(E_n - mgx)}$, the integral can be evaluated

$$\int_0^{x_t} \sqrt{2m(E - mgx)}dx = \sqrt{2}\left(\frac{2E_n\sqrt{E_n m}}{3gm} + \frac{2\sqrt{m}\,(E_n - gmx_t)(-E_n + gmx_t)}{3gm}\right)$$

Since $x_t = E_n/mg$ for the classical turning point, the phase intergral becomes

$$\frac{2\sqrt{2}E_n{}^2}{3g\sqrt{E_n m}} = (n - 1/4)\pi\hbar.$$

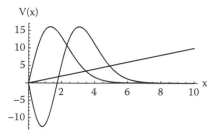

FIGURE 3.3 Quantum bound states in a graviational well.

Solving for E_n yields the semiclassical approximation for the eigenvalues:

$$E_n = \frac{g^{\frac{2}{3}} m^{\frac{1}{3}} \left((1 - 4n)^2\right)^{\frac{1}{3}} (3\pi)^{\frac{2}{3}} \hbar^{\frac{2}{3}}}{4 \, 2^{\frac{1}{3}}} \tag{3.76}$$

In atomic units, the gravitional constant is $g = 1.08563 \times 10^{-22}$ bohr/a.u.^{2}. For $n = 0$, we get for an electron $E_o^{sc} = 2.014 \times 10^{-15}$ hartree or about 12.6 Hz. So, graviational effects on electrons are extremely tiny compared with the typical electronic energy for an atom or molecule. However, quantized gravitational states have been observed in atomic fountains.

3.4   SCATTERING

The collision between two particles plays an important role in the dynamics of reactive molecules. We consider here the collision between two particles interacting via a central force $V(r)$. Working in the center of mass frame, we consider the motion of a point particle with mass μ and position vector $\vec{r}$. We will first examine the process in a purely classical context since it is intuitive and then apply what we know to the quantum and semiclassical case.

3.4.1 CLASSICAL SCATTERING

The angular momentum of the particle about the origin is given by

$$\vec{L} = \vec{r} \times \vec{p} = \mu(\vec{r} \times \dot{\vec{r}}) \tag{3.77}$$

We know that angular momentum is a conserved quantity and it is is easy to show that $\dot{\vec{L}} = 0$, viz,

$$\dot{\vec{L}} = \frac{d}{dt}\vec{r} \times \vec{p} = (\dot{\vec{r}} \times \vec{r} + (\vec{r} \times \dot{\vec{p}}) \tag{3.78}$$

Since $\dot{r} = \dot{p}/\mu$, the first term vanishes; likewise, the force vector, $\dot{\vec{p}} = -dV/dr$, is along $\vec{r}$ so that the second term vanishes. Thus, $L = const$, meaning that angular momentum is a conserved quantity during the course of the collision.

In Cartesian coordinates, the total energy of the collision is given by

$$E = \frac{\mu}{2}(\dot{x}^2 + \dot{y}^2) + V \tag{3.79}$$

To convert from Cartesian to polar coordinates, we use

$$x = r\cos\theta \tag{3.80}$$

$$y = r\sin\theta \tag{3.81}$$

$$\dot{x} = \dot{r}\cos\theta - r\dot{\theta}\sin\theta \tag{3.82}$$

$$\dot{y} = \dot{r}\sin\theta + r\dot{\theta}\cos\theta \tag{3.83}$$

Thus,

$$E = \frac{mu}{2}\dot{r}^2 + V(r) + \frac{L^2}{2\mu r^2} \qquad (3.84)$$

where we use the fact that

$$L = \mu r^2 \dot{\theta}^2 \qquad (3.85)$$

where L is the angular momentum. What we see here is that we have two potential contributions. The first is the physical attraction (or repulsion) between the two scattering bodies. The second is a purely repulsive centrifugal potential that depends upon the angular momentum and ultimately upon the impact parameters. For cases of large impact parameters, this can be the dominant force. The effective radial force is given by

$$\mu\ddot{r} = \frac{L^2}{2r^3\mu} - \frac{\partial V}{\partial r} \qquad (3.86)$$

Again, we note that the centrifugal contribution is always repulsive while the physical interaction $\mathbf{V(r)}$ is typically attractive at long ranges and repulsive at short ranges.

We can derive the solutions to the scattering motion by integrating the velocity equations for r and θ

$$\dot{r} = \pm\left(\frac{2}{\mu}\left(E - V(r) - \frac{L^2}{2\mu r^2}\right)\right)^{1/2} \qquad (3.87)$$

$$\dot{\theta} = \frac{L}{\mu r^2} \qquad (3.88)$$

and taking into account the starting conditions for r and θ. In general, we could solve the equations numerically and obtain the complete scattering path. However, really what we are interested in is the deflection angle χ since this is what is ultimately observed. So, we integrate the last two equations and derive θ in terms of r:

$$\theta(r) = \int_0^\theta d\theta = -\int_\infty^r \frac{d\theta}{dr}dr \qquad (3.89)$$

$$= -\int_\infty^r \frac{L}{\mu r^2}\frac{1}{\sqrt{(2/\mu)(E - V - L^2/2\mu r^2)}}dr \qquad (3.90)$$

where the collision starts at $t = -\infty$ with $r = \infty$ and $\theta = 0$. What we want to do is derive this in terms of an impact parameter, b, and scattering angle χ. These are illustrated in Figure 3.4 and can be derived from basic kinematic considerations. First, energy is conserved throughout, so if we know the asymptotic velocity v, then $E = \mu v^2/2$. Secondly, angular momentum is conserved, so $L = \mu|r \times v| = \mu v b$. Thus the integral above becomes

$$\theta(r) = -b\int_\infty^r \frac{d\theta}{dr}dr \qquad (3.91)$$

$$= -\int_\infty^r \frac{dr}{r^2\sqrt{1 - V/E - b^2/r^2}} \qquad (3.92)$$

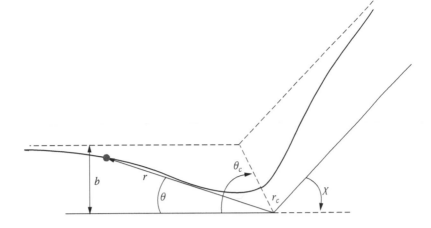

FIGURE 3.4 Elastic scattering trajectory for classical collision.

Finally, the angle of deflection is related to the angle of closest approach by $2\theta_c + \chi = \pi$; hence,

$$\chi = \pi - 2b \int_{r_c}^{\infty} \frac{dr}{r^2 \sqrt{1 - V/E - b^2/r^2}} \qquad (3.93)$$

The radial distance of closest approach is determined by

$$E = \frac{L^2}{2\mu r_c^2} + V(r_c) \qquad (3.94)$$

which can be restated as

$$b^2 = r_c^2 \left(1 - \frac{V(r_c)}{E}\right) \qquad (3.95)$$

Once we have specified the potential, we can compute the deflection angle using Equation 3.95. If $V(r_c) < 0$, then $r_c < b$ and we have an attractive potential; if $V(r_c) > 0$, then $r_c > b$ and the potential is repulsive at the turning point.

If we have a beam of particles incident on some scattering center, then collisions will occur with all possible impact parameters (hence angular momenta) and will give rise to a distribution in the scattering angles. We can describe this by a *differential cross-section*. If we have some incident intensity of particles in our beam I_o, which is the incident flux or the number of particles passing a unit area normal to the beam direction per unit time, then the differential cross-section $I(\chi)$ is defined so that $I(\chi)d\Omega$ is the number of particles per unit time scattered into some solid angle $d\Omega$ divided by the incident flux.

The deflection pattern will be axially symmetric about the incident beam direction due to the spherical symmetry of the interaction potential; thus, $I(\chi)$ depends only upon the scattering angle. Thus, $d\Omega$ can be constructed by the cones defining χ and $\chi + d\chi$, that is, $d\Omega = 2\pi \sin \chi d\chi$. Even if the interaction potential is not spherically symmetric, since most molecules are not spherical, the scattering would be axially symmetric since we would be scattering from a homogeneous distribution of all possible orientations of the colliding molecules. Hence any azimuthal dependency must vanish unless we can orient on the colliding species.

Given an initial velocity v, the fraction of the incoming flux with impact parameters between b and $b+db$ is $2\pi b\,db$. These particles will be deflected between χ and $\chi+d\chi$ if $d\chi/db > 0$ or between χ and $\chi - d\chi$ if $d\chi/db < 0$. Thus, $I(\chi)d\Omega = 2\pi b\,db$ and it follows then that

$$I(\chi) = \frac{b}{\sin \chi \,|d\chi/db|} \tag{3.96}$$

Thus, once we know $\chi(b)$ for a given v, we can get the differential cross-section. The *total cross-section* is obtained by integrating

$$\sigma = 2\pi \int_0^\pi I(\chi) \sin \chi \, d\chi \tag{3.97}$$

This is a measure of the attenuation of the incident beam by the scattering target and has the units of area.

3.4.2 SCATTERING AT SMALL DEFLECTION ANGLES

Our calculations will be greatly simplified if we consider collisions that result in small deflections in the forward direction. If we let the initial beam be along the x axis with momentum p, then the scattered momentum p' will be related to the scattered angle by $p' \sin \chi = p'_y$. Taking χ to be small,

$$\chi \approx \frac{p'_y}{p'} = \frac{\text{momentum transfer}}{\text{momentum}} \tag{3.98}$$

Since the time derivative of momentum is the force, the momentum transfered perpendicular to the incident beam is obtained by integrating the perpendicular force

$$F'_y = -\frac{\partial V}{\partial y} = -\frac{\partial V}{\partial r}\frac{\partial r}{\partial y} = -\frac{\partial V}{\partial r}\frac{b}{r} \tag{3.99}$$

where we used $r^2 = x^2 + y^2$ and $y \approx b$. Thus we find

$$\chi = \frac{p'_y}{\mu(2E/\mu)^{1/2}} \tag{3.100}$$

$$= -b(2\mu E)^{-1/2} \int_{-\infty}^{+\infty} \frac{\partial V}{\partial r}\frac{dt}{r} \tag{3.101}$$

$$= -b(2\mu E)^{-1/2} \left(\frac{2E}{\mu}\right)^{-1/2} \int_{-\infty}^{+\infty} \frac{\partial V}{\partial r}\frac{dx}{r} \tag{3.102}$$

$$= -\frac{b}{E} \int_b^\infty \frac{\partial V}{\partial r}(r^2 - b^2)^{-1/2}dr \tag{3.103}$$

where we used $x = (2E/\mu)^{1/2}t$ and x varies from $-\infty$ to $+\infty$ as r goes from $-\infty$ to b and back.

Let us use this in a simple example of the $V = C/r^s$ potential for $s > 0$. Substituting V into the integral above and solving yields

$$\chi = \frac{sC\pi^{1/2}}{2b^s E} \frac{\Gamma((s+1)/2)}{\Gamma(s/2+1)} \tag{3.104}$$

This indicates that $\chi E \propto b^{-s}$ and $|d\chi/db| = \chi s/b$. Thus, we can conclude by deriving the differential cross-section

$$I(\chi) = \frac{1}{s}\chi^{-(2+2/s)} \left(\frac{sC\pi^{1/2}}{2E}\frac{\Gamma((s+1)/2)}{\Gamma(s/2+1)}\right)^{2/s} \tag{3.105}$$

for small values of the scattering angle. Consequently, a log–log plot of the center of mass differential cross-section as a function of the scattering angle at fixed energy should give a straight line with a slope $-(2+2/s)$ from which one can determine the value of s. For the van der Waals potential, $s = 6$ and $I(\chi) \propto E^{-1/3}\chi^{-7/3}$.

3.4.3 QUANTUM TREATMENT

The quantum mechanical case is a bit more complex. Here we will develop a brief overview of quantum scattering and move on to the semiclassical evaluation. The quantum scattering is determined by the asymptotic form of the wave function,

$$\psi(r, \chi) \stackrel{r \to \infty}{\longrightarrow} A\left(e^{ikz} + \frac{f(\chi)}{r}e^{ikr}\right) \tag{3.106}$$

where A is some normalization constant and $k = 1/\lambda = \mu v/\hbar$ is the initial wave vector along the incident beam direction ($\chi = 0$). The first term represents a plane wave incident upon the scatterer and the second represents an outgoing spherical wave. Notice that the outgoing amplitude is reduced as r increases. This is because the wave function spreads as r increases. If we can collimate the incoming and outgoing components, then the scattering amplitude $f(\chi)$ is related to the differential cross-section by

$$I(\chi) = |f(\chi)|^2 \tag{3.107}$$

What we have is then that the asymptotic form of the wave function carries within it information about the scattering process. As a result, we do not need to solve the wave equation for all of space, we just need to be able to connect the scattering amplitude to the interaction potential. We do so by expanding the wave as a superposition of Legendre polynomials

$$\psi(r, \chi) = \sum_{l=0}^{\infty} R_l(r)P_l(\cos \chi) \tag{3.108}$$

$R_l(r)$ must remain finite as $r = 0$. This determines the form of the solution.

When $V(r) = 0$, then $\psi = A\exp(ikz)$ and we can expand the exponential in terms of spherical waves

$$e^{ikz} = \sum_{l=0}^{\infty}(2l+1)e^{il\pi/2}\frac{\sin(kr-l\pi/2)}{kr}P_l(\cos\chi) \qquad (3.109)$$

$$= \frac{1}{2i}\sum_{l=0}^{\infty}(2l+1)i^l P_l(\cos\chi)\left(\frac{e^{i(kr-l\pi/2)}}{kr} + \frac{e^{-i(kr-l\pi/2)}}{kr}\right) \qquad (3.110)$$

We can interpret this equation in the following intuitive way: The incident plane wave is equivalent to an infinite superposition of incoming and outgoing spherical waves in which each term corresponds to a particular angular momentum state with

$$L = \hbar\sqrt{l(l+1)} \approx \hbar(l+1/2) \qquad (3.111)$$

From our analysis above, we can relate L to the impact parameter, b,

$$b = \frac{L}{\mu v} \approx \frac{l+1/2}{k}\lambda \qquad (3.112)$$

In essence the incoming beam is divided into cylindrical zones in which the lth zone contains particles with impact parameters (and hence angular momenta) between $l\lambda$ and $(l+1)\lambda$.

Problem 3.4 In the collision between hard spheres as described on p. 78, the impact parameter b is treated as continuous; however, in quantum mechanics we allow only discrete values of the angular momentum l. How will this affect our results, since $b = (l+1/2)\lambda$?

If $V(r)$ is short ranged (that is, it falls off more rapidly than $1/r$ for large r), we can derive a general solution for the asymptotic form

$$\psi(r,\chi) \longrightarrow \sum_{l=0}^{\infty}(2l+1)\exp\left(i\left(\frac{l\pi}{2}+\eta_l\right)\right)\frac{\sin(kr-l\pi/2+\eta_l)}{kr}P_l(\cos\chi)$$

$$(3.113)$$

The significant difference between Equation 3.113 and Equation 3.110 for the $V(r) = 0$ case is the addition of a phase shift η_l. This shift *only* occurs in the outgoing part of the wave function and so we conclude that the primary effect of a potential in quantum scattering is to introduce a phase in the asymptotic form of the scattering wave. This phase must be a real number and has the physical interpretation illustrated in Figure 3.5. A repulsive potential will cause a decrease in the relative velocity of the particles at small r resulting in a *longer* de Broglie wavelength. This causes the wave to be "pushed out" relative to that for $V = 0$ and the phase shift is negative. An attractive potential produces a positive phase shift and "pulls" the wave function in a bit. Furthermore, the centrifugal part produces a negative shift of $-l\pi/2$.

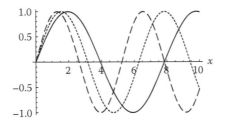

FIGURE 3.5 Form of the radial wave for repulsive (short dashed) and attractive (long dashed) potentials. The form for $V = 0$ is the solid curve for comparison.

Comparing the various forms for the asymptotic waves, we can deduce that the scattering amplitude is given by

$$f(\chi) = \frac{1}{2ik} \sum_{l=0}^{\infty} (2l + 1)(e^{2i\eta_l} - 1)P_l(\cos \chi) \tag{3.114}$$

From this, the differential cross-section is

$$I(\chi) = \lambda^2 \left| \sum_{l=0}^{\infty} (2l + 1)e^{i\eta_l} \sin(\eta_l)P_l(\cos \chi) \right|^2 \tag{3.115}$$

What we see here is the possibility for interference between different angular momentum components.

Moving forward at this point requires some rather sophisticated treatments. However, we can use the semiclassical methods developed in this chapter to estimate the phase shifts.

3.4.4 SEMICLASSICAL EVALUATION OF PHASE SHIFTS

The exact scattering wave is not so important. What is important is the asymptotic extent of the wave function because that is the part that carries the information from the scattering center to the detector. What we want is a measure of the shift in phase between a scattering with and without the potential. From the WKB treatment above, we know that the phase is related to the classical action along a given path. Thus, in computing the semiclassical phase shifts, we are really looking at the difference between the classical actions for a system with the potential switched on and a system with the potential switched off,

$$\eta_l^{SC} = \lim_{R \to \infty} \left(\int_{r_c}^{R} \frac{dr}{\lambda(r)} - \int_{b}^{R} \frac{dr}{\lambda(r)} \right) \tag{3.116}$$

R is the radius a sphere about the scattering center and $\lambda(r)$ is a de Broglie wavelength

$$\lambda(r) = \frac{\hbar}{p} = \frac{1}{k(r)} = \frac{\hbar}{\mu v(1 - V(r) - b^2/r^2)^{1/2}} \tag{3.117}$$

associated with the radial motion. Putting this together:

$$\eta_l^{SC} = \lim_{R \to \infty} k \left(\int_{r_c}^{R} \left(1 - \frac{V(r)}{E} - \frac{b^2}{r^2} \right)^{1/2} dr - \int_{b}^{R} \left(1 - \frac{b^2}{r^2} \right)^{1/2} dr \right) \quad (3.118)$$

$$= \lim_{R \to \infty} \left(\int_{r_c}^{R} k(r) dr - k \int_{b}^{R} \left(1 - \frac{b^2}{r^2} \right)^{1/2} dr \right) \quad (3.119)$$

(k is the incoming wave vector.) The last integral we can evaluate as

$$k \int_{b}^{R} \frac{(r^2 - b^2)^{1/2}}{r} dr = k \left((r^2 - b^2) - b \cos^{-1} \frac{b}{r} \right) \Big|_{b}^{R} = kR - \frac{kb\pi}{2} \quad (3.120)$$

Now, to clean things up a bit, we add and subtract an integral over k (we do this to get rid of the R dependence, which will cause problems when we take the limit $R \to \infty$):

$$\eta_l^{SC} = \lim_{R \to \infty} \left(\int_{r_c}^{R} k(r) dr - \int_{r_c}^{R} k \, dr + \int_{r_c}^{R} k \, dr - \left(kR - \frac{kb\pi}{2} \right) \right) \quad (3.121)$$

$$= \int_{r_c}^{R} (k(r) - k) dr - k(r_c - b\pi/2) \quad (3.122)$$

$$= \int_{r_c}^{R} (k(r) - k) dr - kr_c \pi (l + 1/2)/2 \quad (3.123)$$

This last expression is the standard form of the phase shift.

The deflection angle can be determined in a similar way:

$$\chi = \lim_{R \to \infty} \left[\left(\pi - 2 \int_{actual\ path} d\theta \right) - \left(\pi - \int_{V=0\ path} d\theta \right) \right] \quad (3.124)$$

We transform this into an integral over r:

$$\chi = -2b \left[\int_{r_c}^{\infty} \left(1 - \frac{V(r)}{E} - \frac{b^2}{r^2} \right)^{-1/2} \frac{dr}{r^2} - \int_{b}^{\infty} \left(1 - \frac{b^2}{r^2} \right)^{-1/2} \frac{dr}{r^2} \right] \quad (3.125)$$

Agreed, this is a weird way to express the scattering angle. But let us keep pushing this forward. The last integral can be evaluated as

$$\int_{b}^{\infty} \left(1 - \frac{b^2}{r^2} \right)^{-1/2} \frac{dr}{r^2} = \frac{1}{b} \cos^{-1} \frac{b}{r} \Big|_{b}^{\infty} = -\frac{\pi}{2b} \quad (3.126)$$

which yields the classical result we obtained previously. So, why did we bother? From this we can derive a simple and useful connection between the classical deflection angle and the rate of change of the semiclassical phase shift with angular momentum, $d\eta_l^{SC}/dl$. First, recall the Leibnitz rule for taking derivatives of integrals:

$$\frac{d}{dx} \int_{a(x)}^{b(x)} f(x, y) dy = \frac{b}{dx} f(b(x), y) - \frac{da}{dx} f(a(x), y)$$

$$+ \int_{a(x)}^{b(x)} \frac{\partial f(x, y)}{\partial x} dy \quad (3.127)$$

Taking the derivative of η_l^{SC} with respect to l, using the last equation and the relation that $(\partial b/\partial l)_E = b/k$, we find that

$$\frac{d\eta_l^{SC}}{dl} = \frac{\chi}{2} \qquad (3.128)$$

Next, we examine the differential cross-section, $I(\chi)$. The scattering amplitude

$$f(\chi) = \frac{\lambda}{2i} \sum_{l=0}^{\infty} (2l + 1)e^{2i\eta_l} P_l(\cos \chi) \qquad (3.129)$$

where we use $\lambda = 1/k$ and exclude the singular point where $\chi = 0$ since this contributes nothing to the total flux.

Now, we need a mathematical identity to take this to the semiclassical limit where the potential varies slowly with wavelength. What we do is to first relate the Legendre polynomial, $P_l(\cos\theta)$, to a zeroth-order Bessel function for small values of θ ($\theta \ll 1$),

$$P_l(\cos\theta) = J_0((l + 1/2)\theta) \qquad (3.130)$$

Now, when $x = (l + 1/2)\theta \gg 1$ (that is, large angular momentum), we can use the asymptotic expansion of $J_0(x)$

$$J_0(x) \rightarrow \sqrt{\frac{2}{\pi x}} \sin\left(x + \frac{\pi}{4}\right) \qquad (3.131)$$

Pulling this together,

$$P_l(\cos\theta) \rightarrow \left[\frac{2}{\pi(l + 1/2)\theta}\right]^{1/2} \sin((l + 1/2)\theta + \pi/4)$$

$$\approx \left[\frac{2}{\pi(l + 1/2)}\right]^{1/2} \frac{\sin((l + 1/2)\theta + \pi/4)}{(\sin\theta)^{1/2}} \qquad (3.132)$$

for $\theta(l + 1/2) \gg 1$. Thus, we can write the semiclassical scattering amplitude as

$$f(\chi) = -\lambda \sum_{l=0}^{\infty} \left(\frac{(l + 1/2)}{2\pi \sin \chi}\right)^{1/2} (e^{i\phi^+} + e^{i\phi^-}) \qquad (3.133)$$

where

$$\phi^\pm = 2\eta_l \pm (l + 1/2)\chi \pm \pi/4 \qquad (3.134)$$

The phases are rapidly oscillating functions of l. Consequently, the majority of the terms must cancel and the sum is determined by the ranges of l for which either ϕ^+ or ϕ^- is extremized. This implies that the scattering amplitude is determined almost exclusively by phase shifts that satisfy

$$2\frac{d\eta_l}{dl} \pm \chi = 0 \qquad (3.135)$$

where the $+$ is for $d\phi^+/dl = 0$ and the $-$ is for $d\phi^-/dl = 0$. This demonstrates that only the phase shifts corresponding to impact parameter b can contribute significantly to the differential cross-section in the semiclassical limit. Thus, the classical condition for scattering at a given deflection angle χ is that l be large enough for Equation 3.135 to apply.

3.5   PROBLEMS AND EXERCISES

Problem 3.5   In this problem we will consider the ammonia inversion problem, but this time we will proceed in a semiclassical context.

Recall that the ammonia inversion potential consists of two symmetrical potential wells separated by a barrier. If the barrier were impenetrable, one would find energy levels corresponding to motion in one well or the other. Since the barrier is not infinite, there can be passage between wells via tunneling. This causes the otherwise degenerate energy levels to split. In this problem, we will make life a bit easier by taking

$$V(x) = \alpha(x^4 - x^2)$$

Let ψ_o be the semiclassical wave function describing the motion in one well with energy E_o. Assume that ψ_o is exponentially damped on both sides of the well and that the wave function is normalized so that the integral over ψ_o^2 is unity. When tuning is taken into account, the wave functions corresponding to the new energy levels, E_1 and E_2, are the symmetric and antisymmetric combinations of $\psi_o(x)$ $\langle Q|P|Q'\rangle$ and $\psi_o(-x)$

$$\psi_1 = (\psi_o(x) + \psi_o(-x)/\sqrt{2}$$
$$\psi_2 = (\psi_o(x) - \psi_o(-x)/\sqrt{2}$$

where $\psi_o(-x)$ can be thought of as the contribution from the zeroth-order wave function in the other well. In well 1, $\psi_o(-x)$ is very small, in well 2, $\psi_o(+x)$ is very small, and the product $\psi_o(x)\psi_o(-x)$ is vanishingly small everywhere. Also, by construction, ψ_1 and ψ_2 are normalized.

1. Assume that ψ_o and ψ_1 are solutions of the Schrödinger equations

$$\psi_o'' + \frac{2m}{\hbar^2}(E_o - V)\psi_o = 0$$

and

$$\psi_1'' + \frac{2m}{\hbar^2}(E_1 - V)\psi_1 = 0$$

Multiply the former by ψ_1 and the latter by ψ_o, combine and subtract equivalent terms, and integrate over x from 0 to ∞ to show that

$$E_1 - E_o = -\frac{\hbar^2}{m}\psi_o(0)\psi_o'(0)$$

Perform a similar analysis to show that

$$E_2 - E_o = +\frac{\hbar^2}{m}\psi_o(0)\psi_o'(0)$$

2. Show that the unperturbed semiclassical wave function is

$$\psi_o(0) = \sqrt{\frac{\omega}{2\pi v_o}}\exp\left[-\frac{1}{\hbar}\int_0^a |p|dx\right]$$

and

$$\psi'_o(0) = \frac{mv_o}{\hbar}\psi_o(0)$$

where $v_o = \sqrt{2(E_o - V(0))/m}$ and a is the classical turning point at $E_o = V(a)$.

3. Combining your results, show that the tunneling splitting is

$$\Delta E = \frac{\hbar\omega}{\pi}\exp\left[-\frac{1}{\hbar}\int_{-a}^{+a}|p|dx\right]$$

where the integral is taken between classical turning points on either side of the barrier.

4. Assuming that the potential in the barrier is an upside-down parabola

$$V(x) \approx V_o - kx^2/2$$

What is the tunneling splitting?

5. Now, taking $\alpha = 0.1$, expand the potential about the barrier and determine the harmonic force constant for the upside-down parabola. Use the equations you derived and compute the tunneling splitting for a proton in this well.

Problem 3.6 Use the Bohr–Sommerfield approximation to derive an expression for the number of discrete bound states in the following potentials:

1. $V = mu\omega^2x^2/2$
2. $V = V_o\cot^2(\pi x/a)$ for $0 < x < a$

Problem 3.7 Use the semiclassical approximation to determine the average kinetic energy of a particle in a stationary state.

Problem 3.8 Use the result of the previous problem to determine the average kinetic energy of a particle in the following potentials:

1. $V = mu\omega^2x^2/2$
2. $V = V_o\cot^2(\pi x/a)$ for $0 < x < a$

Problem 3.9 Use the semiclassical approximation to determine the form of the potential $V(x)$ for a given energy spectrum E_n. Assume $V(x)$ to be an even function $V(x) = V(-x)$ increasing monotonically for $x > 0$.

Problem 3.10 Use the Ritz variational principle to show that any purely attractive one-dimensional potential well has at least one bound state.

Problem 3.11 Consider a particle of mass m moving in a potential $\lambda V(x)$ that satisfies the following conditions: For $x < 0$ and $x > a$, $V(x) = 0$ and for $0 \leq x \leq a$,

$$\lambda \int_0^a V(x)dx < 0$$

Show that if λ is small, there exists a bound state with energy

$$E \approx -\frac{m\lambda^2}{2\hbar^2} \left[\int_0^a V(x)dx \right]^2$$

Problem 3.12   Let us revisit the double well tunneling problem by making the following approximation to the tunneling doublet.

$$\psi_\pm = \frac{1}{\sqrt{2}}(\phi_0(x) \pm \phi_0(-x))$$

when $\phi_0(x)$ is a quasi-classical wave function describing motion in the right-hand well.

$$\phi_0'' \frac{2m}{\hbar^2}(V(x) - E_0)\phi_0 = 0.$$

Show that the tunneling splitting between the $\psi_\pm$ states is given by

$$E_- - E_+ = \frac{4\hbar^2}{m}\phi_0(0)\phi_0'(x)$$

4 Quantum Dynamics (and Other Un-American Activities)

Dr. Condon, it says here that you have been at the forefront of a revolutionary movement in physics called *quantum mechanics*. It strikes this hearing that if you could be at the forefront of one revolutionary movement . . . you could be at the forefront of another.

—House Committee on Un-American Activities to
Dr. Edward Condon, 1948

4.1 INTRODUCTION

This chapter is really the heart and soul of this text—not only in a physical sense but also in a scientific sense. In the early days of quantum mechanics and especially chemical physics, we were mostly interested in discerning the energy states or predicting equilibrium structures of a given atomic or molecular system. This provided a good test of quantum theory and deepened our understanding of the nature of the bonding and intermolecular interactions that define a chemical system. With the introduction of time-resolved laser techniques in the 1980s, modern investigations have focused upon pulling apart how an atomic or molecular system undergoes transitions from one state to the next and how the quantum interferences between different pathways influence these transitions. Typically, in a molecular system we treat the electronic degrees of freedom using rigorous quantum theory and allow their energies and states to be parametrized by the instantaneous positions of the nuclei. This is justified through the Born–Oppenheimer approximation, which allows us to separate the fast motion of the electrons from the far slower motions of the nuclei by virtue of their disparity in mass. As we shall see in this chapter, things become interesting when the separation of time scales is no longer valid.

We shall begin with a brief review of the bound states of a coupled two-level system. This is a model problem that captures the essential physics of a wide range of situations. We shall discuss this within first a time-independent perspective and then a time-dependent perspective. Finally, at the end of the chapter we shall discuss what happens when we allow the two-level system to have an additional harmonic degree of freedom that couples the transitions between the two states.

4.2 THE TWO-STATE SYSTEM

A very general problem is to consider two states $\{|1\rangle \& |2\rangle\}$ with energies E_1 and E_2 coupled by some off-diagonal interaction V. We shall refer to these states as the "localized" basis since if we allow V to be parametrized by, say, the spatial separation between the states, then as this distance becomes large, we expect $V \to 0$ and the localized states become the exact states of the system. An elementary example of this is the description of bonding within the hydrogen molecule where the localized states can be taken to be the hydrogenic 1s orbitals localized about each atom center. For the sake of simplicity we assume $\langle 1|2\rangle = 0$.

In the localized basis, we can write the Hamiltonian H as a matrix

$$H = \begin{pmatrix} E_1 & V \\ V & E_2 \end{pmatrix} \tag{4.1}$$

Let us define new energy variables $E_m = (E_1 + E_2)/2$ and $\Delta = (E_1 - E_2)/2$. The Hamiltonian can then be written in these new variables as

$$H = E_m I + \Delta \begin{pmatrix} 1 & \tan(2\theta) \\ \tan(2\theta) & -1 \end{pmatrix} \tag{4.2}$$

where

$$\tan(2\theta) = \frac{V}{\Delta} \tag{4.3}$$

defines a "mixing angle." It is straightforward to determine the energy levels of the coupled system in terms of the mixing angle

$$\varepsilon_{\pm} = E_m + \Delta \sec(2\theta) \tag{4.4}$$

or in terms of the interaction

$$\varepsilon_{\pm} = E_m \mp \sqrt{\Delta^2 + V^2} \tag{4.5}$$

where we have assumed $V < 0$.

The mixing angle θ comes about because H can be brought into diagonal form by introducing the 2×2 rotation matrix

$$T = \begin{pmatrix} \cos(\theta) & \sin(\theta) \\ -\sin(\theta) & \cos(\theta) \end{pmatrix} \tag{4.6}$$

where θ is the mixing angle defined above. What we do is to rotate the initial two orthogonal state vectors, $|1\rangle$ and $|2\rangle$, which lie on a two-dimensional plane, by an angle θ to form new state vectors, $|+\rangle$ and $|-\rangle$,

$$\begin{pmatrix} |+\rangle \\ |-\rangle \end{pmatrix} = \begin{pmatrix} \cos(\theta) & \sin(\theta) \\ -\sin(\theta) & \cos(\theta) \end{pmatrix} \begin{pmatrix} |1\rangle \\ |2\rangle \end{pmatrix} \tag{4.7}$$

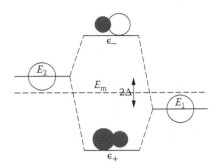

FIGURE 4.1 Energy levels of the two-state system. The superimposed circles are representative of the initial (localized) and final (delocalized) states of the system.

These new states, which we shall term the "delocalized" basis, are linear combinations of the original localized states as illustrated in Figure 4.1. When the energy gap $V \ll \Delta$, $\tan(2\theta) = V/\Delta$ becomes small and $\theta \to 0$. In this limit, the eigenstates become more and more like the initial localized states. In the other limit, as $\Delta \to 0$ and the initial states become degenerate, $\tan(2\theta)$ diverges and $\theta = \pi/4$. In this case, the true eigenstates of the system are the totally delocalized states:

$$|\pm\rangle = \frac{1}{\sqrt{2}}(|1\rangle \pm |2\rangle) \tag{4.8}$$

Let us briefly examine the impact of these two limits on the final energies of the system. From above, the exact energy levels are given by

$$\varepsilon_\pm = E_m \mp \sqrt{\Delta^2 + V^2} \tag{4.9}$$

We can use the binomial theorem to expand the exact energies either in terms of the initial energy gap Δ or in terms of the coupling. When $V/\Delta \ll 1$, then we can expand

$$\varepsilon_\pm = E_m \mp \Delta\sqrt{1 + (V/\Delta)^2}$$

$$= E_m \mp \Delta\left(1 + \frac{1}{2}\frac{V^2}{\Delta^2} + \cdots\right) \tag{4.10}$$

to obtain a lowest-order correction to the energy levels:

$$\varepsilon_+ \approx = E_1 - \frac{1}{2}\frac{V^2}{\Delta} \tag{4.11}$$

$$\varepsilon_- \approx = E_2 + \frac{1}{2}\frac{V^2}{\Delta} \tag{4.12}$$

(Note, we have assumed that $E_1 < E_2$.)

In the opposite limit where $\Delta/V \ll 1$, we pull V out from under the square root and perform the binomial expansion

$$\varepsilon_\pm = E_m \mp V\left(1 + \frac{1}{2}\frac{\Delta^2}{V^2} + \cdots\right) \tag{4.13}$$

In this limit (as $\Delta \to 0$) the exact energies are

$$\varepsilon_\pm = E_1 \mp |V| \qquad (4.14)$$

We can perform a similar analysis on the wave functions. In the weak coupling limit, $V/\Delta \ll 1$; hence, $\theta \approx 0$. Thus, we can expand the coefficients of the $|\pm\rangle$ about $\theta = 0$ to obtain the lowest-order corrections to the states:

$$|+\rangle = \cos(\theta)|1\rangle + \sin(\theta)|2\rangle$$

$$= \left(1 - \frac{\theta^2}{2} + \cdots\right)|1\rangle + (\theta + \cdots)|2\rangle \qquad (4.15)$$

$$\approx |1\rangle + \frac{V}{2\Delta}|2\rangle \qquad (4.16)$$

where we have used $\tan(2\theta) \approx 2\theta$ for small values of θ. Similarly for $|-\rangle$,

$$|-\rangle = \cos(\theta)|2\rangle - \sin(\theta)|1\rangle$$

$$= \left(1 - \frac{\theta^2}{2} + \cdots\right)|2\rangle - (\theta + \cdots)|1\rangle \qquad (4.17)$$

$$\approx |2\rangle - \frac{V}{2\Delta}|1\rangle \qquad (4.18)$$

In short, within the weak coupling limit, both the energies and states resemble their parent uncoupled energies and states.

4.3 PERTURBATIVE SOLUTIONS

One of the most important techniques we will use thoughout this text will be the use of perturbative expansions whereby we consider the response or reaction of some reference system to some sort of applied or additional interaction. In most cases, it is simply impossible to obtain the exact solution to the Schrödinger equation. In fact, the vast majority of problems that are of physical interest cannot be resolved exactly and one is forced to make a series of well-posed approximations. The simplest approximation is to say that the system we want to solve looks a lot like a much simpler system that we can solve with some additional complexity (which hopefully is quite small). In other words, we want to be able to write our total Hamiltonian as

$$H = H_o + V \qquad (4.19)$$

where H_o represents that part of the problem we can solve exactly and V some extra part that we cannot. This we take as a correction or *perturbation* to the exact problem.

Perturbation theory can be formuated in a variery of ways, but we begin with what is typically termed *Rayleigh–Schrödinger* perturbation theory. This is the typical approach and used most commonly. Let $H_o|\phi_n\rangle = W_n|\phi_n\rangle$ and $(H_o + \lambda V)|\psi\rangle = E_n|\psi\rangle$ be the Schrödinger equations for the uncoupled and perturbed systems.

In what follows, we take λ as a small parameter and expand the exact energy in terms of this parameter. Clearly, we write E_n as a function of λ and write

$$E_n(\lambda) = E_n^{(0)} + \lambda E_n^{(1)} + \lambda^2 E_n^{(2)} \dots \qquad (4.20)$$

Likewise, we can expand the exact wave function in terms of λ:

$$|\psi_n\rangle = \left|\psi_n^{(0)}\right\rangle + \lambda\left|\psi_n^{(1)}\right\rangle + \lambda^2\left|\psi_n^{(2)}\right\rangle \dots \qquad (4.21)$$

Since we require that $|\psi\rangle$ be a solution of the exact Hamiltonian with energy E_n, then

$$H|\psi\rangle = (H_o + \lambda V)\left(\left|\psi_n^{(0)}\right\rangle + \lambda\left|\psi_n^{(1)}\right\rangle + \lambda^2\left|\psi_n^{(2)}\right\rangle \dots\right) \qquad (4.22)$$

$$= \left(E_n^{(0)} + \lambda E_n^{(1)} + \lambda^2 E_n^{(2)} \dots\right)\left(\left|\psi_n^{(0)}\right\rangle + \lambda\left|\psi_n^{(1)}\right\rangle + \lambda^2\left|\psi_n^{(2)}\right\rangle \dots\right) \qquad (4.23)$$

Now, we collect terms order by order in λ:

- $\lambda^0 : H_o\left|\psi_n^{(0)}\right\rangle = E_n^{(0)}\left|\psi_n^{(0)}\right\rangle$
- $\lambda^1 : H_o\left|\psi_n^{(1)}\right\rangle + V\left|\psi_n^{(0)}\right\rangle = E_n^{(0)}|\psi^{(1)}\rangle + E_n^{(1)}|\psi^{(0)}\rangle$
- $\lambda^2 : H_o\left|\psi_n^{(2)}\right\rangle + V|\psi^{(1)}\rangle = E_n^{(0)}\left|\psi_n^{(2)}\right\rangle + E_n^{(1)}\left|\psi_n^{(1)}\right\rangle + E_n^{(2)}\left|\psi_n^{(0)}\right\rangle$

and so on.

The λ^0 problem is just the unperturbed problem we can solve. Taking the λ^1 terms and multiplying by $\langle\psi_n^{(0)}|$ we obtain

$$\left\langle\psi_n^{(0)}\left|H_o\right|\psi_n^{(0)}\right\rangle + \left\langle\psi_n^{(0)}|V|\psi^{(0)}\right\rangle = E_n^{(0)}\left\langle\psi_n^{(0)}|\psi_n^{(1)}\right\rangle + E_n^{(1)}\left\langle\psi_n^{(0)}|\psi_n^{(0)}\right\rangle \quad (4.24)$$

In other words, we obtain the first-order correction for the nth eigenstate:

$$E_n^{(1)} = \left\langle\psi_n^{(0)}|V|\psi^{(0)}\right\rangle \qquad (4.25)$$

This is easy to check by performing a similar calculation, except by multiplying by $\langle\psi_m^{(0)}|$ for $m \neq n$ and noting that $\langle\psi_n^{(0)}|\psi_m^{(0)}\rangle = 0$ are orthogonal states.

$$\left\langle\psi_m^{(0)}\left|H_o\right|\psi_n^{(0)}\right\rangle + \left\langle\psi_m^{(0)}|V|\psi^{(0)}\right\rangle = E_n^{(0)}\left\langle\psi_m^{(0)}|\psi_n^{(1)}\right\rangle \qquad (4.26)$$

Rearranging things a bit, one obtains an expression for the overlap between the unperturbed and perturbed states:

$$\left\langle\psi_m^{(0)}|\psi_n^{(1)}\right\rangle = \frac{\left\langle\psi_m^{(0)}|V|\psi_n^{(0)}\right\rangle}{E_n^{(0)} - E_m^{(0)}} \qquad (4.27)$$

Now, we use the resolution of the identity to project the perturbed state onto the unperturbed states:

$$\left|\psi_n^{(1)}\right\rangle = \sum_m \left|\psi_m^{(0)}\right\rangle\left\langle\psi_m^{(0)}|\psi_n^{(1)}\right\rangle$$

$$= \sum_{m \neq n} \frac{\left\langle\psi_m^{(0)}|V|\psi_n^{(0)}\right\rangle}{E_n^{(0)} - E_m^{(0)}}\left|\psi_m^{(0)}\right\rangle \qquad (4.28)$$

where we explictly exclude the $n = m$ term to avoid the singularity. Thus, the first-order correction to the wave function is

$$|\psi_n\rangle \approx |\psi_n^{(0)}\rangle + \sum_{m \neq n} \frac{\langle \psi_m^{(0)}|V|\psi_n^{(0)}\rangle}{E_n^{(0)} - E_m^{(0)}} |\psi_m^{(0)}\rangle \qquad (4.29)$$

This also justifies our assumption above.

4.3.1 Dipole Molecule in Homogenous Electric Field

Here we take the example of ammonia inversion in the presence of an electric field. The NH_3 molecule can tunnel between two equivalent C_{3v} configurations and, as a result of the coupling between the two configurations, the unperturbed energy levels E_o are split by an energy A. Defining the unperturbed states as $|1\rangle$ and $|2\rangle$, we can define the tunneling Hamiltonian as

$$H = \begin{pmatrix} E_o & -A \\ -A & E_o \end{pmatrix} \qquad (4.30)$$

or in terms of Pauli matrices:

$$H = E_o\sigma_o - A\sigma_x \qquad (4.31)$$

Now we apply an electric field. When the dipole moment of the molecule is aligned parallel with the field, the molecule is in a lower energy configuration, whereas for the antiparallel case, the system is in a higher energy configuration. This lifts the degeneracy between the two otherwise equivalent states. We can denote the contribution to the Hamiltonian from the electric field as

$$H' = \mu_e\mathscr{E}\sigma_z \qquad (4.32)$$

The total Hamiltonian in the $\{|1\rangle, |2\rangle\}$ basis is thus

$$H = \begin{pmatrix} E_o + \mu_e\mathscr{E} & -A \\ -A & E_o - \mu_e\mathscr{E} \end{pmatrix} \qquad (4.33)$$

Solving the eigenvalue problem

$$|H - \lambda I| = 0 \qquad (4.34)$$

we find two eigenvalues:

$$\lambda_\pm = E_o \pm \sqrt{A^2 + \mu_e^2\mathscr{E}^2} \qquad (4.35)$$

These are the exact eigenvalues. In Figure 4.2 we show the variation of the energy levels as a function of the field strength.

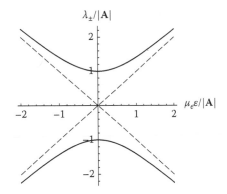

FIGURE 4.2 Variation of energy levels ($\lambda_\pm$) as a function of the applied field for a polar molecule in an electric field.

4.3.1.1 Weak Field Limit

If $\mu_e \mathscr{E}/A \ll 1$, then we can use the binomial expansion

$$\sqrt{1+x^2} \approx 1 + x^2/2 + \cdots \tag{4.36}$$

to write

$$\sqrt{A^2 + \mu_e^2 \mathscr{E}^2} = A \left(1 + \left(\frac{\mu_e \mathscr{E}}{A}\right)^2\right)^{\!\!1}\!\!/2$$

$$\approx A \left(1 + \frac{1}{2}\left(\frac{\mu_e \mathscr{E}}{A}\right)^2\right) \tag{4.37}$$

Thus in the weak field limit, the system can still tunnel between configurations, and the energy splittings are given by

$$E_\pm \approx (E_o \mp A) \mp \frac{\mu_e^2 \mathscr{E}^2}{A} \tag{4.38}$$

To understand this a bit further, let us use perturbation theory in which the tunneling dominates and treat the external field as a perturbing force. The unperturbed Hamiltonian can be diagonalized by taking symmetric and antisymmetric combinations of the $|1\rangle$ and $|2\rangle$ basis functions. This is exactly what we did above with the time-dependent coefficients. Here the stationary states are

$$|\pm\rangle = \frac{1}{\sqrt{2}}(|1\rangle \pm |2\rangle) \tag{4.39}$$

with energies $E_\pm = E_o \mp A$, so in the $|\pm\rangle$ basis, the unperturbed Hamiltonian becomes

$$H = \begin{pmatrix} E_o - A & 0 \\ 0 & E_o + A \end{pmatrix} \tag{4.40}$$

The first-order correction to the ground-state energy is given by

$$E^{(1)} = E^{(0)} + \langle +|H'|+\rangle \tag{4.41}$$

To compute $\langle +|H'|+\rangle$ we need to transform H' from the $\{|1\rangle, |2\rangle\}$ uncoupled basis to the new $|\pm\rangle$ coupled basis. This is accomplished by inserting the identity on either side of H' and collecting terms:

$$\langle +|H'|+\rangle = \langle +|(|1\rangle < 1| + |2\rangle\langle 2|)H'(|1\rangle < 1| + |2\rangle\langle 2|)) \tag{4.42}$$

$$= \frac{1}{2}((\langle 1| + \langle 2|)H'(|1\rangle + |2\rangle)) \tag{4.43}$$

$$= 0 \tag{4.44}$$

This also applies to $\langle -|H'|-\rangle = 0$. Thus, the first-order correction vanishes. However, since $\langle +|H'|-\rangle = \mu_e \mathscr{E}$ does not vanish, we can use second-order perturbation theory to find the energy correction:

$$W_+^{(2)} = \sum_{m \neq i} \frac{H'_{mi} H'_{im}}{E_i - E_m} \tag{4.45}$$

$$= \frac{\langle +|H'|-\rangle\langle -|H'|+\rangle}{E_+^{(0)} - E_-^{(0)}} \tag{4.46}$$

$$= \frac{(\mu_e \mathscr{E})^2}{E_o - A - E_o - A} \tag{4.47}$$

$$= -\frac{\mu_e^2 \mathscr{E}^2}{2A} \tag{4.48}$$

This also applies to $W_-^{(2)} = +\mu_e^2 \mathscr{E}^2/A$. So we get the same variation as we estimated above by expanding the exact energy levels when the field was weak.

Now let us examine the wave functions. Remember the first-order correction to the eigenstates is given by

$$|+^{(1)}\rangle = \frac{\langle -|H'|-\rangle}{E_+ - E_-}|-\rangle \tag{4.49}$$

$$= -\frac{\mu \mathscr{E}}{2A}|-\rangle \tag{4.50}$$

Thus,

$$|+\rangle = |+^{(0)}\rangle - \frac{\mu \mathscr{E}}{2A}|-\rangle \tag{4.51}$$

$$|-\rangle = |-^{(0)}\rangle + \frac{\mu \mathscr{E}}{2A}|+\rangle \tag{4.52}$$

So we see that by turning on the field, we begin to mix the two tunneling states. However, since we have assumed that $\mu \mathscr{E}/A \ll 1$, the final state is not too unlike our initial tunneling states.

4.3.1.2   Strong Field Limit

In the strong field limit, we expand the square-root term such that $\left(\frac{A}{\mu_e\mathscr{E}}\right)^2 \ll 1$

$$\sqrt{A^2 + \mu_e^2\mathscr{E}^2} = \mathscr{E}\mu_e\left(\left(\frac{A}{\mu_e\mathscr{E}}\right)^2 + 1\right)^{1/2}$$

$$= \mathscr{E}\mu_e\left[1 + \frac{1}{2}\left(\frac{A}{\mu_e\mathscr{E}}\right)^2\right]\cdots$$

$$\approx \mathscr{E}\mu_e + \frac{1}{2}\frac{A^2}{\mu_e\mathscr{E}} \tag{4.53}$$

For very strong fields, the first term dominates and the energy splitting becomes linear in the field strength. In this limit, the tunneling has been effectively suppressed.

Let us analyze this limit using perturbation theory. Here we will work in the $|1, 2\rangle$ basis and treat the tunneling as a perturbation. Since the electric field part of the Hamiltonian is diagonal in the $1, 2$ basis, our unperturbed strong-field Hamiltonian is simply

$$H = \begin{pmatrix} E_o - \mu_e\mathscr{E} & 0 \\ 0 & E_o - \mu_e\mathscr{E} \end{pmatrix} \tag{4.54}$$

and the perturbation is the tunneling component. As stated previously, the first-order corrections to the energy vanish and we are forced to resort to second-order perturbation theory to get the lowest-order energy correction. The result is

$$W^{(2)} = \pm\frac{A^2}{2\mu\mathscr{E}} \tag{4.55}$$

which is exactly what we obtained by expanding the exact eigenenergies above. Likewise, the lowest-order correction to the state vectors are

$$|1\rangle = |1^0\rangle - \frac{A}{2\mu\mathscr{E}}|2^0\rangle \tag{4.56}$$

$$|2\rangle = |2^0\rangle + \frac{A}{2\mu\mathscr{E}}|1^0\rangle \tag{4.57}$$

So, for large $\mathscr{E}$ the second-order correction to the energy vanishes, the correction to the wave function vanishes, and we are left with the unperturbed (that is, localized) states. We also find that the perturbative results exactly agree with the series expansion results we obtained above. Thus, perturbative approaches work in the limit that the coupling remains small compared with the energy gap of the unperturbed system.

4.4   DYSON EXPANSION OF THE SCHRÖDINGER EQUATION

The Rayleigh–Schrödinger approach is useful for discrete spectra. However, it is not very useful for scattering or for systems with continuous spectra. On the other hand, the Dyson expansion of the wave function can be applied to both cases. Its development is similar to the Rayleigh–Schrödinger case. We begin by writing the Schrödinger equation as usual:

$$(H_o + V)|\psi\rangle = E|\psi\rangle \tag{4.58}$$

where we define $|\phi\rangle$ and W to be the eigenvectors and eigenvalues of part of the full problem. We shall call this the "uncoupled" problem and assume it is something we can easily solve:

$$H_o|\phi\rangle = W|\phi\rangle \tag{4.59}$$

We want to write the solution of the fully coupled problem in terms of the solution of the uncoupled problem. First we note that

$$(H_o - E)|\psi\rangle = V|\psi\rangle \tag{4.60}$$

Using the "uncoupled problem" as a "homogeneous" solution and the coupling as an inhomogeneous term, we can solve the Schrödinger equation and obtain $|\psi\rangle$ *exactly* as

$$|\psi\rangle = |\phi\rangle + \frac{1}{H_o - E}V|\psi\rangle \tag{4.61}$$

This may seem a bit circular. But we can iterate the solution:

$$|\psi\rangle = |\phi\rangle + \frac{1}{H_o - E}V|\phi\rangle + \frac{1}{H_o - E}V\frac{1}{H_o - W}V|\psi\rangle \tag{4.62}$$

Taking this out to all orders, one obtains:

$$|\psi\rangle = |\phi\rangle + \sum_{n=1}^{\infty}\left(\frac{1}{H_o - E}V\right)^n|\phi\rangle \tag{4.63}$$

Assuming that the series converges rapidly (true for $V \ll H_o$ weak coupling case), we can truncate the series at various orders and write

$$|\psi^{(0)}\rangle = |\phi\rangle \tag{4.64}$$

$$|\psi^{(1)}\rangle = |\phi\rangle + \left(\frac{1}{H_o - E}V\right)|\phi\rangle \tag{4.65}$$

$$|\psi^{(2)}\rangle = |\psi^{(1)}\rangle + \left(\frac{1}{H_o - E}V\right)^2|\phi\rangle \tag{4.66}$$

and so on. Let us look at $|\psi^{(1)}\rangle$ for a moment. We can insert one in the form of $\sum_n |\phi_n\rangle\langle\phi_n|$:

$$|\psi_n^{(1)}\rangle = |\phi_n\rangle + \sum_n \left(\frac{1}{H_o - W_m}\right)|\phi_m\rangle\langle\phi_n|V|\phi_m\rangle \tag{4.67}$$

that is,

$$\left|\psi_n^{(1)}\right\rangle = |\phi_n\rangle + \sum_n \left(\frac{1}{W_n - W_m}\right) |\phi_m\rangle\langle\phi_n|V|\phi_m\rangle \tag{4.68}$$

Likewise,

$$\left|\psi_n^{(2)}\right\rangle = \left|\psi_n^{(1)}\right\rangle + \sum_{lm} \left(\frac{1}{(W_m - W_l)(W_n - W_m)}\right)^2 V_{lm}V_{mn}|\phi_n\rangle \tag{4.69}$$

where

$$V_{lm} = \langle\phi_l|V|\phi_m\rangle \tag{4.70}$$

is the matrix element of the coupling in the uncoupled basis. These last two expressions are the first- and second-order corrections to the wave function.

Note that we can actually solve the perturbation series exactly by noting that the series has the form of a geometric progression, for $x < 1$ converge uniformly to

$$\frac{1}{1-x} = 1 + x + x^2 + \cdots = \sum_{n=0}^{\infty} x^n \tag{4.71}$$

Thus, we can write

$$|\psi\rangle = \sum_{n=0}^{\infty} \left(\frac{1}{H_o - E}V\right)^n |\phi\rangle \tag{4.72}$$

$$= \sum_{n=0}^{\infty} (G_o V)^n |\phi\rangle \tag{4.73}$$

$$= \frac{1}{1 - G_o V}|\phi\rangle \tag{4.74}$$

where $G_o = (H_o - E)^{-1}$ (this is the "time-independent" form of the propagator for the uncoupled system). This particular analysis is particularly powerful in deriving the propagator for the fully coupled problem.

We now calculate the first-order and second-order corrections to the energy of the system. To do so, we make use of the wave functions we just derived and write

$$E_n^{(1)} = \left\langle\psi_n^{(0)}\big|H\big|\psi_n^{(0)}\right\rangle = W_n + \langle\phi_n|V|\phi_n\rangle = W_n + V_{nn} \tag{4.75}$$

So the lowest-order correction to the energy is simply the matrix element of the perturbation in the uncoupled or unperturbed basis. That was easy. What about the next order corrections? Using the same procedure as previously (assuming the states are normalized)

$$E_n^{(2)} = \left\langle\psi_n^{(1)}\big|H\big|\psi_n^{(1)}\right\rangle$$

$$= \langle\phi_n|H|\phi_n\rangle$$

$$+ \sum_{m\neq n}\langle\phi_n|H|\phi_m\rangle \left(\frac{1}{W_n - W_m}\right)\langle\phi_m|V|\phi_n\rangle + \mathcal{O}[V^3]$$

$$= W_n + V_{nn} + \sum_{m\neq n} \frac{|V_{nm}|^2}{W_n - W_m} \tag{4.76}$$

Notice that I am avoiding the case where $m = n$ as that would cause the denominator to be zero, leading to an infinity. This must be avoided. The "degenerate case" must be handled via explicit matrix diagonalization. Closed forms can be obtained for the doubly degenerate case easily. Also note that the successive approximations to the energy require one less level of approximation to the wave function. Thus, second-order energy corrections are obtained from first-order wave functions.

4.4.1 Van der Waals Forces: Origin of Long-Range Attractions

One of the underlying principles in chemistry is that molecules at long range are attracted toward each other. This is clearly true for polar and oppositely charged species. It is also true for nonpolar and neutral species, such as methane, noble gases, and so on. These forces are due to polarization forces or van der Waals forces, which are attractive and decrease as $1/R^7$; that is, the attractive part of the potential goes as $-1/R^6$. In this section we will use perturbation theory to understand the origins of this force, restricting our attention to the interaction between two hydrogen atoms separated by some distance R.

Let us take the two atoms to be motionless and separated by distance R with $\vec{n}$ being the vector pointing from atom A to atom B. Now let $\vec{r}_a$ be the vector connecting nuclei A to its electron and likewise for $\vec{r}_b$. Thus each atom has an *instantaneous* electric dipole moment

$$\vec{\mu}_a = q\vec{R}_a \tag{4.77}$$

$$\vec{\mu}_b = q\vec{R}_b \tag{4.78}$$

We will assume that $R \gg r_a$ & r_b so that the electronic orbitals on each atom do not come into contact.

Atom A creates an electrostatic potential U for atom B in which the charges in B can interact. This creates an interaction energy W. Since both atoms are neutral, the most important source for the interactions will come from the dipole–dipole interactions. Thus, the dipole of A interacts with an electric field $E = -\nabla U$ generated by the dipole field about B and vice versa. To calculate the dipole–dipole interaction, we start with the expression for the electrostatic potential created by μ_a at B,

$$U(R) = \frac{1}{4\pi\varepsilon_o} \frac{\mu_a \cdot R}{R^3} \tag{4.79}$$

Thus,

$$\vec{E} = -\nabla U = -\frac{q}{4\pi\varepsilon_o} \frac{1}{R^3} \left(\vec{r}_a - 3(\vec{r}_a \cdot \vec{n})\vec{n} \right) \tag{4.80}$$

Thus, the dipole–dipole interaction energy is

$$W = -\vec{\mu}_b \cdot \vec{E}$$

$$= \frac{e^2}{R^3} \left(\vec{r}_a \cdot \vec{r}_b - 3(\vec{r}_a \cdot \vec{n})(\vec{r}_b \cdot \vec{n}) \right) \tag{4.81}$$

where $e^2 = q^2/4\pi\varepsilon_o$. Now, we set the z axis to be along $\vec{n}$, so we can write

$$W = \frac{e^2}{R^3}(x_a x_b + y_a y_b - 2z_a z_b) \tag{4.82}$$

This will be our perturbing potential that we add to the total Hamiltonian:

$$H = H_a + H_b + W \tag{4.83}$$

where H_a are the unperturbed Hamiltonians for the atoms. Let us take, for example, the interaction between two hydrogen atoms each in the 1s (ground) state. The unperturbed system has energy

$$H|1s_1; 1s_2\rangle = (E_1 + E_2)|1s_1; 1s_2\rangle = -2E_I|1s_1; 1s_2\rangle \tag{4.84}$$

where E_I is the ionization energy of the hydrogen 1s state ($E_I = 13.6$ eV). The first order vanishes because it involves integrals over odd functions. This we can anticipate since the 1s orbitals are spatially isotropic, so the time-averaged value of the dipole moments is zero. So, we have to look toward second-order corrections. The second-order energy correction is

$$E^{(2)} = \sum_{nlm} \sum_{n'l'm'} \frac{|\langle nlm; n'l'm'|W|1s_a; 1s_b\rangle|^2}{-2E_I - E_n - E_{n'}} \tag{4.85}$$

where we restrict the summation to avoid the $|1s_a; 1a_b\rangle$ state. Since $W \propto 1/R^3$ and the denominator is negative, we can write

$$E^{(2)} = -\frac{C}{R^6} \tag{4.86}$$

which explains the origin of the $1/R^6$ attraction.

Now we evaluate the proportionality constant C. Written explicitly,

$$C = e^4 \sum_{nml} \sum_{n'l'm'} \frac{|\langle nlm'n'l'm'|(x_a x_b + y_a y_b - 2z_a z_b)|1s_a; 1s_b\rangle|^2}{2E_I + E_n + E_{n'}} \tag{4.87}$$

Since n and $n' \geq 2$ and $|E_n| = E_I/n^2 < E_I$, we can replace E_n and $E_{n'}$ with 0 without appreciable error. Now, we can use the resolution of the identity

$$1 = \sum_{nml} \sum_{n'l'm'} |nlm; n'l'm'\rangle\langle nlm; n'l'm'| \tag{4.88}$$

to remove the summation and we get

$$C = \frac{e^4}{2E_I}\langle 1s_a; 1a_b|(x_a x_b + y_a y_b - 2z_a z_b)^2|1s_a; 1s_b\rangle \tag{4.89}$$

where E_I is the ionization potential of the 1s state ($E_I = 1/2$). Surprisingly, this is simple to evaluate because we can use symmetry to our advantage. Since the 1s orbitals are spherically symmetric, any terms involving cross-terms of the sort

$$\langle 1s_a|x_a y_a|1s\rangle = 0 \tag{4.90}$$

vanish. This leaves only terms of the sort

$$\langle 1s | x^2 | 1s \rangle \tag{4.91}$$

all of which are equal to 1/3 of the mean value of $R_A = x_a^2 + y_a^2 + z_a^2$. Thus,

$$C = 6 \frac{e^2}{2E_I} \left| \left\langle 1s \left| \frac{R}{3} \right| 1s \right\rangle \right|^2 = 6e^2 a_o \tag{4.92}$$

where a_o is the Bohr radius. Thus,

$$E^{(2)} = -6e^2 \frac{a_o}{R^6} \tag{4.93}$$

What does all this mean? We stated at the beginning that the *average* dipole moment of a H 1s atom is zero. That does not mean that every single measurement of μ_a will yield zero. What it means is that the probability of finding the atom with a dipole moment μ_a is the same for finding the dipole vector pointed in the opposite direction. Adding the two together produces a net zero dipole moment. So it is the fluctuation about the mean that gives the atom an instantaneous dipole field. Moreover, the fluctuations in A are independent of the fluctuations in B, so first-order effects must be zero since the average interaction is zero.

Just because the fluctuations are independent does not mean they are not correlated. Consider the field generated by A as felt by B. This field is due to the fluctuating dipole at A. This field induces a dipole at B. This dipole field is in turn felt by A. As a result, the fluctuations become correlated and explain why this is a second-order effect. In a sense, A interacts with its own dipole field through "reflection" off B.

4.4.2 Attraction between an Atom and a Conducting Surface

The interaction between an atom or molecule and a surface is a fundamental physical process in surface chemistry. In this example, we will use perturbation theory to understand the long-range attraction between an atom, again taking a H 1s atom as our species for simplicity, and a conducting surface. We will take the z axis to be normal to the surface and assume that the atom is high enough off the surface that its altitude is much larger than its atomic dimensions. Furthermore, we will assume that the surface is a metal conductor and we will ignore any atomic level of detail on the surface. Consequently, the atom can only interact with its dipole image on the opposite side of the surface.

We can use the same dipole–dipole interaction as previously with the following substitutions

$$e^2 \longrightarrow -e^2 \tag{4.94}$$

$$R \longrightarrow 2d \tag{4.95}$$

$$x_b \longrightarrow x_a' = x_a \tag{4.96}$$

$$y_b \longrightarrow y_a' = y_a \tag{4.97}$$

$$z_b \longrightarrow z_a' = -z_a \tag{4.98}$$

where the sign change reflects the sign difference in the image charges. So we get

$$W = -\frac{e^2}{8d^3}\left(x_a^2 + y_a^2 + 2z_a^2\right) \tag{4.99}$$

as the interaction between a dipole and its image. Taking the atom to be in the 1s ground state, the first-order term is nonzero:

$$E^{(1)} = \langle 1s|W|1s\rangle. \tag{4.100}$$

Again, using spherical symmetry to our advantage,

$$E^{(1)} = -\frac{e^2}{8d^3}4\langle 1s|r^2|1s\rangle = -\frac{e^2 a_o^2}{2d^3} \tag{4.101}$$

Thus an atom is attracted to the wall with an interaction energy that varies as $1/d^3$. This is a first-order effect since there is perfect correlation between the two dipoles.

4.5 TIME-DEPENDENT SCHRÖDINGER EQUATION

Our discussion of time-dependent quantum mechanics begins with a brief overview of the time-dependent Schrödinger equation (TDSE) that governs the time evolution of the system:

$$i\hbar \frac{\partial}{\partial t}|\psi(t)\rangle = H|\psi(t)\rangle \tag{4.102}$$

where H is the Hamiltonian operator for the system. If we assume that H is independent of time and $\psi(r, t)$ can be separated into time-dependent and time-independent (spatial or otherwise) components

$$\psi(r, t) = \phi(r)f(t) \tag{4.103}$$

then

$$i\hbar \frac{1}{f(t)} \frac{\partial}{\partial t} f(t) = \frac{1}{\phi(r)} H\phi(r) \tag{4.104}$$

Since the left-hand side is a function of only t and the right-hand side is a function of only r, both sides must be equal to the same constant, E. Hence,

$$H\phi(r) = E\phi(r) \tag{4.105}$$

which is the *time-independent* Schrödinger equation. Turning to the other equation,

$$i\hbar \frac{1}{f(t)} \frac{\partial}{\partial t} f(t) = E \tag{4.106}$$

Solving this yields

$$f(t) = A\exp(-i\omega t) \tag{4.107}$$

where $\omega = E/\hbar$ is the (angular) frequency. Functions $\phi_n(r)$ satisfying Equation 4.105 are eigenfunctions of H with eigenvalue E_n. Consequently, the complete wave function can be written as

$$\psi(r, t) = e^{-i\omega_n t}\phi_n(r) \qquad (4.108)$$

This is *stationary* because the probability density $P(r)|\psi(r, t)|^2$ is independent of time. More generally, since the complete set of eigenfunctions forms a suitable space for representing any arbitrary function, Ψ,

$$\Psi(r) = \sum_m \langle \phi_m | \Psi \rangle \phi_m(r) \qquad (4.109)$$

it is not stationary since

$$i\hbar \frac{\partial}{\partial t} \Psi = \sum_m \langle \phi_m | \Psi \rangle e^{-i\omega_m t} \phi_m(r) \qquad (4.110)$$

leads to a probability distribution that evolves in time.

The formal solution of the time-dependent Schrödinger equation is given by

$$\psi(t) = e^{-iHt/\hbar}\psi(0) = U(t, 0)\psi(0) \qquad (4.111)$$

where $\psi(0)$ is the state at time $t = 0$ and $\psi(t)$ is the state evolved forward in time. The operator $U(t, 0)$ is the time-evolution operator. It is formally defined via the expansion

$$U(t, 0) = 1 - \frac{it}{\hbar}H + \frac{1}{2!}\left(\frac{i}{\hbar}H\right)^2 - \cdots \qquad (4.112)$$

The time-evolution operator has a number of useful properties:

1. It is unitary, $I = U^\dagger U$ and $U(t', t) = U^*(t, t') = U^\dagger(t, t')$
2. U obeys the semigroup property that $U(t, t_o) = U(t, t')U(t', t_o)$ for $t \geq t' \geq t_o$
3. U itself is a solution of the time-dependent Schrodinger equation

$$i\hbar \frac{\partial}{\partial t}U(t, t') = HU(t, t') \qquad (4.113)$$

4. $U(t, t') = U(t - t')$
5. $U(0) = 1$

Notice that U is a polynomial function of an operator H. Hence if we know the eigenvectors and eigenvalues of H, we know that $f(H)\phi_n = f(a_n)\phi_n$. Thus, we can write U in an eigenbasis representation as

$$U = \sum_n e^{-i\omega_n}|\phi_n\rangle\langle\phi_n| \qquad (4.114)$$

This form is especially convenient when we have at hand the eigenvalues and eigenvectors of the system.

4.6 TIME EVOLUTION OF A TWO-LEVEL SYSTEM

As in the time-independent case, a great deal can be learned by examining what happens to a two-state system subject to some sort of coupling. Again, we write our Hamiltonian in the $\{|1\rangle, |2\rangle\}$ basis

$$H = \begin{pmatrix} E_1 & V \\ V & E_2 \end{pmatrix} \qquad (4.115)$$

and this time we consider the solutions of the time-dependent Schrödinger equation:

$$i\hbar \frac{\partial}{\partial t}|\psi\rangle = H|\psi\rangle \qquad (4.116)$$

We can write $|\psi\rangle$ in terms of either the $|\pm\rangle$ eigenstates of H or in terms of the localized basis states

$$|\psi\rangle = c_1(t)|1\rangle + c_2(t)|2\rangle$$
$$= c_+(t)|+\rangle + c_-(t)|-\rangle \qquad (4.117)$$

where the c's are time-dependent coefficients. Either representation will work and we can transform between the two easily enough.

From our discussion above, the time evolution of $|\psi\rangle$ is generated by

$$|\psi(t)\rangle = U(t, 0)|\psi(0)\rangle \qquad (4.118)$$

We can write the time-evolution operator in terms of the eigenstates:

$$U(t, 0) = \begin{pmatrix} e^{-i\omega_+ t} & 0 \\ 0 & e^{-i\omega_- t} \end{pmatrix} \qquad (4.119)$$

where $\omega_\pm = \varepsilon_\pm/\hbar$. If our initial state, however, is one of the states of the uncoupled system, say $|\psi(0)\rangle = |1\rangle$, then we need to write this in terms of the $|\pm\rangle$. Using the rotation matrix above,

$$|1\rangle = \cos\theta|+\rangle - \sin\theta|-\rangle \qquad (4.120)$$

Thus, our time-evolved state is

$$|\psi(t)\rangle = U(t, 0)|1\rangle = e^{-i\omega_1 t}\cos\theta|+\rangle - e^{-i\omega_2 t}\sin\theta|-\rangle \qquad (4.121)$$

We now ask, what is the probability that at time $t > 0$ the system will be found in the other state? It is straightforward to show that

$$P_{12}(t) = \frac{V^2}{V^2 + \Delta^2}\sin^2\omega_R t \qquad (4.122)$$

where $\omega_R = \sqrt{\Delta^2 + V^2}/\hbar$ is the Rabi frequency, which gives the frequency at which the system oscillates between the two states.

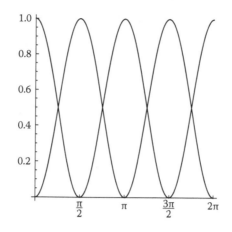

FIGURE 4.3 Rabi oscillation between two degenerate states starting off in $|1\rangle$.

In Figure 4.3 we show the $P_{12}(t)$ and $P_{11}(t)$ for the case of a degenerate system $\Delta = 0$. Here the $\omega_R = V/\hbar$ and the system oscillates between the two localized states with a period $\tau = \pi/\omega_R$. In other words, if we prepare the system in state $|1\rangle$ then for every $t = n\tau$ we have a 100% likelihood of finding the system in state 1 and for every $t = n\tau/2$ we have a 100% likelihood of finding the system in state 2. The amount of amplitude transferred depends upon both the coupling V and the energy gap Δ. For the degenerate case, $\Delta = 0$ and 100% of the initial population in 1 is transferred to 2 and back every π/ω_R. For nondegenerate cases, a maximum of $V^2/(V^2 + \Delta^2)$ is transferred every Rabi period. Ultimately, in the weak coupling limit, the population remains localized in the the initial state.

4.7 TIME-DEPENDENT PERTURBATIONS

Time-dependent perturbation theory is a powerful tool for deriving approximate solutions and theories whenever the complete solution is impossible to obtain and the coupling between the unperturbed system and coupling is weak. One of the advantages of the perturbative approach is that it allows us to discuss quantum transitions as a series of state-to-state interactions followed by free propagation of the unperturbed system by writing

$$H = H_o + \lambda V(t) \tag{4.123}$$

where H_o represents the Hamiltonian for the uncoupled system and λV is some coupling. We begin by writing the state in the basis of unperturbed states as

$$\psi(t)\rangle = \sum_n c_n(t)|\phi_n\rangle \tag{4.124}$$

where the expansion coefficients $c_n(t) = \langle \psi(t)|\phi_n\rangle$ are simply the projection of the evolving state onto the unperturbed basis. In this representation, the time-dependent equation for the coefficients is given by

$$i\hbar \dot{c}_n(t) = \varepsilon_n c_n(t) + \lambda \sum_n V_{nm}(t) c_m(t) \tag{4.125}$$

where $V_{nm}(t) = \langle \phi_n | V(t) | \phi_m \rangle$ is the matrix element of the coupling in the ϕ_n basis. As such, this is a set of coupled linear differential equations to first order in time, and, in principle at least, we can determine the coefficients for the time-evolved state. The coupling between the equations comes from the fact that the operator $V(t)$ is nondiagonal in this basis representation. When $\lambda V = 0$, our system of equations becomes totally decoupled and the solutions are simply

$$c_n(t) = b_n e^{-i\varepsilon_n t/\hbar} \tag{4.126}$$

where b_n depends entirely upon the choice of initial condition. We can also make a simple change of variables by writing the general solution for the coefficients as

$$c_n(t) = b_n(t) e^{-i\varepsilon_n t/\hbar} \tag{4.127}$$

and determine the equations that govern the evolution of the new coefficients. The advantage here is that this will eliminate any rapidly evolving phase terms $e^{-i\varepsilon_n t/\hbar}$, and we expect the $b_n(t)$ to be slowly varying functions of time. Upon substitution into the TDSE and introducing the Bohr frequency $\omega_{nm} = (\varepsilon_n - \varepsilon_m)/\hbar$,

$$i\hbar \dot{b}_n(t) = \lambda \sum_m V_{nm}(t) e^{-i\omega_{nm} t} b_m(t) \tag{4.128}$$

Again, this is a system of linear equations first order in time, but the $b_n(t)$ coefficients are now slowly varying in time.

Next, let us expand the $b_n(t) = b_n^{(0)}(t) + \lambda^1 b_n^{(1)}(t) + \lambda^2 b_n^{(2)}(t) + \cdots$ and substitute this into Equation 4.128. Equating terms on each side to the equation with equal orders in λ^α, one finds for λ^0

$$i\hbar \dot{b}_n^{(0)}(t) = 0 \tag{4.129}$$

However, for $\alpha > 0$

$$i\hbar \dot{b}_n^{(\alpha+1)}(t) = \sum_m e^{-i\omega_{nm} t} V_{nm}(t) b_m^{(\alpha)}(t) \tag{4.130}$$

one obtains a recursive solution whereby lower-order solutions serve as input to the next higher-order term.

Suppose at time $t = 0$ our initial state was prepared in the state $|\phi_i\rangle$. Hence, $b_i(t = 0) = 1$ and all other $b_{i \neq 1}(0) = 0$ (that is, $b_n(t = 0) = \delta_{ni}$). Prior to $t = 0$, we assume that the interaction is turned off $V(t < 0) = 0$ and at time $t = 0$, it is instantly switched on (but remains finite). To first order in the perturbation series,

$$i\hbar \dot{b}_n^{(1)}(t) = e^{-i\omega_{ni} t} V_{ni}(t) b_i^{(0)}(t) \tag{4.131}$$

This can be easily integrated

$$b_n^{(1)}(t) = \frac{1}{i\hbar} \int_0^t dt' V_{ni}(t') e^{i\omega_{ni} t'} \tag{4.132}$$

to give the first-order probability amplitude for starting in state i and finding the system in state n some time t later. Notice that this is the partial Fourier transform of the coupling operator. (Partial in the sense that we integrate only out to intermediate times.) The transition probability is then found by $P_{ni}(t) = |b_n(t)|^2$.

4.7.1 Harmonic Perturbation

Take, for example, a harmonic perturbation $V_{ni}(t) = V_{ni}\sin(\omega t)$. Integrating Equation 4.132 and writing the corresponding transition probability yields

$$P_{ni}(t) = \frac{|V_{ni}|^2}{4\hbar^2} \left| \frac{1 - e^{i(\omega_{ni} - \omega)t}}{\omega_{ni} - \omega} + \frac{1 - e^{i(\omega_{ni} + \omega)t}}{\omega_{ni} + \omega} \right|^2 \tag{4.133}$$

In the limit of a slowly varying perturbation, ω can be set to zero and we find

$$P_{ni}(t) = \frac{|V_{ni}|^2}{\hbar^2} \frac{4\sin^2(\omega_{ni}t/2)}{\omega_{ni}^2} \tag{4.134}$$

that is,

$$P_{ni}(t) = \frac{4|V_{ni}|^2}{\hbar^2} f(t, \omega_{ni}) \tag{4.135}$$

where $f(t, \omega_{ni})$ is shown in Figure 4.4 as a function of the transition frequency ω_{ni} for fixed t. Notice that $f(t, \omega_{ni})$ has a sharp peak at $\omega_{ni} = 0$ with a height proportional to $t^2/4$ while its width (at half maximum) is given by $2\pi/t$. A straightforward application of the residue theorem indicates that

$$\int_{-\infty}^{\infty} d\omega f(t, \omega) = \frac{t\pi}{2} \tag{4.136}$$

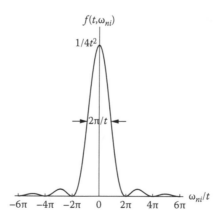

FIGURE 4.4 $f(t, \omega_{ni})$, a function of transition frequency for fixed t.

so that the total area under the curve is proportional to t. Also, one has

$$\lim_{t \to \infty} f(t, \omega_{ni}) \propto \frac{\pi t}{2} \delta(\omega) \qquad (4.137)$$

This tells us that transitions occur mainly between states whose final energies E_n do not differ from the initial energy E_i by more than

$$\delta E = 2\pi\hbar/t \qquad (4.138)$$

hence, energy is *approximately* conserved with the spread in energy given by $2\pi\hbar/t$.

We can relate this result with the so-called time-energy uncertainty relationship $\delta E \delta t \geq \hbar$. In a sense, the perturbation is akin to making a measurement of the energy of the system by inducing a transition from the initial state i to the final state n. Since the time associated with making this observation is t, the associated uncertainty with the observation should be approximately $\hbar/t$, which is good agreement with the estimate above.

For processes that occur to a *continuum* of final states whose energies lie within a given interval $(E_f - \varepsilon, E_f + \varepsilon)$ about some central energy E_f, it should be apparent from this discussion that we need to consider transitions from the initial state to groups of states. Let us denote by $\rho_f(E_f)$ the density of levels so that $\rho_f(E_f)dE_f$ is the number of states with energy levels in the interval $(E_f, E_f + dE_f)$. Integrating our result for a single state over a continuum of final states yields

$$P_{fi}(t) = \int_{E_f - \varepsilon}^{E_f + \varepsilon} dE_f P_{fi}(t, E_f) \rho_f(E_f) \qquad (4.139)$$

If we assume that both $|V_{fi}|$ and $\rho_f(E_f)$ are slowly varying over a narrow integration range,

$$P_{fi}(t) = \frac{4}{\hbar} |V_{fi}|^2 \rho_f(E_f) \int_{\omega_{fi} - \varepsilon'}^{\omega_{fi} + \varepsilon'} f(t, \omega_{fi}) d\omega_{fi} \qquad (4.140)$$

In this last expression, we make a change of integration variable to ω_{fi} and $\varepsilon' = \varepsilon/\hbar$. Clearly, the overwhelming contribution to the integral here comes from those transitions that *do in fact conserve energy* (within $\delta E \approx 2\pi/t$). Moreover, if we set the range of integration so that $\varepsilon' \gg 2\pi/t$ (that is, a long enough time), then the entire peak will fall within the integration range and all transitions will conserve energy.

Since $\varepsilon' \gg 2\pi/t$, we can take the limits of integration out to infinity and we arrive at

$$P_{fi}(t) = \frac{2\pi}{\hbar} |V_{fi}|^2 \rho_f(E_f) t \qquad (4.141)$$

with $E_f = E_i$.

We are now in the position to write perhaps the most physically interesting and important result of this discussion, namely, the transition probability per unit time:

$$k_{fi} = \frac{d}{dt} P_{fi}(t) \qquad (4.142)$$

that is,

$$k_{fi} = \frac{2\pi}{\hbar}|V_{fi}|^2 \rho_f(E) \qquad (4.143)$$

This is often referred to as Fermi's *Golden Rule*[4] (even though it was first obtained by Paul Dirac)[5] since it plays an important role in many physical processes.

4.7.2 CORRELATION FUNCTIONS

We can gain some deeper understanding of the transition rate by doing some further analysis. For this, let us write the golden rule expression as

$$k_{fi} = \frac{2\pi}{\hbar}|\langle f|V|i\rangle|^2 \delta(E_f - E_i) \qquad (4.144)$$

This is the expression for the transition rate that we previously derived for the limiting case where the external field varied slowly with time. It is applicable when the initial and final energies are the same. For nondegenerate systems, we obtain for the transition between states 1 and 2

$$k_{21} = \frac{2\pi}{\hbar}|\langle 2|V|1\rangle|^2 \delta(E_2 - E_1 - \hbar\omega) \qquad (4.145)$$

corresponding to a transition from the initial to final state that involves the absorption of a quantum of energy $\hbar\omega$. We can also write the transition rate for the reverse $2 \rightarrow 1$ process as

$$k_{12} = \frac{2\pi}{\hbar}|\langle 1|V|2\rangle|^2 \delta(E_1 - E_2 + \hbar\omega) \qquad (4.146)$$

Because we are dealing with Hermitian operators, $|\langle 1|V|2\rangle|^2 = |\langle 2|V|1\rangle|^2$, we can conclude that

$$k_{12} = k_{21} \qquad (4.147)$$

This is an example of *microscopic reversibility* and stems from the fact that our equations of motion are symmetric in time.

In general, however, we rarely encounter isolated systems. Typically in chemical dynamical systems we deal with an ensemble of identically prepared systems. Thus, to compute a statistical transition rate for the ensemble we need to sum over all initial conditions, weighted by their respective Boltzmann probability, and sum over all possible final states. Let us write this as

$$P(\omega) = \sum_{f,i} k_{fi}(\omega)w_i \qquad (4.148)$$

where

$$w_i = \frac{e^{-\beta E_i}}{Z} \qquad (4.149)$$

is the canonical density. Next, let us write the microscopic rate as

$$k_{fi} = \frac{2\pi}{\hbar} |\mathscr{F}(\omega)|^2 |\langle f|B|i\rangle|^2 \delta(E_f - E_i - \hbar\omega) \tag{4.150}$$

where we have written the time-dependent driving field in terms of a frequency spectrum $\mathscr{F}(\omega)$ and an operator coupling the initial and final states. Thus, we can write $P(\omega)$ as

$$P(\omega) = \frac{2\pi}{\hbar} |\mathscr{F}(\omega)|^2 \sum_{f,i} w_i |\langle f|B|i\rangle|^2 \delta(E_f - E_i - \hbar\omega) \tag{4.151}$$

for an absorption process where $E_f = E_i + \hbar\omega$ and

$$P(-\omega) = \frac{2\pi}{\hbar} |\mathscr{F}(\omega)|^2 \sum_{f,i} w_i |\langle f|B|i\rangle|^2 \delta(E_f - E_i + \hbar\omega) \tag{4.152}$$

for the emission process where $E_f = E_i - \hbar\omega$. In order to relate these two expressions, let us now assume that $E_f > E_i$ so that i and f now serve as state indices rather than simply referring to the initial and final states. Thus, the sum over initial states in $P(-\omega)$ is really a sum over f's, so we need to write

$$w_f = \frac{e^{-\beta E_f}}{Z} = \frac{e^{-\beta(E_i + \hbar\omega)}}{Z} = e^{-\beta\hbar\omega} w_i \tag{4.153}$$

It should be clear that the emission and absorption rates are related by

$$P(-\omega) = e^{-\beta\hbar\omega} P(+\omega) \tag{4.154}$$

In essence, the rate for stimulated emission is statistically lower than that for stimulated absorption. It makes sense because at thermal equilibrium, we are less likely to find the system in a higher energy state than in a lower energy state. The relation also assumes that the transition occurs from a thermally populated distribution of initial states at time $t = 0$. Since $P(\omega) > P(-\omega)$, we have lost microscopic reversibility. This is the principle of *detailed balance*. Reversibility is lost the moment we place the system in contact with a thermal bath.

Let us consider the part of $P(\omega)$ that only involves the summations:

$$C_>(\omega) = \sum_{f,i} w_i |\langle f|B|i\rangle|^2 \delta(E_f - E_i - \hbar\omega) \tag{4.155}$$

and

$$C_<(\omega) = \sum_{f,i} w_i |\langle f|B|i\rangle|^2 \delta(E_f - E_i + \hbar\omega) \tag{4.156}$$

so that

$$P(\omega) = \frac{2\pi}{\hbar} |\mathscr{F}(\omega)|^2 C_>(\omega) \tag{4.157}$$

and

$$P(-\omega) = \frac{2\pi}{\hbar} |\mathscr{F}(\omega)|^2 C_<(\omega) \tag{4.158}$$

Clearly from the discussion above, $C_<(\omega) = \exp(-\beta\hbar\omega)C_>(\omega)$.

For the moment, consider just $C_>(\omega)$ and recast this using the integral form of $\delta(E)$:

$$\delta(E) = \frac{1}{2\pi\hbar} \int_{-\infty}^{\infty} dt\, e^{-iEt/\hbar} \tag{4.159}$$

Using this we can write

$$C_>(\omega) = \int_{-\infty}^{\infty} dt \sum_{i,f} w_i |B_{if}|^2 e^{i(E_f - E_i - \hbar\omega)t/\hbar} \tag{4.160}$$

Now, we break up the matrix element

$$C_>(\omega) = \int_{-\infty}^{\infty} dt\, e^{i\omega t} \sum_{i,f} w_i \langle i|B|f\rangle \langle f|B|i\rangle e^{i(E_f - E_i)t/\hbar} \tag{4.161}$$

and use

$$\langle i|B|f\rangle e^{i(E_f - E_i)t/\hbar} = \langle i|e^{+iHt/\hbar} B e^{-iHt/\hbar}|f\rangle \tag{4.162}$$

to write this as

$$C_>(\omega) = \int_{-\infty}^{\infty} dt\, e^{i\omega t} \sum_{i,f} w_i \langle i|e^{+iHt/\hbar} B e^{-iHt/\hbar}|f\rangle \langle f|B|i\rangle \tag{4.163}$$

We can now eliminate the sum over the final states since this is simply a resolution of the identity

$$C_>(\omega) = \int_{-\infty}^{\infty} dt\, e^{i\omega t} \sum_{i} w_i \langle i|B(t)B(0)|i\rangle \tag{4.164}$$

Lastly, we can condense our notation by letting the sum over the initial conditions be written as the trace over the thermal density

$$\sum_{i} w_i \langle i|B(t)B(0)|i\rangle = Tr[\rho e^{+iHt/\hbar} B e^{-iHt/\hbar} B(0)] = \langle B(t)B(0)\rangle \tag{4.165}$$

Thus, we can write the $C_>(\omega)$ and $C_<(\omega)$ in terms of Fourier transform of correlation functions:

$$C_>(\omega) = \int_{-\infty}^{\infty} dt\, e^{i\omega t} \langle B(t)B(0)\rangle \tag{4.166}$$

and

$$C_<(\omega) = \int_{-\infty}^{\infty} dt\, e^{i\omega t} \langle B(0)B(t)\rangle \tag{4.167}$$

It is vitally important to note that $C_>(\omega) \neq C_<(\omega)$. This is because $B(t)$ and $B(0)$ are quantum mechanical operators and do not necessarily commute. Also, while $B(t)$ and $B(0)$ are Hermitian operators, their product is not Hermitian.

Symmetry properties: It is important to take a close look at the properties of time-correlation functions. Consider the time-correlation function

$$C(t) = \langle B(t)B(0) \rangle \tag{4.168}$$

discussed above. If B is a real operator, then we can immediately recognize that

$$C(t) = C^*(-t) \tag{4.169}$$

which implies that

$$Re(C(t)) = Re(C(-t)) \tag{4.170}$$

and

$$Im(C(t)) = -Im(C(t)) \tag{4.171}$$

In other words, the real part of $C(t)$ is an even function of time and the imaginary part must be an odd function of time. Thus, we can write

$$P(\omega) = |\mathscr{F}(\omega)|^2 I(\omega) \tag{4.172}$$

and consider

$$I(\omega) = \int_{-\infty}^{\infty} e^{i\omega t} C(t) dt \tag{4.173}$$

If we were measuring the absorption cross-section of a molecular species, $P(\omega)$ would be the rate of energy absorption given an external driving frequency ω. Thus, we can understand the physical origin of its two components. The part involving $|\mathscr{F}(\omega)|^2$ will depend upon the specific nature of the driving field itself and can be referred to as an "instrument" function. The other part, $I(\omega)$, depends upon only the internal details of the system being investigated. This we term the "line-shape" function and we can write this as

$$
\begin{aligned}
I(\omega) &= \left(\int_{-\infty}^{0} e^{i\omega t} C(t) dt + \int_{0}^{\infty} e^{i\omega t} C(t) dt \right) \\
&= \left(\int_{0}^{-\infty} e^{-i\omega t} C(-t) dt + \int_{0}^{\infty} e^{i\omega t} C(t) dt \right) \\
&= \left(\int_{0}^{-\infty} e^{+i\omega t} C^*(t) dt + \int_{0}^{\infty} e^{i\omega t} C(t) dt \right) \\
&= 2Re \int_{0}^{\infty} C(t) e^{+i\omega t} dt \tag{4.174}
\end{aligned}
$$

Typical forms for correlation functions: In general, $C(t)$ is an oscillatory function that decays in time. Since the real part of $C(t)$ must be even, $dC(0)/dt = 0$. Thus, we anticipate the general form

$$C(t) = C(0)e^{-(t/\tau_g)^2}\cos(\omega_o t) \tag{4.175}$$

where t_g is a Gaussian decay time and ω_o is a characteristic frequency. One can also encounter correlation functions that decay exponentially with time:

$$C(t) = C(0)e^{-t/\tau}\cos(\omega_o t) \tag{4.176}$$

While formally incorrect, this can occur whenever there is a loss of time reversibility within the system being probed. This can occur either through coarse-graining over some intermediate time scale or through the presence of a dissipative (that is, velocity-dependent) force. The line shapes corresponding to these two correlation functions are easy to obtain:

$$I_g(\omega) = \frac{1}{2}C(0)\sqrt{\pi}\tau(e^{-\tau^2(\omega-\omega_o)^2/4} + e^{-\tau^2(\omega+\omega_o)^2/4}) \tag{4.177}$$

for the Gaussian case and

$$I_l(\omega) = \left(\frac{\tau}{\tau^2(\omega+\omega_o)^2+1} + \frac{\tau}{\tau^2(\omega-\omega_o)^2+1}\right) \tag{4.178}$$

for the exponential decaying case. Here, the line shape is the characteristic Lorentzian. In the limit that $\omega_o = 0$ we have

$$I_g(\omega) = C(0)\sqrt{\pi}\tau e^{-\tau^2\omega^2/4} \tag{4.179}$$

and

$$I_l(\omega) = \frac{2\tau}{\tau^2(\omega+\omega_o)^2+1} \tag{4.180}$$

Finally, we can define the correlation time as

$$\tau_c = \int_0^\infty \frac{C(t)}{C(0)}dt \tag{4.181}$$

For the Gaussian decay: $\tau_c = \sqrt{\pi}\tau$ while for the exponential decay: $\tau_c = \tau$.

Lastly, we note that for a classical system, the time-correlation function is symmetric in time with $C(t) = C(-t)$. This is the result of the time-reversal symmetry that arises in Newtonian mechanics.

Example: A Brownian particle. To gain some understanding and practice in computing correlation functions, we consider the time-correlation function of the position for a particle with unit mass ($m = 1$) undergoing Brownian motion. This can be described via the Langevin equation:

$$\ddot{x}(t) = -\gamma\dot{x}(t) + R(t) \tag{4.182}$$

where $R(t)$ is a random force with $\langle R(t) \rangle = 0$ and $\langle R(t)R(0) \rangle = 2\pi kT\gamma\delta(t)$. In essence, each random kick is uncorrelated with the previous one that occurred an instant earlier in time. Multiply both sides on the right by $x(0)$,

$$\ddot{x}(t)x(0) = -\gamma\dot{x}(t)x(0) + R(t)x(0) \qquad (4.183)$$

then perform the thermal average,

$$\langle \ddot{x}(t)x(0) \rangle = -\gamma\langle \dot{x}(t)x(0) \rangle + \langle R(t)x(0) \rangle \qquad (4.184)$$

The last term vanishes since $\langle R(t) \rangle = 0$. For the other terms, the time derivative can be pulled out in front of the thermal average:

$$\frac{d^2}{dt^2}\langle x(t)x(0) \rangle = -\gamma\frac{d}{dt}\langle x(t)x(0) \rangle \qquad (4.185)$$

This gives us a simple ordinary differential equation for our correlation function:

$$\frac{d^2}{dt^2}C(t) = -\gamma\frac{d}{dt}C(t) \qquad (4.186)$$

This we can easily solve to find

$$C(t) = C(0)e^{-\gamma t} \qquad (4.187)$$

In other words, the line-shape function for a randomly kicked particle is a Lorentzian

$$I(\omega) = \frac{2\gamma}{\omega^2 + \gamma^2}C(0) \qquad (4.188)$$

4.8 INTERACTION BETWEEN MATTER AND RADIATION

Much of what we shall discuss in this book hinges upon what happens once a molecule has been promoted into one of its excited states. However, the dynamics that do occur depend upon just how the molecule found itself in this excited state. Moreover, the ultimate fate of the excitation depends upon how strongly coupled the initial excited state is to other excited states or to the ground state. Since much of this book is concerned with electronic processes in an excited state, we review here briefly the basic interactions between molecules and the electromagnetic field. We take a traditional, semiclassical approach to describe the coupling between matter and radiation rather than a fully quantum mechanical treatment in which the radiation field is treated entirely within the context of Maxwell's equations. This will allow us to describe the coupling entirely in terms of time-dependent operators involving position and momentum operators acting on the material degrees of freedom.

4.8.1 FIELDS AND POTENTIALS OF A LIGHT WAVE

An electromagnetic wave consists of two oscillating vector field components that are perpendicular to each other and oscillate at an angular frequency $\omega = ck$ where k is the magnitude of the wave vector that points in the direction of propagation and c is the speed of light. For such a wave, we can always set the scalar part of its potential to zero with a suitable choice in gauge and describe the fields associated with the wave in terms of a vector potential $\vec{A}$, given by

$$\vec{A}(r, t) = A_o \mathbf{e}_z e^{iky - i\omega t} + A_o^* \mathbf{e}_z e^{-iky + i\omega t} \tag{4.189}$$

Here, the wave vector points in the $+y$ direction, the electric field E is polarized in the yz plane, and the magnetic field B is in the xy plane. Using Maxwell's relations

$$\vec{E}(r, t) = -\frac{\partial A}{\partial t} = i\omega \mathbf{e}_z (A_o e^{i(ky - \omega t)} - A_o^* e^{-i(ky - \omega t)}) \tag{4.190}$$

and

$$\vec{B}(r, t) = \nabla \times \vec{A} = ik \mathbf{e}_x (A_o e^{i(ky - \omega t)} - A_o^* e^{-i(ky - \omega t)}) \tag{4.191}$$

We are free to choose the time origin, so we will choose it so as to make A_o purely imaginary, and set

$$i\omega A_o = \mathscr{E}/2 \tag{4.192}$$

$$ik A_o = \mathscr{B}/2 \tag{4.193}$$

where $\mathscr{E}$ and $\mathscr{B}$ are real quantities such that

$$\frac{\mathscr{E}}{\mathscr{B}} = \frac{\omega}{k} = c \tag{4.194}$$

Thus

$$E(r, t) = \mathscr{E}\mathbf{e}_z \cos(ky - \omega t) \tag{4.195}$$

$$B(r, t) = \mathscr{B}\mathbf{e}_z \sin(ky - \omega t) \tag{4.196}$$

where $\mathscr{E}$ and $\mathscr{B}$ are the magnitudes of the electric and magnetic field components of the plane wave.

Lastly, we define what is known as the Poynting vector, which is parallel to the direction of propagation:

$$\vec{S} = \varepsilon_o c^2 \vec{E} \times \vec{B} \tag{4.197}$$

Using the expressions for $\vec{E}$ and $\vec{B}$ above and averaging over several oscillation periods:

$$\vec{S} = \varepsilon_o c^2 \frac{\mathscr{E}}{2} \mathbf{e}_y \tag{4.198}$$

4.8.2 Interactions at Low Light Intensity

The electromagnetic wave we just discussed can interact with an atomic electron. The Hamiltonian of this electron can be given by

$$H = \frac{1}{2m}(\mathbf{P} - q\mathbf{A}(\mathbf{r}, t))^2 + V(r) - \frac{q}{m}\mathbf{S} \cdot \mathbf{B}(\mathbf{r}, t) \qquad (4.199)$$

where the first term represents the interaction between the electron and the electrical field of the wave and the last term represents the interaction between the magnetic moment of the electron and the magnetic moment of the wave. In expanding the kinetic energy term, we have to remember that momentum and position do not commute. However, in the present case, $\mathbf{A}$ is parallel to the z axis, and P_z and y commute. So, we wind up with the following:

$$H = H_o + W \qquad (4.200)$$

where

$$H_o = \frac{\mathbf{P}^2}{2m} + V(r) \qquad (4.201)$$

is the unperturbed (atomic) Hamiltonian and

$$W = -\frac{q}{m}\mathbf{P} \cdot \mathbf{A} - \frac{q}{m}\mathbf{S} \cdot \mathbf{B} + \frac{q^2}{2m}A^2 \qquad (4.202)$$

The first two terms depend linearly upon A and the second is quadratic in A. So, for low intensity we can take

$$W = -\frac{q}{m}\mathbf{P} \cdot \mathbf{A} - \frac{q}{m}\mathbf{S} \cdot \mathbf{B} = W_E + W_B \qquad (4.203)$$

Before moving on, we need to evaluate the relative importance of each term by orders of magnitude for transitions between bound states. In the second term, the contribution of the spin operator is on the order of $\hbar$ and the contribution from B is on the order of kA. Thus,

$$\frac{W_B}{W_E} = \frac{\frac{q}{m}\mathbf{S} \cdot \mathbf{B}}{\frac{q}{m}\mathbf{P} \cdot \mathbf{A}} \approx \frac{\hbar k}{p} \qquad (4.204)$$

where $h/p = \Lambda_{dB}$ is the de Broglie wavelength of the particle. For an electron in an atom, Λ_{dB} is on the order of an atomic radius, a_o The wave number is related to the wavelength via $k = 2\pi/\lambda$. For electronic excitations in the UV or visible spectral range, λ is on the order of $1000a_o$ to $10\,000\,a_o$. Thus,

$$\frac{W_B}{W_E} \approx \frac{a_o}{\lambda} \ll 1 \qquad (4.205)$$

We can safely conclude then that the magnetic interaction is not at all important for ordinary optical transitions and we focus only upon the coupling to the electric field.

Using the expressions we derived previously, the coupling to the electric field component of the light wave is given by

$$W_E = -\frac{q}{m} p_z \left(A_o e^{iky} e^{-i\omega t} + A_o^* e^{-iky} e^{+i\omega t} \right) \tag{4.206}$$

Now, we expand the exponential in powers of y:

$$e^{\pm iky} = 1 \pm iky - \frac{1}{2}k^2 y^2 + \cdots \tag{4.207}$$

Since $ky \approx a_o/\lambda \ll 1$, we can, with good approximation, keep only the first term. Thus we get the dipole operator

$$W_D = \frac{q\mathcal{E}}{m\omega} p_z \sin(\omega t) \tag{4.208}$$

In the electric dipole approximation, $W(t) = W_D(t)$.

Note that we might expect that W_D should have been written as

$$W_D = -q\mathcal{E}z \cos(\omega t) \tag{4.209}$$

since we are, after all, talking about a dipole moment associated with the motion of the electron about the nucleus. Actually, the two expressions are equivalent because we can always choose a different gauge to represent the physical problem without changing the physical result. In electrodynamics, the electric and magnetic fields are described in terms of a vector potential $\vec{A}$ and a scalar potential U. To get the present result, we used

$$\mathbf{A} = \frac{\mathcal{E}}{\omega} \mathbf{e}_z \sin(\omega t) \tag{4.210}$$

and set the scalar potential

$$U(r) = 0 \tag{4.211}$$

But this is completely arbitrary. We can always choose another vector potential and scalar potential to describe the fields and require that, in the end, the physics be invariant to how we choose to describe these potentials. Formally, when we choose the potential we make a particular choice of *gauge*. We can transform from one gauge to another by taking a function f and defining a new vector potential and a new scalar potential as

$$\mathbf{A}' = \mathbf{A} + \nabla f \tag{4.212}$$

$$U' = U - \frac{\partial f}{\partial t} \tag{4.213}$$

We are free to choose f. Let us take $f = z\mathcal{E}\sin(\omega t)/\omega$ so that

$$\mathbf{A}' = \mathbf{e}_z \frac{\mathcal{E}}{\omega}(\sin(ky - \omega t) + \sin(\omega t)) \tag{4.214}$$

is the new vector potential and

$$U' = -z\mathscr{E}\cos\omega t \tag{4.215}$$

is the new scalar potential. In the electric dipole approximation, ky is small, so we set $ky = 0$ everywhere and obtain $A' = 0$. Thus, the total Hamiltonian becomes

$$H = H_o + qU'(r, t) \tag{4.216}$$

with perturbation

$$W'_D = -qz\mathscr{E}\cos(\omega t) \tag{4.217}$$

Now our perturbation depends upon the displacement operator rather than the momentum operator. This is the usual form of the dipole coupling operator.

Next, let us consider the matrix elements of the dipole operator between two stationary states of H_o: $|\psi_i\rangle$ and $|\psi_f\rangle$ with eigenenergy E_i and E_f, respectively. The matrix elements of W_D are given by

$$W_{fi}(t) = \frac{q\mathscr{E}}{m\omega}\sin(\omega t)\langle\psi_f|p_z|\psi_i\rangle \tag{4.218}$$

We can evaluate this by noting that

$$[z, H_o] = i\hbar\frac{\partial H_o}{\partial p_z} = i\hbar\frac{p_z}{m} \tag{4.219}$$

Thus,

$$\langle\psi_f|p_z|\psi_i\rangle = im\omega_{fi}\langle\psi_f|z|\psi_i\rangle \tag{4.220}$$

Consequently,

$$W_{fi}(t) = iq\mathscr{E}\omega_{fi}\frac{\sin(\omega t)}{\omega}z_{fi} \tag{4.221}$$

Thus, the matrix elements of the dipole operator are those of the position operator. This determines the selection rules for the transition.

Before going through any specific details, let us consider what happens if the frequency ω does not coincide with ω_{fi}. Specifically, we limit ourselves to transitions originating from the ground state of the system, $|\psi_o\rangle$. We will assume that the field is weak and that in the field the atom acquires a time-dependent dipole moment that oscillates at the same frequency as the field via a forced oscillation. To simplify matters, assume that the electron is harmonically bound to the nucleus with a classical Hooke's law force,

$$V(r) = \frac{1}{2}m\omega_o r^2 \tag{4.222}$$

where ω_o is the natural frequency of the electron.

The classical motion of the electron is given by the equations of motion (via the Ehrenfest theorem)

$$\ddot{z} + \omega^2 z = \frac{q\mathscr{E}}{m}\cos(\omega t) \tag{4.223}$$

This is the equation of motion for a harmonic oscillator subject to a periodic force. This inhomogeneous differential equation can be solved (using Fourier transform methods) and the result is

$$z(t) = A\cos(\omega_o t - \phi) + \frac{q\mathscr{E}}{m\left(\omega_o^2 - \omega^2\right)}\cos(\omega t) \tag{4.224}$$

where the first term represents the harmonic motion of the electron in the absence of the driving force. The two coefficients, A and ϕ, are determined by the initial condition. If we have a very slight damping of the natural motion, the first term disappears after a while, leaving only the second, forced oscillation, so we write

$$z = \frac{q\mathscr{E}}{m\left(\omega_o^2 - \omega^2\right)}\cos(\omega t) \tag{4.225}$$

Thus, we can write the classical induced electric dipole moment of the atom in the field as

$$D = qz = \frac{q^2\mathscr{E}}{m\left(\omega_o^2 - \omega^2\right)}\cos(\omega t) \tag{4.226}$$

Typically this is written in terms of a susceptibility, χ, where

$$\chi = \frac{q^2}{m\left(\omega_o^2 - \omega^2\right)} \tag{4.227}$$

Now we look at this from a quantum mechanical point of view. Again, take the initial state to be the ground state and $H = H_o + W_D$ as the Hamiltonian. Since the time-evolved state can be written as a superposition of eigenstates of H_o,

$$|\psi(t)\rangle = \sum_n c_n(t)|\phi_n\rangle \tag{4.228}$$

To evaluate this we can use the results derived previously in our derivation of the golden rule,

$$|\psi(t)\rangle = |\phi_o\rangle + \sum_{n\neq 0} \frac{q\mathscr{E}}{2im\hbar\omega}\langle n|p_z|\phi_o\rangle$$
$$\times \left\{ \frac{e^{-i\omega_{no}t} - e^{i\omega t}}{\omega_{no} + \omega} - \frac{e^{-i\omega_{no}t} - e^{-i\omega t}}{\omega_{no} - \omega} \right\}|\phi_n\rangle \tag{4.229}$$

where we have removed a common phase factor. We can then calculate the dipole moment expectation value, $\langle D(t)\rangle$, as

$$\langle D(t)\rangle = \frac{2q^2}{\hbar}\mathscr{E}\cos(\omega t)\sum_n \frac{\omega_{on}|\langle\phi_n|z|\phi_o\rangle|^2}{\omega_{no}^2 - \omega^2} \tag{4.230}$$

From this we can begin to clearly appreciate the physics behind the absorption or emission of light by an atom or molecule. When an oscillating dipole field is applied to an atom or molecule, the electrons in the atom respond by oscillating with the applied field. Ordinarily, this oscillation is not very significant when $\omega \neq \omega_{no}$. However, at the resonance condition, the induced dipole moment atom or molecule (due to its interaction with the field) oscillates readily at the transition frequency and the atom or molecule readily absorbs or emits energy in the form of electromagnetic radiation.

4.8.2.1   Oscillator Strength

We can now notice the similarity between a driven harmonic oscillator and the expectation value of the dipole moment of an atom in an electric field. We can define the *oscillator strength* as a dimensionless and real number characterizing the transition between $|\phi_o\rangle$ and $|\phi_n\rangle$,

$$f_{no} = \frac{2m\omega_{no}}{\hbar}|\langle\phi_n|z|\phi_o\rangle|^2 \tag{4.231}$$

The term *oscillator strength* comes from the analysis of a harmonically bound electron. In a sense, such an electron is a perfect absorber since its motion is perfectly harmonic and as such it can maintain a perfect phase relationship with the external driving field.

Summing over all possible transitions from the original state yields the Thomas–Reiche–Kuhn (TRK) sum rule:

$$\sum_n f_{no} = 1 \tag{4.232}$$

which can be written in a very compact form:

$$\frac{m}{\hbar^2}\langle\phi_o|[z,[H,z]]|\phi_o\rangle = 1 \tag{4.233}$$

From this, one concludes that the highest possible oscillator strength for a transition is 1. In fact, strong electronic transitions can have oscillator strengths on the order of unity and often are greater than 1. For example, polyacetylenes can have f's as large as 5. This seems like a contradiction until we realize that the above sum rule is for a single particle system. If we extend this for all electrons in the system, then the sum adds up to the total number of electrons in the system. So, in a sense, one can think of an $f > 1$ as a measure of the total number of electronic states that are involved in the transition.

4.8.3   SPONTANEOUS EMISSION OF LIGHT

The emission and absorption of light by an atom or molecule is perhaps the most spectacular and important phenomena in the universe. It happens when an atom or molecule undergoes a transition from one state to another due to its interaction with the electromagnetic field. Because the electromagnetic field cannot be entirely eliminated from any so-called isolated system (except for certain quantum confinement experiments), no atom or molecule is ever really isolated. Thus, even in the absence of an explicitly applied field, an excited system can spontaneously emit a photon and relax to a lower energy state. Since we have all done spectroscopy experiments at one point or another in our education, we all know that the transitions are between *discrete* energy levels. In fact, it was in the examination of light passing through glass and light emitted from flames that people in the nineteenth century began to speculate that atoms can absorb and emit light only at specific wavelengths.

We will use the golden rule to deduce the probability of a transition under the influence of an applied light field (laser or otherwise). We will argue that the system is in equilibrium with the electromagnetic field and that the laser drives the system out of equilibrium. From this we can deduce the rate of spontaneous emission in the absence of the field.

The electric field associated with a monochromatic light wave of average intensity I is

$$\langle I \rangle = c \langle \rho \rangle \tag{4.234}$$

$$= c \left(\varepsilon_o \frac{\langle \vec{\mathscr{E}}_o^2 \rangle}{2} + \frac{1}{\mu_o} \frac{\langle B_o^2 \rangle}{2} \right) \tag{4.235}$$

$$= \left(\frac{\varepsilon_o}{\mu_o} \right)^{1/2} \frac{\mathscr{E}_o^2}{2} \tag{4.236}$$

$$= c \varepsilon_o \frac{\mathscr{E}_o^2}{2} \tag{4.237}$$

where ρ is the energy density of the field, $|\vec{\mathscr{E}}|$ and $|B_o| = (1/c)|\vec{\mathscr{E}}|$ are the maximum amplitudes of the E and B fields of the wave, and we are using meter-kilogram-second (mks) units. The electromagnetic wave in reality contains a spread of frequencies, so we must also specify the intensity density over a definite frequency interval:

$$\frac{dI}{d\omega} d\omega = c u(\omega) d\omega \tag{4.238}$$

where $u(\omega)$ is the energy density per unit frequency at ω.

Within the "semiclassical" dipole approximation, the coupling between a molecule and the light wave is

$$\vec{\mu} \cdot \vec{\mathscr{E}}(t) = \vec{\mu} \cdot \vec{\varepsilon} \frac{\mathscr{E}_o}{2} \cos(\omega t) \tag{4.239}$$

where $\vec{\mu}$ is the dipole moment vector and $\vec{\varepsilon}$ is the polarization vector of the wave. Using this result, we can plug directly into the golden rule and deduce that

$$P_{fi}(\omega, t) = 4|\langle f|\vec{\mu} \cdot \vec{\varepsilon}|i \rangle|^2 \frac{\mathscr{E}_o^2}{4} \frac{\sin^2((E_f - E_i - \hbar\omega)t/(2\hbar))}{(E_f - E_i - \hbar\omega)^2} \tag{4.240}$$

Now, we can take into account the spread of frequencies of the electromagnetic wave around the *resonant value* of $\omega_o = (E_f - E_i)/\hbar$. To do this we note

$$\mathscr{E}_o^2 = 2 \frac{\langle I \rangle}{c \varepsilon_o} \tag{4.241}$$

and replace $\langle I \rangle$ with $(dI/d\omega)d\omega$.

$$P_{fi}(t) = \int_0^\infty d\omega P_{fi}(t, \omega) \tag{4.242}$$

$$= \frac{2}{c\varepsilon_o} \left(\frac{dI}{d\omega}\right)_{\omega_o} |\langle f|\vec{\mu} \cdot \vec{\varepsilon}|i\rangle|^2 \int_0^\infty \frac{\sin^2((\hbar\omega_o - \hbar\omega)(t/(2\hbar)))}{(\hbar\omega_o - \hbar\omega)^2} d\omega \tag{4.243}$$

To get this we assume that $dI/d\omega$ and the matrix element of the coupling vary slowly with frequency as compared to the $\sin^2(x)/x^2$ term. Thus, as far as doing integrals are concerned, they are both constants. With ω_o so fixed, we can do the integral over dw and get $\pi t/(2\hbar^2)$, and we obtain the golden rule transition rate:

$$k_{fi} = \frac{\pi}{c\varepsilon_o\hbar^2} |\langle f|\vec{\mu} \cdot \vec{\varepsilon}|i\rangle|^2 \left(\frac{dI}{d\omega}\right)_{\omega_o} \tag{4.244}$$

Notice also that this equation predicts that the rate for excitation is identical to the rate for de-excitation. This is because the radiation field contains both $+\omega$ and $-\omega$ terms (unless the field is circularly polarized), and the transition rate from a state of lower energy to a higher energy is the same as that of the transition from a higher energy state to a lower energy state. However, we know that systems can emit spontaneously in which a state of higher energy can go to a state of lower energy in the absence of an external field. This is difficult to explain in the present framework since we have assumed that $|i\rangle$ is stationary. Let us assume that we have an ensemble of atoms in a cavity containing electromagnetic radiation and the system is in thermodynamic equilibrium. (Thought you could escape thermodynamics, eh?) Let E_1 and E_2 be the energies of two states of the atom with $E_2 > E_1$. When equilibrium has been established, the number of atoms in the two states is determined by the Boltzmann equation:

$$\frac{N_2}{N_1} = \frac{Ne^{-E_2\beta}}{Ne^{-E_1\beta}} = e^{-\beta(E_2-E_1)} \tag{4.245}$$

where $\beta = 1/kT$. The number of atoms (per unit time) undergoing the transition from 1 to 20 is proportional to k_{21} induced by the radiation and to the number of atoms in the initial state N_1:

$$\frac{dN}{dt}(1 \rightarrow 2) = N_1 k_{21} \tag{4.246}$$

The number of atoms going from 2 to 1 is proportional to N_2 and to $k_{21} + A$ where A is the *spontaneous transition rate*

$$\frac{dN}{dt}(2 \rightarrow 1) = N_2(k_{21} + A) \tag{4.247}$$

At equilibrium, these two rates must be equal. Thus,

$$\frac{k_{21} + A}{k_{21}} = \frac{N_1}{N_2} = e^{\hbar\omega\beta} \tag{4.248}$$

Now, let us refer to the result for the induced rate k_{21} and express it in terms of the energy density per unit frequency of the cavity, $u(\omega)$,

$$k_{21} = \frac{\pi}{\varepsilon_o \hbar^2} |\langle 2|\vec{\mu} \cdot \vec{\varepsilon}|1\rangle|^2 u(\omega) = B_{21} u(\omega) \tag{4.249}$$

where

$$B_{21} = \frac{\pi}{\varepsilon_o \hbar^2} |\langle 2|\vec{\mu} \cdot \vec{\varepsilon}|1\rangle|^2 \tag{4.250}$$

For electromagnetic radiation in equilibrium at temperature T, the energy density per unit frequency is given by Planck's law:

$$u(\omega) = \frac{1}{\pi^2 c^3} \frac{\hbar \omega^3}{e^{\hbar \omega \beta} - 1} \tag{4.251}$$

Combining the results we obtain

$$\frac{B_{12}}{B_{21}} + \frac{A}{B_{21}} \frac{1}{u(\omega)} = e^{\hbar \omega \beta} \tag{4.252}$$

$$\frac{B_{21}}{B_{12}} + \frac{A}{B_{21}} \frac{\pi^2 c^3}{\hbar \omega^3} (e^{\hbar \omega \beta} - 1) = e^{\hbar \omega \beta} \tag{4.253}$$

$$\tag{4.254}$$

which must hold for all temperatures. Since

$$\frac{B_{21}}{B_{12}} = 1 \tag{4.255}$$

we get

$$\frac{A}{B_{21}} \frac{\pi^2 c^3}{\hbar \omega^3} = 1 \tag{4.256}$$

and thus, the spontaneous emission rate is

$$A = \frac{\hbar \omega^3}{\pi^2 c^3} B_{12} \tag{4.257}$$

$$= \frac{\omega^3}{\varepsilon_o \pi \hbar c^3} |\langle 2|\vec{\mu} \cdot \vec{\varepsilon}|1\rangle|^2 \tag{4.258}$$

This is a key result in that it determines the probability for the emission of light by atomic and molecular systems. We can use it to compute the intensity of spectral lines in terms of the electric dipole moment operator. The lifetime of the excited state is then inversely proportional to the spontaneous decay rate,

$$\tau = \frac{1}{A} \tag{4.259}$$

To compute the matrix elements, we can make a rough approximation that $\langle \mu \rangle \propto \langle x \rangle e$ where e is the charge of an electron and $\langle x \rangle$ is on the order of atomic dimensions. We also must include a factor of 1/3 for averaging over all orientations of $(\vec{\mu} \cdot \vec{\varepsilon})$. Since at any given time the moments are not all aligned,

$$\frac{1}{\tau} = A = \frac{4}{3} \frac{\omega^3}{\hbar c^3} \frac{e^2}{4\pi\varepsilon_o} |\langle x \rangle|^2 \tag{4.260}$$

The factor

$$\frac{e^2}{4\pi\varepsilon_o \hbar c} = \alpha \approx \frac{1}{137} \tag{4.261}$$

is the *fine structure constant*. Also, $\omega/c = 2\pi/\lambda$. So, setting $\langle x \rangle \approx 1$ Å,

$$A = \frac{4}{3} \frac{1}{137} c \left(\frac{2\pi}{\lambda}\right)^3 (1\text{Å})^2 \approx \frac{6 \times 10^{18}}{[\lambda(\text{Å})]^3} \sec^{-1} \tag{4.262}$$

So, for a typical wavelength, $\lambda \approx 4 \times 10^3$ Å,

$$\tau = 10^{-8} \sec \tag{4.263}$$

which is consistent with observed lifetimes.

We can also compare with classical radiation theory. The power radiated by an accelerated particle of charge e is given by the Larmor formula (cf. Jackson)

$$P = \frac{2}{3} \frac{e^2}{4\pi\varepsilon_o} \frac{(\dot{v})^2}{c^3} \tag{4.264}$$

where $\dot{v}$ is the acceleration of the charge. Assuming the particle moves in a circular orbit of radius r with angular velocity ω, the acceleration is $\dot{v} = \omega^2 r$. Thus, the time required to radiate energy $\hbar\omega/2$ is equivalent to the lifetime τ

$$\frac{1}{\tau_{\text{class}}} = \frac{2P}{\hbar\omega} \tag{4.265}$$

$$= \frac{1}{\hbar\omega} \frac{4}{3} \frac{e^2}{4\pi\varepsilon_o} \frac{\omega^4 r^2}{c^3} \tag{4.266}$$

$$= \frac{4}{3} \frac{\omega^3}{\hbar c^3} \frac{e^2}{4\pi\varepsilon_o} r^2 \tag{4.267}$$

This qualitative agreement between the classical and quantum results is a manifestation of the correspondence principle. However, it must be emphasized that the *MECHANISM* for radiation is entirely different. The classical result will never predict a discrete spectrum. This was in fact a very early indication that something was certainly amiss with the classical electromagnetic field theories of Maxwell and others.

4.9 APPLICATION OF GOLDEN RULE: PHOTOIONIZATION OF HYDROGEN 1S

We consider here the photoionization of the hydrogen 1s orbital to illustrate how the golden rule formalism can be used to calculate photoionization cross-sections as a function of the photon frequency. We already have an expression for dipole coupling:

$$W_D = \frac{q\mathcal{E}}{m\omega} p_z \sin(\omega t) \tag{4.268}$$

and we have derived the golden rule rate for transitions between states:

$$k_{if} = \frac{2\pi}{\hbar} |\langle f|V|i\rangle|^2 \delta(E_i - E_f + \hbar\omega) \tag{4.269}$$

For transitions to the continuum, the final states are the plane waves,

$$\psi(\mathbf{k}) = \frac{1}{\Omega^{1/2}} e^{i\mathbf{k}\cdot\mathbf{r}} \tag{4.270}$$

where Ω is the volume element. Thus the matrix element $\langle 1s|V|\mathbf{k}\rangle$ can be written as

$$\langle 1s|p_z|\mathbf{k}\rangle = \frac{\hbar k_z}{\Omega^{1/2}} \int \psi_{1s}(\mathbf{r}) e^{i\mathbf{k}\cdot\mathbf{r}} d\mathbf{r} \tag{4.271}$$

To evaluate the integral, we need to transform the plane-wave function into spherical coordinates. This can be done via the expansion

$$e^{i\mathbf{k}\cdot\mathbf{r}} = \sum_l i^l (2l+1) j_l(kr) P_l(\cos(\theta)) \tag{4.272}$$

where $j_l(kr)$ is the spherical Bessel function and $P_l(x)$ is a Legendre polynomial, which we can also write as a spherical harmonic function,

$$P_l(\cos(\theta)) = \sqrt{\frac{4\pi}{2l+1}} Y_{l0}(\theta, \phi) \tag{4.273}$$

Thus, the integral we need to perform is

$$\langle 1s|\mathbf{k}\rangle = \frac{1}{\sqrt{\pi\Omega}} \sum_l \left(\int Y_{00}^* Y_{l0} d\Omega \right) i^l \sqrt{4\pi(2l+1)} \int_0^\infty r^2 e^{-r} j_l(kr) dr \tag{4.274}$$

The angular integral we do by orthogonality and this produces a delta function that restricts the sum to $l = 0$ only leaving

$$\langle 1s|\mathbf{k}\rangle = \frac{1}{\sqrt{\Omega}} \int_0^\infty r^2 e^{-r} j_0(kr) dr \tag{4.275}$$

The radial integral can be easily performed using

$$j_0(kr) = \frac{\sin(kr)}{kr} \tag{4.276}$$

leaving

$$\langle 1s|\mathbf{k}\rangle = \frac{4}{k}\frac{1}{\Omega^{1/2}}\frac{1}{(1+k^2)^2} \tag{4.277}$$

Thus, the matrix element is given by

$$\langle 1s|V|k\rangle = \frac{q\mathcal{E}\hbar}{m\omega}\frac{1}{\Omega^{1/2}}\frac{2}{(1+k^2)^2} \tag{4.278}$$

This we can insert directly into the golden rule formula to get the photoionization rate to a given k-state:

$$R_{0k} = \frac{2\pi\hbar}{\Omega}\left(\frac{q\mathcal{E}}{m\omega}\right)^2\frac{4}{(1+k^2)^4}\delta(E_o - E_k + \hbar\omega) \tag{4.279}$$

which we can manipulate into reading as

$$R_{0k} = \frac{16\pi}{\hbar\Omega}\left(\frac{q\mathcal{E}}{m\omega}\right)^2 m\frac{\delta(k^2 - K^2)}{(1+k^2)^4} \tag{4.280}$$

where we write $K^2 = 2m(E_1 + \hbar\omega)/\hbar^2$ to make our notation a bit more compact. Eventually, we want to know the rate as a function of the photon frequency, so let us put everything except the frequency and the volume element into a single constant $\mathcal{I}$, which is related to the intensity of the incident photon,

$$R_{0k} = \frac{\mathcal{I}}{\Omega}\frac{1}{\omega^2}\frac{\delta(k^2 - K^2)}{(1+k^2)^4} \tag{4.281}$$

Now, we sum over all possible final states to get the total photoionization rate. To do this, we need to turn the sum over final states into an integral, and this is done by

$$\sum_k = \frac{\Omega}{(2\pi)^3}4\pi\int_0^\infty k^2 dk \tag{4.282}$$

Thus,

$$\begin{aligned}
R &= \frac{\mathcal{I}}{\Omega}\frac{1}{\omega^2}\frac{\Omega}{(2\pi)^3}4\pi\int_0^\infty k^2\frac{\delta(k^2 - K^2)}{(1+k^2)^4}dk \\
&= \frac{\mathcal{I}}{\omega^2}\frac{1}{2\pi^2}\int_0^\infty k^2\frac{\delta(k^2 - K^2)}{(1+k^2)^2}dk
\end{aligned}$$

Now we do a change of variables, $y = k^2$ and $dy = 2kdk$, so that the integral becomes

$$\int_0^\infty k^2\frac{\delta(k^2 - K^2)}{(1+k^2)^2}dk = \frac{1}{2}\int_0^\infty \frac{y^{1/2}}{(1+y^2)^4}\delta(y - K^2)dy$$

$$= \frac{K}{2(1+K^2)^4} \tag{4.283}$$

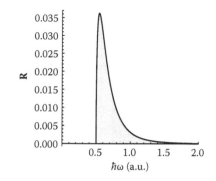

FIGURE 4.5 Photoionization spectrum for hydrogen atom. Note that the vertical axis is scaled by the incident photon flux.

Pulling everything together, we see that the total photoionization rate is given by

$$R = \frac{\mathscr{I}}{\omega^2} \frac{1}{2\pi^2} \frac{K}{(1 + K^2)^4}$$

$$= \frac{\mathscr{I}\sqrt{\frac{m}{\hbar^2}} \sqrt{\omega\hbar - \varepsilon_o}}{\sqrt{2}\,\pi^2\,\omega^2 \left(1 + \frac{2m\,(\omega\hbar - \varepsilon_o)}{\hbar^2}\right)^4}$$

$$= \mathscr{I}\frac{\sqrt{2\omega - 1}}{32\,\pi^2\,\omega^6} \qquad\qquad (4.284)$$

where in the last line we have converted to atomic units to clean things up a bit. This expression is clearly valid only when $\hbar\omega > E_I = 1/2$ hartree (13.6 eV); a plot of the photoionization rate is given in Figure 4.5.

4.10   COUPLED ELECTRONIC/NUCLEAR DYNAMICS

We conclude this chapter with a brief discussion of the coupling between nuclear and electronic degrees of freedom in a molecular system. For the sake of connecting to the rest of this chapter, let us take the nuclear degrees of freedom to be a time-dependent driving field for the electronic degrees of freedom. However, unlike the electromagnetic field, there will be a considerable back reaction since a change in the electronic state will result in a force acting on the nuclei. We begin by writing the Hamiltonian describing this as

$$H = H_e(r(t)) + \frac{p^2}{2m} \qquad\qquad (4.285)$$

where the first term is the electronic part that depends parametrically upon the nuclear coordinate r and the second is the nuclear kinetic energy. If $\psi(t)$ is the electronic state at time t, its total time derivative contains two terms:

$$i\hbar\frac{d\psi}{dt} = i\hbar\frac{\partial\psi}{\partial t} + i\hbar\dot{r}\frac{\partial\psi}{\partial r} \tag{4.286}$$

where the first term gives the contribution from the explicit dependency on time while the second gives the implicit dependency. In the language of fluid dynamics, this is the *advective* derivative since we can imagine that ψ is being carried along some path $r(t)$. If we expand ψ in terms of the eigenstates of $H_e(r)$ at a given value of r,

$$\psi(r) = \sum_n c_n(r)\phi_n(r) \tag{4.287}$$

then

$$i\hbar\frac{dc_n}{dt} = \varepsilon_n c_n + i\hbar\dot{r}\sum_m c_m\left\langle\phi_n|\frac{\partial}{\partial r}|\phi_m\right\rangle \tag{4.288}$$

First, in the limit of slow nuclear motion, $\dot{r}\approx 0$ or if the electronic wave function varies slowly along r, then the second term gives no contribution to the dynamics. If our initial electronic state is prepared in an eigenstate of H_e, then under these conditions it will remain in the same eigenstate even as the nuclei move. This is the adiabatic approximation whereby the nuclear motion is slow enough such that the electronic state instantly responds to any small change. Within this approximation, we can use the Hellmann–Feynman theorem to compute the forces exerted on the nuclei. The resulting equations of motion for the nuclei then read:

$$m\ddot{r} = -\frac{\partial\varepsilon_n(r)}{\partial r} = -\left\langle\phi_n(r)\left|\frac{\partial H_e(r)}{\partial r}\right|\phi_n(r)\right\rangle \tag{4.289}$$

In general, we do not need to assume that ψ is initially an eigenstate of H_e; it can be a superposition of eigenstates, in which case we need to take a weighted average over forces

$$m\ddot{r} = -\left\langle\psi\left|\frac{\partial H_e(r)}{\partial r}\right|\psi\right\rangle$$

$$= -\sum_n |c_n|^2\frac{\partial\varepsilon_n(r)}{\partial r} \tag{4.290}$$

In other words, in this expression, the nuclear degrees of freedom (represented by r) experience an *average* force weighted by the population in each state. This is a very compelling picture since one can effectively partition a very large system with many degrees of freedom into one consisiting of two interacting subsystems, one of which behaves classically and the other quantum mechanically with a time-dependent Hamiltonian that depends parametrically upon the classical variables

$$i\hbar\dot{\psi}(t) = H_e(r(t))\psi(t) \tag{4.291}$$

At first glance, there does not seem to be any problem with this description. However, consider the case where the system evolves into two very different config- urations, perhaps one corresponding to the case in Figure 4.6 where an electron is

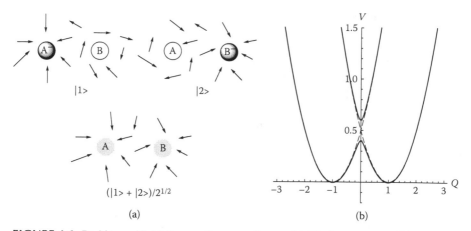

FIGURE 4.6 Problem with Hellmann–Feynman forces: (a) We have two possible electron transfer states. One ($|1\rangle$) has the electron localized on site A with B being neutral and the other ($|2\rangle$) has the electron on site B with A being neutral. The arrows indicate the dipole moments of surrounding solvent molecules. Since $|1\rangle$ and $|2\rangle$ are coupled, the state will naturally evolve into a linear combination of the two possible outcomes. Consequently, the Hellmann–Feynman forces will see an average of the two. (b) We have a potential well representing the two states with Q being an order parameter. For $Q = -1$, the dipoles are oriented about A and for $Q = +1$ the dipoles are oriented about B. $Q = 0$ corresponds to the unstable case of neither A nor B being fully solvated.

localized on the left-hand molecule and the other corresponding to where the electron is localized on the right-hand molecule. If the final populations are such that there is a 1:1 mixture between the left and right configurations, the solvent molecules following the transfer of the electron from the left to the right will "see" an averaged case and will not fully solvate either side. The problem stems from the fact that when we partition the full system into interacting subsystems and then make the mean-field assumption, we essentially "trap" quantum coherence within the two separate subspaces and do not allow for the mixing of phase coherence. Energy can flow between the two subspaces, but phase information cannot. Consequently, the system is forced to remain too coherent and never resolves itself into either state. A number of "fixes" have been proposed [6–14] to kill off coherence and force the system to localize in one state or the other. We shall pick up with this discussion of coherence and decoherence in detail in a later chapter.

In effect, we are really solving the two-level system problem posed earlier in this chapter. For the sake of discussion, we limit ourselves to two electronic states, labeled a and b, and write our $H_e(r)$ as

$$H_e(r) = \begin{pmatrix} E_a(r) & \lambda \\ \lambda & E_b(r) \end{pmatrix}$$

(4.292)

where $E_a(r)$ and $E_b(r)$ define two potential energy surfaces and λ is some constant coupling. When $\lambda = 0$, the two potential energy curves will cross at some point.

However, when $\lambda \neq 0$, the two curves avoid each other as seen in Figure 4.6. Let us center our frame of reference at the point of intersection where $E_a(0) = E_b(0)$. We know from our previous analysis that if the coupling is much weaker than the energy difference (in the uncoupled representation), then the probability to make a transition between the two states will be vanishingly small. Also for the sake of discussion, let us consider this as a scattering problem whereby the nuclear motion is from $r \to -\infty$ to $r \to +\infty$ and appears at the crossing point at $t = 0$. Also at $t \to -\infty$, we presume the system is prepared in one of the two states ϕ_a or ϕ_b, which are eigenstates of the uncoupled system (that is, with $\lambda = 0$). We shall refer to this representation as the diabatic representation. As the system progresses from left to right, the diabatic states will mix, leading to a superposition of states

$$\psi = c_a\phi_a + c_b\phi_b \tag{4.293}$$

At $t \to \infty$ the coefficients $|c_a|^2$ and c_b give the probability for either remaining in the initial state or making a transition from state a to state b.

We can equally well picture a representation where $H_e(r)$ is diagonal at each point along r. Although the physics (that is, what we eventually compute or observe) will not depend upon our choice of representation, our description of the physics may be quite different. In this *adiabatic* representation, the electronic coupling is described by Equation 4.288.

$$\hat{V}_{12} = \dot{r} \cdot \langle \phi_1 | \nabla_r | \phi_2 \rangle \tag{4.294}$$

We can estimate this using the "off-diagonal" Hellmann–Feynman theorem

$$\frac{d}{dr}\langle \phi_1 | H | \phi_2 \rangle = \langle \phi_1' | H | \phi_2 \rangle + \langle \phi_1 | H' | \phi_2 \rangle + \langle \phi_1 | H | \phi_2' \rangle = 0. \tag{4.295}$$

Rearranging, we find

$$\langle \phi_1 | \nabla_r | \phi_2 \rangle = \frac{\langle \phi_1 | H' | \phi_2 \rangle}{E_1 - E_2} \tag{4.296}$$

Again, when we are far from the point of intersection, $\dot{r} \cdot \langle \phi_1 | H' | \phi_2 \rangle \ll E_1 - E_2$ and the coupling can be ignored. In this limit, both the adiabatic states ϕ_1 and ϕ_2 and the diabatic states ϕ_a and ϕ_b are equivalent. To the left, the lower state $\phi_1 = \phi_a$ and the upper state $\phi_2 = \phi_b$. However, to the right as $r \to +\infty$, the lower state becomes $\phi_1 = \phi_b$ while the upper adiabatic state becomes $\phi_2 = \phi_a$. In other words, as our state evolves, it becomes a superposition of adiabatic states:

$$\psi = a_1\phi_1 + a_2\phi_2 \tag{4.297}$$

with $|a_1|^2$ being the probability for the system to be found in the lower adiabatic state. Again starting off at t in the distant past with the system prepared to the left in the lower adiabatic state $|a_1(r \to -\infty)|^2 = 1$, we find at long time and when the nuclear coordinate has progressed though the intersection $|a_1(r \to \infty)|^2 = |c_b|^2$ and $|a_2(r \to \infty)|^2 = |c_a|^2$. Thus, the probability to remain on the lower adiabatic surface is given by $P_{1\to 1} = |c_b(r \to \infty)|^2 = |a_1(r \to \infty)|^2 = P_{a\to b}$ and the probability for making a transition to the other surface is $P_{1 \to 2} = |c_a(r \to \infty)|^2 = |a_2(r \to \infty)|^2 = P_{a \to a}$.

We can approximate the probability for making the transition using the Landau–Zener approach [15–17]:

$$P_{1 \to 1} = P_{a \to b} = \left(1 - \exp\left(-\frac{2\pi |V_{ab}|^2}{\hbar \partial_t (E_a(r) - E_b(r))} \right) \right)_{r=r_c} \tag{4.298}$$

where the time dependence of the energy gap is due to the motion along r. Taking the derivative,

$$\frac{d}{dt}(E_a(r) - E_b(r)) = \dot{r}(F_b - F_a) \tag{4.299}$$

where $F_a = -\nabla_r E_a$ is the force acting on the nuclear coordinate at r from either the lower (F_a) or upper (F_b) adiabatic surfaces. Note that all of these quantities are to be evaluated at the point of crossing r_c, and $\dot{r}$ is the velocity at the point of crossing. Hence, F_a and F_b are the slopes of the diabatic curves at the point of crossing.

Again, in the limit of weak coupling or high velocity through the coupling region, $2\pi |V_{ab}|^2 \ll \hbar \dot{r}(F_b - F_a)$ and

$$P_{1 \to 1} \approx \left[\frac{2\pi |V_{ab}|^2}{\hbar \dot{r} |F_b - F_a|} \right]_{r=r_c} \tag{4.300}$$

Thus, the probability to remain in the original adiabatic state is very small. This is referred to as the *nonadiabatic* limit. On the other hand, in the limit of large coupling or slow motion, the exponential term in the Landau–Zener equation (Equation 4.298) vanishes and the system remains on the original adiabatic surface throughout the scattering process.

4.10.1 ELECTRONIC TRANSITION RATES

As an application of the Landau–Zener treatment, let us consider the simple model for charge transfer suggested by Figure 4.6. For the sake of discussion, let $|a\rangle$ represent the quantum state where species "A" is an electron donor and "B" is an electron acceptor. The polarization field (represented by the arrows) is generated by polar solvent molecules around the two species. We assume that the A and B are fixed in space. The reaction then is for A to pass its electron over to B. For example, in the ferric-ferrous *self-exchange reaction*,

$$Fe^{3+} + Fe^{2+} \to Fe^{2+} + Fe^{3+}$$

Thus, $|a\rangle$ corresponds to the reactant state and $|b\rangle$ is the product state. We can also have a more general *cross-electron transfer* if the species are different, for example,

$$Fe^{2+} + Ce^{4+} \to Fe^{3+} + Ce^{3+}$$

In these chemical reactions, no bonds are broken, we simply have a rearrangement of the electronic charge density about the two ion sites. If these reactions were to be carried out in a polar medium, the dipoles within the medium would respond by reorganizing themselves to minimize the electrostatic interactions. As suggested by Figure 4.6 and Figure 4.7, we have two minima corresponding to the cases where the solvent polarization fields are organized to minimize these interactions. If we take Q to be some collective polarization coordinate, then we can easily arrive at the parabolic curves:

$$V_a(Q) = E_a + \frac{k}{2}(Q - Q_a)^2 \tag{4.301}$$

$$V_b(Q) = E_b + \frac{k}{2}(Q - Q_b)^2 \tag{4.302}$$

Furthermore, since the electronic coupling is only significant close to the crossing point, we shall assume it is independent of Q and equal to V_{ab}.* These curves cross at Q_c where $V_a(Q_c) = V_b(Q_c)$. Simple algebra yields

$$Q_c = \frac{(E_a - E_b) + k\left(Q_a^2 - Q_b^2\right)/2}{k(Q_a - Q_b)} \tag{4.303}$$

For the forward reaction, the activation energy is the energy difference between E_1 and the energy at the crossing:

$$E_A = \frac{(E_a - E_b) - \lambda}{4\lambda} \tag{4.304}$$

where

$$\lambda = \frac{k}{2}(Q_a - Q_b)^2 \tag{4.305}$$

This last term carries an important physical meaning. It is the energy required to reorganize the polarization following the transfer of a charge from A to B. These terms are shown in Figure 4.7 along with a simple sketch of the parabolic potentials.

To get to a transition rate, we need to compute the expectation value that our system will arrive at the crossing at an appropriate velocity,

$$k_{a \to b} = \int_0^\infty d\dot{Q} P(Q_c, \dot{Q}) P_{a \to b}(\dot{Q}) \tag{4.306}$$

$P_{a \to b}(\dot{Q})$ we get from the Landau–Zener expression above. $P(Q_c, \dot{Q})$ we get by taking the Boltzmann probability that the system will have the appropriate velocity at the crossing point

$$P(Q_c, \dot{Q}) = \sqrt{\frac{\beta m}{2\pi}} e^{-\beta m \dot{Q}^2/2} e^{-\beta E_A} \int_{-\infty}^{Q_c} e^{-\beta(V_a - E_a)} \tag{4.307}$$

---

* In keeping with our notation above, V_a and V_b will represent the uncoupled (diabatic) potentials and V_1 and V_2 will denote the adiabatic potentials.

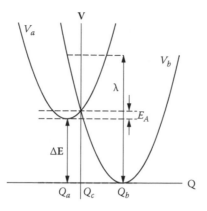

FIGURE 4.7 Sketch of parabolic free-energy potentials arising from Marcus' treatment of electron transfer. In this figure, E_A is the activation energy, ΔE is the driving force taken as the (free) energy difference between the initial and final states. λ is the reorganization energy.

Taking the integration limit to $+\infty$, we arrive at

$$P(Q_c, \dot{Q}) = \frac{\beta}{2\pi} \sqrt{mk} e^{-\beta E_A} e^{-\beta m \dot{Q}^2 / 2} \qquad (4.308)$$

where $\beta = 1/k_B T$. In the adiabatic limit, $P_{a \to b} \approx 1$ and we arrive at an expression for the transition rate very much like what we expect from transition state theory

$$k_{ad} = \frac{\omega_c}{2\pi} e^{-\beta \tilde{E}_A}$$

where $\omega_c / 2\pi$ is the transmission frequency, $\omega_c = \sqrt{k/m}$, and $\tilde{E}_A$ is the activation energy. Since the electronic coupling opens an energy gap of $2V_{ab}$ at the crossing point, we have written $\tilde{E}_A = E_A - V_{ab}$ to account for the electronic coupling.

In the nonadiabatic limit, we use the weakcoupling form for $P_{a \to b}$ and obtain

$$k_{na} = \sqrt{\frac{\pi k \beta}{2}} \frac{|V_{ab}|^2}{\hbar |F_b - F_a|} e^{-\beta E_A}$$

The $\Delta F = |F_b - F_a|$ is simply the difference in forces at the crossing point. Since our potentials are parabolic $F_a = k(Q_c - Q_a)$ and $F_b = k(Q_c - Q_b)$, $\Delta F = k \Delta Q$ where ΔQ is the distance between the two potential minima. The nonadiabatic rate is similar to the adiabatic rate in that it depends upon both the activation energy and the force constant. However, it does not depend upon the mass. The fact that the electron coupling appears as $|V_{ab}|^2 / \hbar$ reminds us that this is the first-order term in the perturbation expansion.

While the Landau–Zener model does account for the nuclear motion in a semi-classical way, it does not treat the motion fully quantum mechanically nor does it fully account for the electronic coherences between the two electronic states. These coherences and correlations are vitally important in the weak coupling (nonadiabatic limit) and cannot be entirely ignored. Finally, there are a number of parameters—k, Q_a, Q_b, and so on—that need to be inferred from experiments and as such may be difficult to obtain.

4.10.2 MARCUS' TREATMENT OF ELECTRON TRANSFER

In the 1950s, Rudy Marcus examined the problem of electron transfer in a polar medium and gave a solid physical significance to the parabolic potentials we used above. In particular, we used a generic coordinate Q to characterize the reorganization of the medium about the two charge distributions about initial and final states.[18,19] In his original work, he treats the molecular solvent as a dielectric continuum rather than an explicit solvent model. Although this lacks the molecularity of an explicit solvent, it does allow one to simplify response of the solvent by its characteristic dielectric time scales. Within the dielectric continuum model, one typically assumes that the total response is characterized by two distinct time scales: a fast one characterizing the electronic dynamics and a much slower time scale characterizing the nuclear (molecular) dynamics. The static dielectric constant ε_s contains contributions from both. The electronic contribution (termed the *optical response*) is related to the index of refraction via $\varepsilon_o = n^2$. The nuclear component ε_s contains contributions from the translational, rotational, and vibrational motions of the solvent species. Unfortunately, one cannot write ε_s as $\varepsilon_n + \varepsilon_e$, because the electronic degrees of freedom of the solvent depend strongly upon its instantaneous nuclear arrangement.

The critical assumption underlying Marcus' approach is that change in the electronic density associated with the transfer of an electron occurs on a time scale much faster than the time scale for the nuclear charges to respond *but* it is slow on the time scale for the electronic motions that determine ε_e. As such, electron-transfer events occur at constant nuclear *polarization* as determined by fixed nuclear positions. This is more or less a statement of the Franck–Condon principle.

The essential feature of the Marcus treatment is that the rate constant can be expressed in a very compact form:

$$k_{et} = \frac{2\pi}{\hbar} |V_{ab}|^2 FCF \qquad (4.309)$$

The Franck–Condon factor (FCF) can be written as the overlap integral between a vibrational state in the V_a potential centered about Q_a and vibrational eigenstate in the V_b potential centered about Q_b. Assuming the two wells have the same vibrational frequency, this is the overlap between two displaced Gauss–Hermite wave functions. In Marcus' approach, the FCF term is computed using a semiclassical approximation and takes the form:

$$FCF = \frac{1}{\sqrt{4\pi\lambda kT}} \exp\left[-\frac{(\Delta E + \lambda)^2}{4\pi\lambda kT}\right] \qquad (4.310)$$

Since the potentials are parabolic, the displacements between the wells can be expressed in terms of specific energy differences, namely, the driving force $\Delta E = E_b - E_a$, which is the difference between the energy minima of the initial and final states, and the reorganization energy λ, which is the energy required to change the nuclear coordinates without changing the electronic state. In other words, this is the energy of the initial electronic state evaluated at the equilibrium geometry of the final electronic state. Both ΔE and λ can be obtained from the emission and absorption spectra of the system, respectively.

The coupling V_{ab} can also be determined spectroscopically by examining the transition moment between the two *adiabatic* states. Recall, that a and b label the localized or diabatic states and 1 and 2 label the adiabatic eigenstates of a model two-level system with Hamiltonian

$$H = \begin{pmatrix} E_a & V_{ab} \\ V_{ab} & E_b \end{pmatrix} \tag{4.311}$$

Electronic transitions occur between energy eigenstates of H, not between the diabatic states. It is important that we make this distinction. Recall our discussion of light absorption earlier in this chapter. We assumed initially that the system was at equilibrium and perturbed only by the electromagnetic field of the incident photon. Consequently, the initial state for optical absorption must be an eigenstate of H. Fortunately, far from the crossing region, $\psi_a \approx \psi_1$ close to Q_a and $\psi_b \approx \psi_1$ close to Q_b.

To find an expression for the coupling, begin by writing the transition moment between ψ_1 and ψ_2 as

$$\mu_{12} = e\langle \psi_1 | r | \psi_2 \rangle$$

and then expand the eigenstates in terms of the diabatic states

$$|\psi_1\rangle = \cos\theta |\psi_a\rangle + \sin\theta |\psi_b\rangle \tag{4.312}$$

$$|\psi_2\rangle = -\sin\theta |\psi_a\rangle + \cos\theta |\psi_b\rangle \tag{4.313}$$

where θ is the mixing angle. If we assume that the transition moment between the diabatic state vanishes, $\mu_{ab} = e\langle \psi_a | r | \psi_b \rangle = 0$, and let μ_a and μ_b be the *static* dipole moments of the donor and acceptor species, then we can write

$$V_{ab}|\Delta\mu| = \hbar\omega_{max}\mu_{12}$$

where we have written $\hbar\omega_{max} = E_2 - E_1$ is the optical absorption maximum. As we showed earlier as well, the oscillator strength is related to the transition moment by

$$f_{osc} = \frac{2m_e\omega_{no}}{e^2\hbar}|\mu_{12}|^2 \tag{4.314}$$

where e and m_e are the charge and mass of an electron. Thus, measuring the oscillator strength gives the transition moment. Furthermore, if we assume that a single charge is being passed between the donor and acceptor and they are separated by some distance R_{ab}, then $\Delta\mu = eR_{ab}$.

Predictions of the Marcus' theory: The Marcus rate equation makes an interesting prediction. If we consider the rate as a function of the driving force,

$$\log k \propto -\frac{(\Delta E + \lambda)^2}{\lambda}$$

as we increase the driving force and keep the reorganization energy roughly constant, then at some point we reach a maximum rate where $\Delta E = \lambda$. Increasing ΔE beyond this will actually lead to a decrease in the electron transfer rate. Experimental verification of this turnover in the rate did not occur until nearly 30 years after Marcus made

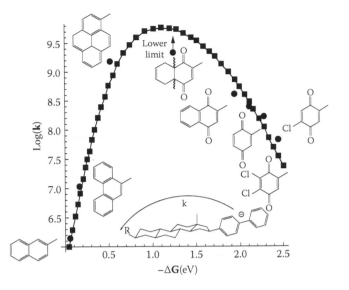

FIGURE 4.8 Comparison between predicted and experimental electron transfer rates between a series of donor/acceptor species. Here, k is in units of 1/sec. Note that $-\Delta G$ in this figure is equivalent to ΔE in our discussion. (Adopted from Refs. 19 and 20.)

this prediction. In a true tour de force of synthesis and spectroscopy, Gerhard Closs' group produced a series of donor-acceptor molecules in the form of a bi-phenyl radical anion separated from an organic acceptor held a fixed distance away by a linking chain.[20] A plot of the observed rates vs. the driving force is shown in Figure 4.8. Here we see that the rate constant increases up to a maximum value of about $2 \times 10^9 \mathrm{s}^{-1}$ with increasing $|\Delta G^\circ|$. Note, that the ΔE used in our discussion should be taken as free energy change between final and initial states and the activation free energy $\Delta G^{\dagger\dagger} = (\lambda + \Delta G^\circ)^2/4\lambda$.

From the rate expression, $\Delta G^{\dagger\dagger}$ decreases as ΔG° becomes increasingly negative and the reaction becomes more and more exothermic. When $\Delta G^\circ = -\lambda$, the activation energy vanishes and any further increase in the exothermicity causes $\Delta G^{\dagger\dagger}$ to increase, leading to a decrease in the rate constant. Looking at Figure 4.8, we can identify these three regimes. First, the normal regime where $-\Delta G^\circ < \lambda$. Here increasing $|\Delta G^\circ|$ leads to an increase in the rate since the barrier for the reaction steadily decreases. At the point $-\Delta G = \lambda \approx 1$ eV, the barrier vanishes and we achieve the maximum electron transfer rate. Making the reaction increasingly exothermic only serves to increase the energetic barrier for the reaction and hence leads to a decrease in the rate. This regime is termed the "inverted regime." Sketches of the energy parabolas corresponding to each of these regimes are given in Figure 4.9.

4.10.3 INCLUDING VIBRATIONAL DYNAMICS

Now that we have established the basic physical picture, let us expand on it a bit and apply the golden rule technology we have just developed to analyze this problem.

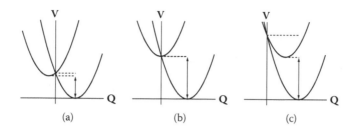

FIGURE 4.9 Potential energy curves corresponding to the normal (a), barrierless (b), and inverted (c) regimes for electron transfer. The vertical arrow denotes the driving force ΔE and the gap between the dashed lines is the activation energy.

Clearly, one of the problems with the Landau–Zener approach is that we have neglected the quantum motion of the nuclear degree of freedom. As such, we consider here a semiclassical approach developed by Neria and Nitzan.[21] The idea here is that the nuclear vibrational motion on the potential energy surface of the initial electronic state drive transitions to vibrational states on the potential energy surface of the final electronic state. We begin by writing the golden rule expression for the transition between electronic states 1 and 2 as

$$k_{12} = \frac{2\pi}{\hbar} \sum_i \frac{e^{-\beta E_{1i}}}{Z_1} \sum_f |\langle 1i|V|2f\rangle|^2 \delta(E_{1i} - E_{2f}) \tag{4.315}$$

where $\beta = 1/kT$, Z is the vibrational partition function for the initial state, and $|i\rangle$ and $|f\rangle$ are nuclear states associated with the initial and final electronic states. Integrating over electronic degrees of freedom, we can write $V_{12} = \langle 1|V|2\rangle$ as it is still an operator acting on the nuclear degrees of freedom. Moreover, following our discussion above, the rate constant can be expressed as a correlation function as

$$k_{12} = \int_{-\infty}^{\infty} dt\, e^{i\Delta Et/\hbar} C(t) \tag{4.316}$$

where ΔE is the difference between the energy origin of the two potential energy surfaces. The correlation function is given by

$$C(t) = \frac{1}{\hbar^2 Z} \sum_i e^{-\beta E_i} \langle i\, V_{12} e^{i H_2 t/\hbar} V_{21} e^{-i H_1 t/\hbar} |i\rangle \tag{4.317}$$

$$= \frac{1}{\hbar^2} \langle V_{12} e^{i H_2 t/\hbar} V_{21} e^{-i H_1 t/\hbar} \rangle_T \tag{4.318}$$

where the T subscript denotes a thermal averaging over initial conditions. H_1 and H_2 denote the Hamiltonians for nuclear motion on the *adiabatic* potential energy curves and V_{12} is the nonadiabatic coupling as given above.

As a model, we consider the crossing of two diabatic potential curves: one, a harmonic well describing a bound molecular state, and the other a linear potential representing an unbound or dissociative state:

$$V_a(x) = x^2/2 \tag{4.319}$$

$$V_b(x) = \alpha x + E_o \tag{4.320}$$

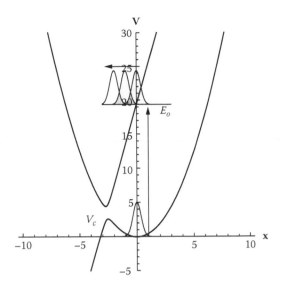

FIGURE 4.10 Schematic view of Gaussian wave packet propagation scheme in computing the correlation function in Equation 4.318. The shaded Gaussians denote the initial wavepackets with the arrow indicating evolution on the upper potential energy curve. (Figure adopted from Ref. 22.)

where α is the slope and E_o determines the vertical excitation energy. The crossing energy is given by $V_c = \alpha \pm \sqrt{\alpha^2 + 2E_o}$. These curves are shown in Figure 4.10 for the case of $\alpha = 6$ and $E_o = 19.65$ following Ref. 22. Also, for the sake of convenience, we use scaled units for position ($\sqrt{\hbar/(m\omega)}$) and momentum ($\sqrt{m\hbar\omega}$) so that the energy is in units of the harmonic oscillator frequency, ($\hbar\omega$). Also, as above, we take the diabatic coupling $V_{ab} = \lambda$ to be a constant. The resulting adiabatic curves are given in Figure 4.10.

As a first approximation, let us adopt a purely diabatic viewpoint and factor completely the electronic coupling so that the correlation function can be written as

$$C(t) = \frac{|V_{ab}|^2}{\hbar^2 Z_1} \langle J(t) \rangle \qquad (4.321)$$

where $J(t)$ is a time-dependent overlap between a Gaussian wave packet starting at position $x = 0$ on the lower potential and a Gaussian wave packet starting at $x = 0$ on the the upper potential surface. In doing so, we assume that the electronic transition occurs instantaneously on the time scale of nuclear motion. Given the disparity between the forces on either surface, the two wave packets will rapidly move apart and $C(t)$ will rapidly converge to zero. Consequently, over this time frame, the wave packets will more or less retain their original shape and only their centroids will change. Thus, we can estimate $J(t)$ by taking the time-dependent overlap between two Gaussians, one moving in the lower harmonic well weighted by the initial thermal populations and the other moving on the upper linear slope.

We take the lower state to be a harmonic oscillator eigenstate

$$\psi_n(x) = \frac{1}{\sqrt{n!2^n a \sqrt{\pi}}} e^{-x^2/(2a^2)} H_n(x/a) \tag{4.322}$$

where $a = \sqrt{\hbar/(m\omega)}$. The upper state is not stationary and evolves according to

$$i\hbar\dot{\psi} = \left(\frac{p^2}{2m} + \alpha x + E_o \right)\psi \tag{4.323}$$

The exact solution was originally derived by de Broglie[1,23–25]

$$\psi(x,t) = R(t)e^{iS(t)/\hbar} \tag{4.324}$$

$$R(t) = \frac{1}{(\sqrt{2\pi}\sigma(t))^{1/2}} e^{-\frac{(\frac{\alpha t^2}{2m} - ut + x)^2}{4\sigma(t)^2}} \tag{4.325}$$

$$S(t) = -\frac{\alpha^2 t^3}{6m} + \frac{\left(\frac{\alpha t^2}{2m} - ut + x\right)^2 \hbar^2 t}{8m\sigma^2\sigma(t)^2} + \left(x - \frac{tu}{2}\right)(mu - t\alpha)$$
$$- \frac{3}{2}\hbar \tan^{-1}\left(\frac{t\hbar}{2m\sigma^2}\right) \tag{4.326}$$

with the time-dependent width given by $\sigma(t) = \sqrt{\sigma + (\hbar^2 t^2/(2m)2\sigma}$ with σ being the initial width and u being the initial group velocity. Notice that this is the same as what one finds for the spread of a free particle. We thus construct the integral

$$J(t) = \int_{-\infty}^{\infty} \psi_0(x)R(t)e^{iS(t)/\hbar}dx \tag{4.327}$$

choosing the initial width of the upper state to match that of the initial harmonic wave function. (See Figure 4.10.) The resulting integral can be tediously worked out by hand; however, numerical evaluation can be readily done. The resulting $J(t)$ decay curve for the problem at hand is shown in Figure 4.11c.

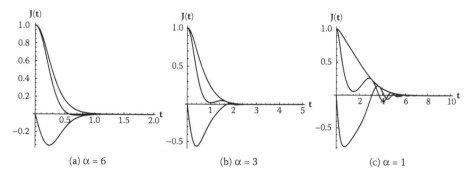

FIGURE 4.11 Time-dependent Franck–Condon overlap integral between harmonic oscillator ground state and a Gaussian moving on a linear ramp. Plotted are the real, imaginary, and absolute values.

The rate of decay of $\langle J(t)\rangle$ depends upon how rapidly the wave packet on the linear potential moves away from the initial state. The steeper the potential, the faster the upper wave packet loses overlap with the lower wave function. This is an indication of how long it takes the system to lose memory of its initial condition. Once this memory has been lost, the correlation between the initial and final states is zero, and the Fourier integral required to compute the golden rule transfer rate will converge.

The final rate constant will depend upon two factors. The transfer will be more efficient if in fact the vibrational spectrum of the final states has a significant overlap with the initial state. For example, if we write as the overlap of two vibrational wave packets evolving on electronic potentials i and f,

$$\langle J(t)\rangle = \langle \psi_f(t)|\psi_i(0)\rangle = \langle \psi_f(0)|\exp[+iH_ft/\hbar]|\psi_i(0)\rangle \tag{4.328}$$

where $\langle \psi_f(0)|\psi_i(0)\rangle = 1$ and H_f is the Hamiltonian for nuclear motion on the final surface, then by inserting a complete set of vibrational states

$$\langle J(t)\rangle = \sum_{n_f} |\exp[+iE_{n_f}t/\hbar]|\langle n_f|\psi_i(0)\rangle|^2 \tag{4.329}$$

we see that $\langle J(t)\rangle$ involves the projection of the initial state onto all possible vibrational eigenstates on the final electronic potential surface and E_f is relative to a common energy origin. Since we are dealing with a continuum of final energy states, the sum must be converted to an integral

$$\sum_{n_f} \rightarrow \int dEg(E) \tag{4.330}$$

where $g(E)$ is normalized to 1:

$$\langle J(t)\rangle = \int dEg(E)\exp[+iEt/\hbar]|\langle E|\psi_i(0)\rangle|^2 \tag{4.331}$$

Thus, upon taking the Fourier transform we can write the transition rate from the nth vibrational eigenstate on the initial electronic state as

$$k_{if} = \frac{2\pi}{\hbar}|V|^2 \int dEg(E)|\langle E|\psi_{ni}\rangle|^2 \delta(E + Eo - E_i) \tag{4.332}$$

The overlap integral we can evaluate exactly since we know the energy eigenstates for a particle under the influence of a constant force $V = x\alpha$

$$\langle x|E\rangle = \frac{1}{2^{1/3}\alpha^{1/6}} Ai\left[(2\alpha)^{1/3}(x - E/\alpha)\right] \tag{4.333}$$

$$= \frac{1}{2\pi\sqrt{\alpha}} \int_{-\infty}^{\infty} dk\,\exp[ik^3/6\alpha + ik(x - E/\alpha)] \tag{4.334}$$

where $Ai(x)$ is the Airy function chosen to be regular at the origin. The overlap integral is then computed using

$$\langle E|\psi_{ni}\rangle = \int_{-\infty}^{\infty} dx\langle E|x\rangle\langle x|\psi_{ni}\rangle \tag{4.335}$$

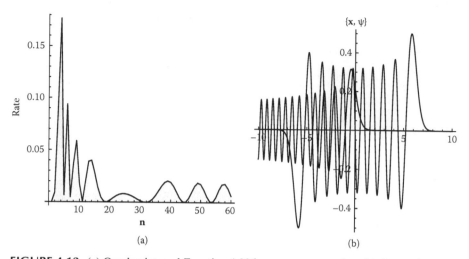

FIGURE 4.12 (a) Overlap integral Equation 4.336 vs. quantum number. (b) Comparison between continuum wavefunction with $E = E_o$ and eigenstate #19.

which can be recast as

$$\langle E | \psi_{ni} \rangle = \frac{1}{\sqrt{2\pi\alpha}} \int_{-\infty}^{\infty} dk \exp[-ik^3/6\alpha + ikE/a] \tilde{\psi}_{ni}(k) \qquad (4.336)$$

where

$$\tilde{\psi}_{ni}(k) = (-i)^n \psi_n(x \to k) \qquad (4.337)$$

is the Fourier transform of the nth harmonic oscillator eigenstate. This last integral over k can be evaluated numerically.

For the case of $\alpha = 6$, $E_o = 19.36$, and $\lambda = 0.5$, we show the rate as a function of the initial quantum number in the lower harmonic well. Notice that it undergoes a series of oscillations corresponding to constructive and destructive overlaps between oscillations in the initial and final vibrational wave functions. Remarkably, even though the two diabatic potential curves cross at $E_o = 19.36$, there is very poor integrated overlap between the $n \approx 19$ vibrational state and an Airy function at that energy due to the oscillations in both wave functions as seen in Figure 4.12b.

4.11 PROBLEMS AND EXERCISES

Problem 4.1 A simple analysis of the two-well model can be performed using the golden rule techniques developed thus far. Consider the case of two identical wells, one displaced from the other by x_o. We can also add an energy shift to the problem E_b.

$$V_a(x) = m\Omega^2 x^2/2 \qquad (4.338)$$

$$V_b(x) = m\Omega^2(x - x_o)^2/2 + E_b \qquad (4.339)$$

Show that the time-dependent overlap between the harmonic oscillator ground-state wave function in state a and an initially identical Gaussian evolving in state b is given

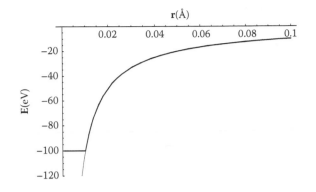

FIGURE 4.13 Coulomb potential for H atom including a cutoff approximating the finite radius of the proton.

where

$$J(t) = \exp\left(-\frac{\Delta^2}{2}(1 - e^{-i\Omega t}) - i\Omega t/2\right) \qquad (4.340)$$

and $\Delta = x_o\sqrt{m\Omega/\hbar}$ is a dimensionless displacement (Huang–Rhys parameter). Show this by taking the Fourier transform of $J(t)$ that the spectral function is given by

$$\sigma(\omega) = \sum_{n=0} e^{-\Delta^2/2}\left[\frac{\Delta^{2n}}{2^n n!}\delta(\hbar\omega - \Omega(n + 1/2))\right] \qquad (4.341)$$

Finally, evaluate and plot the transition rate from state a to state b as a function of temperature.

Problem 4.2 A one-dimensional harmonic oscillator, with frequency ω, in its ground state is subjected to a perturbation of the form

$$H'(t) = C\hat{p}e^{-\alpha|t|}\cos(\Omega t) \qquad (4.342)$$

where $\hat{p}$ is the momentum operator and C, α, and Ω are constants. What is the probability that as $t \to \infty$ the oscillator will be found in its first excited state in first-order perturbation theory? Discuss the result as a function of Ω, ω, and α.

Problem 4.3 A particle is in a one-dimensional infinite well of width $2a$. A time-dependent perturbation of the form

$$H'(t) = T_o V_o \sin\left(\frac{\pi x}{a}\right)\delta(t) \qquad (4.343)$$

acts on the system, where T_o and V_o are constants. What is the probability that the system will be in the first excited state at time t afterwards?

Problem 4.4 Because of the finite size of the nucleus, the actual potential seen by the electron is more like what is seen in Figure 4.12.

1. Calculate this effect on the ground-state energy of the H atom using first-order perturbation theory with

$$H' = \begin{cases} \frac{e^2}{r} - \frac{e^3}{R} & \text{for } r \leq R \\ 0 & \text{otherwise} \end{cases} \tag{4.344}$$

2. Explain this choice for H'.
3. Expand your results in powers of $R/a_o \ll 1$. (Be careful!)
4. Evaluate numerically your result for $R = 1$ fm and $R = 100$ fm.
5. Give the fractional shift of the energy of the ground state.
6. A more rigorous approach is to take into account the fact that the nucleus has a homogeneous charge distribution. In this case, the potential energy experienced by the electron goes as

$$V(r) = -\frac{Ze^2}{r} \tag{4.345}$$

when $r > R$ and

$$V(r) = -\frac{Ze^2}{r} \left(\frac{1}{2R} \left(\left(\frac{r}{R}\right)^2 + 2\frac{R}{r} - 3 \right) - 1 \right) \tag{4.346}$$

for $r \leq R$. What is the perturbation in this case? Calculate the energy shift for the H (1s) energy level for $R = 1$ fm and compare to the result you obtained above.

Note that this effect is the "isotope shift" and can be observed in the spectral lines of the heavy elements.

Problem 4.5   As a good exercise in commutation relations and identity insertions, derive the Thomas–Reiche–Kuhn sum rule. Show that for a harmonic oscillator, the result is exact. Finally, apply the sum rule to a linear rotor. Does the rule still hold?

Problem 4.6   Adiabatic vs. Sudden Approximations. There are two essential limits for time-dependent problems: first, where the perturbation or coupling varies slowly in time and the other when the coupling is suddenly switched on. In the chapter we discussed the case where the reference system was coupled to some time-dependent field. In this problem, we consider the case where the boundary conditions are changed. Consider the case of an electron trapped in an infinite well of length L. The energy levels are discrete and we shall assume that the electron is initially prepared in the lowest energy level. The twist here is that we shall allow L to change with time, something like a piston that can compress and expand the electron's box.

1. What outside pressure must be exerted on the electron in order for L to be fixed at some length L_{eq}?
2. Show that by expanding the electron's wave function in terms of the time-dependent states

$$\psi(t) = \sum_j u_j(t)|j(t)\rangle$$

where

$$|j(t)\rangle = \sqrt{\frac{2}{L(t)}} \cos(\pi(2j+1)x/L(t))$$

are the particle-in-a-box states with x measured from the center of the well and with energy

$$\varepsilon_j(t) = \frac{\hbar^2 2\pi(2j+1)}{2mL(t)^2} = \hbar\omega_j(t)$$

and substituting this into the time-dependent Schrödinger equation,

$$\hbar i \frac{\partial}{\partial t} u_j e^{-\omega_j(t)t} = \hbar i \left\langle j \left| \frac{\partial}{\partial t} \right| 0 \right\rangle e^{-\omega_0(t)t}$$

3. Evaluate $\langle j|\frac{\partial}{\partial t}|0\rangle$ and show that

$$\frac{\partial u_j}{\partial t} = -\lambda_j \pi \frac{\partial \log L(t)}{\partial t} \exp[i(\omega_j(t) - \omega_0(t)]$$

where

$$\lambda_j = 2 \int_{-1/2}^{1/2} \cos[\pi(2j+1)u] u \sin[\pi u] du$$

4. To proceed, we need to specify $L(t)$. Consider a sudden change in L_o to $L_o + \Delta L$ over a time t_o. In other words $L(t) = L_o + \Delta L(\exp(t/t_o) - 1)$. In doing so, $\partial \log L/\partial t \approx -(\Delta L/L)\exp(-t/t_o)/t_o$. Use this and show that the transition probability from the initial state to some final state $|j\rangle$ is given by

$$|u_j|^2 = \left(\frac{\pi\lambda_j\Delta L}{L}\right)^2 \frac{1}{1 + (\omega_j - \omega_o)^2 t_o^2}$$

5. Show that if the rate of change in the boundary is large compared to the transition frequency, then this last expression approaches unity, indicating an abrupt transition. Show also if $1/t_o$ is small (slowly changing boundary), then $|u_j|^2 \to 0$, indicating the electron remains in its initial state.

Problem 4.7   The potential function for an anharmonic oscillator of mass m is given by

$$V(x) = \frac{k}{2}x^2 + cx^4$$

where the second term is small compared with the first. First, show that the first-order correction to the ground-state energy is given by

$$E - E_o = 3c(\hbar/2m\omega)^2$$

What would the first-order energy correction be if there were an x^3 term in V?

Problem 4.8   Consider a particle of mass m in a harmonic well with force constant k. A small perturbation is applied that changes the force constant by δk. Show that the first- and second-order corrections to the ground-state energy are given by

$$E^{(1)} = \frac{1}{4}\frac{\delta k}{k}\hbar\omega$$

and

$$E^{(2)} = -\frac{1}{16}\left(\frac{\delta k}{k}\right)^2\hbar\omega$$

How do these expressions relate to the exact expression for the energy?

Problem 4.9   Taking a trial wave function of the form $\phi = \exp(-\beta r^2)$ where β is an adjustable parameter, use the variational procedure to obtain an estimate of the ground-state energy for the hydrogen atom in terms of atomic constants. How does this compare to the exact answer? Also, use your optimized wave function to compute $\langle r \rangle$, $< 1/r >$, and $\langle p^2 \rangle$. Compare your results with the exact values for a hydrogenic system.

Problem 4.10   For an attractive 1D square well potential, it is possible to show that there is always at least one bound state. Does this hold true for any one-dimensional attractive potential of arbitrary shape?

Problem 4.11   Consider a one-dimensional harmonic oscillator with mass m and angular frequency ω. At time $t = 0$, the following state is prepared:

$$|\psi(0)\rangle = \frac{1}{\sqrt{2s}}\sum_{n=N-s}^{N+s}|n\rangle$$

where $|n\rangle$ is an eigenstate of the Hamiltonian with energy $E_n = \hbar\omega(n + 1/2)$. The summation runs from $n = N - s$ to $n = N + s$ with $N \gg s \gg 1$. Show that the expectation value of $\langle x(t) \rangle$ is oscillatory with amplitude $(2\hbar N/m\omega)^{1/2}$. How does this compare to the time variation of the displacement for a classical oscillator?

Problem 4.12   At $t = -\infty$ an oscillator is in its ground state $|0\rangle$. Determine the probability that at $t = +\infty$ the oscillator will be in the nth excited state if it is acted upon by a force $f(t)$, where $f(t)$ is an arbitrary even function of time with $f = 0$ at $t = \pm\infty$. Evaluate the expression for the following choices of $f(t)$:

1. $f(t) = f_o e^{-t^2/\tau^2}$
2. $f(t) = f_o/((t/\tau)^2 + 1)$

Problem 4.13   The S states for an electron in a spherical cavity of radius R are given by

$$\psi_n(r) = A_n \sin(n\pi r/R)/r$$

where n is the radial quantum number $n = 1, 2, 3, \ldots$ and A_n is chosen to ensure normalization. The first two of these are shown in the graphic below:

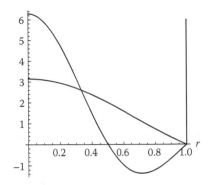

At time $t = 0$, the radius of the cavity is rapidly (and instantly) expanded to $R_f = 1.1 R_i$. Plot as a function of n the probability of finding the electron in each of the eigenstates of the new (expanded) potential.

What would be the R_f if the probability for finding the electron in the $n = 1$ state was exactly 0.5?

Problem 4.14 Consider the case of the vibrational motion of a linear triatomic molecule such as $C = N - H$ where the harmonic stretching frequency of one bond is much higher than the stretching frequency of the other bond so that the low-frequency mode can treated essentially classically. A suitable Hamiltonian for this is

$$H = \hbar \omega a^\dagger a + \lambda (a^\dagger + a)x + \frac{1}{2}(p^2 + \Omega^2 x^2) \qquad (4.347)$$

where a and $a^\dagger$ are the anhiliation and creation operators for a quantum harmonic well with frequency ω, p and x are the classical momentum and position for an oscillator with frequency Ω, and λ the coupling between the two systems. If the low-frequency mode is described by a classical harmonic oscillator, what is the golden rule transition rate for the high-frequency part to relax from its first excited state to the ground state? What happens if $\Omega \ll \omega$?

Problem 4.15 Along a reaction coordinate, $R(t)$, the harmonic frequency of a molecule can change. For the sake of building a model, let us consider the case where the harmonic frequency of a diatomic molecule is increased then decreased,

$$\omega(t) = \omega_o + \Delta \omega \exp(-t^2/\tau)$$

with time scale τ due to a collision with an atom. Derive an expression for finding the molecule in its lowest vibrational state at $t \to \infty$ given that it was in its ground vibrational state at $t \to -\infty$. Since τ is related to the collision energy (that is, the speed of the colliding atom), what happens if the collision is very slow or very fast on the time scale of ω? What collisional time scale is needed for the survival probability to be exactly 50%?

5 Representations and Dynamics

In this chapter we shall examine different ways in which one can represent the evolution of a quantum state. While mathematically different, the various representations are equivalent in that, at the end of the day, one obtains the same physical prediction. This is good because physics and physical measurements should not depend upon how one chooses to represent the quantum state. Depending upon the problem at hand, each representation has its unique advantages and disadvantages.

5.1 SCHRÖDINGER PICTURE: EVOLUTION OF THE STATE FUNCTION

By its very nature, quantum mechanics is a linear theory since all results are ultimately derived by starting with the time-dependent Schrödinger equation

$$i \frac{\bar{\partial}}{\partial t} \psi(t) = \hat{H} \psi(t) \tag{5.1}$$

As we know from the postulates of quantum mechanics, the state of the system at time t is described by $\psi(t)$, which is a solution of the TDSE where $\hat{H}$ is the Hamiltonian operator derived from the classical Hamiltonian function by the substitution of

$$r \rightarrow \hat{r} \tag{5.2}$$

$$V(r) \rightarrow V(\hat{r}) \tag{5.3}$$

$$p \rightarrow -i\hbar \frac{\partial}{\partial r} \tag{5.4}$$

in the coordinate representation or equivalently in the momentum representation

$$p \rightarrow \hat{p} \tag{5.5}$$

$$r \rightarrow i\hbar \frac{\partial}{\partial r} \tag{5.6}$$

$$V(r) \rightarrow V\left(i\hbar \frac{\partial}{\partial p}\right) \tag{5.7}$$

with

$$V\left(i\hbar \frac{\partial}{\partial p}\right) = \sum_{n=0}^{\infty} \frac{i^n \hbar^n}{n!} \left.\frac{\partial^n V}{\partial r^n}\right|_{r=0} \frac{\partial^n}{\partial p^n} \tag{5.8}$$

Consider the time evolution of an observable associated with an operator $\mathscr{A}_S$. The subscript S indicates that we will be working in the Schrödinger representation, which is more or less the de facto representation of quantum mechanics as specified by the postulates. In the Schrödinger representation (or picture), the time evolution is carried by the state vector, which in turn is a solution of the TDSE. The expectation value of $\mathscr{A}_S$ is given by

$$\langle \mathscr{A}_S \rangle (t) = \langle \psi_S(t) | \mathscr{A}_S | \psi_S(t) \rangle \tag{5.9}$$

Taking the time derivative,

$$i\hbar \frac{d}{dt} \langle \mathscr{A}_S \rangle (t) = i\hbar \langle \dot{\psi}_S(t) | \mathscr{A}_S | \psi_S(t) \rangle i\hbar \langle \psi_S(t) | \mathscr{A}_S | \psi_S(t) \rangle i\hbar \langle \psi_S(t) | \mathscr{A}_S | \dot{\psi}_S(t) \rangle$$

$$= -\langle \psi_S(t) | [H, \mathscr{A}_S] | \psi_S(t) \rangle + i\hbar \langle \psi_S(t) | \dot{\mathscr{A}}_S | \psi_S(t) \rangle \tag{5.10}$$

If the operator itself carries no explicit time dependency, then the time evolution of the expectation value of an observable is specified by

$$i\hbar \frac{d}{dt} \langle \mathscr{A}_S \rangle = -\langle \psi_S(t) | [H, \mathscr{A}] | \psi_S(t) \rangle \tag{5.11}$$

If, in fact, $[\mathscr{A}, H] = 0$, then the observable associated with $\mathscr{A}$ is a constant of the motion.

For completion, the time evolution of the Schrödinger state is given by

$$\psi_S(t) = U_S(t, t_o) \psi_S(t_o) \tag{5.12}$$

5.1.1 Properties of the Time-Evolution Operator

Let us consider briefly the mathematical properties of the time-evolution operator. Expanding the Schrödinger state as a polynomial in time,

$$\psi_S(t) = \psi(0) + t\dot{\psi}(0) + \frac{t^2}{2}\ddot{\psi}(0) + \cdots \tag{5.13}$$

$$= \left(1 + \frac{t}{i\hbar} H + t^2 \frac{1}{2(i\hbar^2)} H^n \cdots \right) \psi(0) \tag{5.14}$$

$$= \sum_{n=0}^{\infty} t^n \frac{1}{n!(i\hbar)^n} H^n \tag{5.15}$$

$$= e^{-iHt/\hbar} \psi(0) \tag{5.16}$$

$$= U_S(t, 0) \psi(0). \tag{5.17}$$

The fact that U_S can be expressed as a polynomial in time has a number of advantages. We can choose any one of a number of polynomial bases for this expansion since we can take advantage of various recurrence relations. Later on, when dealing with numerical solutions, we will compare different ways of approximating the evolution operator over short periods of time.

First, we note that U_S is in fact a unitary operator since $U^\dagger U = UU^\dagger = I$ and $U^\dagger = U^{-1}$ where the $\dagger$ denotes the Hermitian conjugate. Written in a basis, U_S has matrix elements

$$\langle a|U|b \rangle = \langle b|U|a \rangle^* \tag{5.18}$$

U_S is also a solution of the time-dependent Schrodinger equation since

$$i\hbar \frac{\partial U_s}{\partial t} = HU_S \tag{5.19}$$

and

$$\frac{\partial U_S^\dagger}{\partial t} = U_S^\dagger H \tag{5.20}$$

subject to the initial condition that $U(0) = 1$.

Notice that $U(t) = U^\dagger(-t)$ so that when operating on a ket $U^\dagger(t)$ affects evolution backwards in time

$$U_S^\dagger(t)|\psi(t)\rangle = |\psi(0)\rangle \tag{5.21}$$

while it has the effect of evolving a bra forward in time.

$$\langle \psi(0)|U^\dagger(t) = \langle \psi(t)| \tag{5.22}$$

Of particular importance is the semigroup property

$$U_S(t_2, t_o) = U_S(t_2, t_1)U_S(t_1, t_o) \tag{5.23}$$

where $t_2 \geq t_1 \geq t_o$, since this allows us to approximate the long-time evolution operator as a product of short-time evaluations

$$\langle b|U(t)|a \rangle = \sum_{i_1, i_2, \dots, i_n} \langle b|U(\delta t_n)|i_n \rangle \langle i_n|U(\delta t_{n-1})|i_{n-1} \rangle \cdots \langle i_1|U(\delta t_1)|a \rangle \tag{5.24}$$

where $t = \sum_i \delta t_i$ and we have inserted n complete sets of states at various intermediate times.

5.2   HEISENBERG PICTURE: EVOLUTION OF OBSERVABLES

In Heisenberg's viewpoint, one never directly observes the state of the system; it is simply a mathematical abstraction that you use to compute observables and make predictions regarding the outcome of specific experiments. What we do, in fact, observe is the spectrum of eigenvalues associated with a given physical operator:

$$\mathscr{A}|a_n\rangle = \alpha_n|a_n\rangle \tag{5.25}$$

If the state of the system is given by ψ and a measurement is made to determine $\mathscr{A}$, then the probability of finding the *value* α_n is given by $|\langle a_n|\psi\rangle|^2$. Because the observation should not depend upon *how* we choose to represent the state vector ψ, any new picture or representation of quantum mechanics must satisfy the following two criteria:

1. The eigenspectrum of an operator must not change upon moving to the new representation.
2. The probability amplitude for a given observation $\langle a_n | \psi \rangle$ must not change.

Both of these criteria are satisfied by unitary transformations

$$A|x\rangle = |y\rangle \tag{5.26}$$

$$A'|x'\rangle = |y'\rangle \tag{5.27}$$

Let U be a unitary transformation such that

$$|x'\rangle = U|x\rangle \quad \langle x'| = \langle x|U^\dagger \tag{5.28}$$

and

$$|y'\rangle = U|y\rangle \quad \langle y'| = \langle y|U^\dagger \tag{5.29}$$

Thus,

$$A'U|x\rangle = A'|x'\rangle = U|y\rangle = UA|x\rangle \tag{5.30}$$

From this we can conclude that

$$U^\dagger A'U = A \tag{5.31}$$

Applying this to the eigenvalue equation $A|a_n\rangle = \alpha_n |a_n\rangle$ we see that

$$UAU^\dagger U|a_n\rangle = \alpha_n U|a_n\rangle = \alpha_n |a_n\rangle \tag{5.32}$$

that is,

$$A'|a_n'\rangle = \alpha_n |a_n'\rangle \tag{5.33}$$

In other words, the eigenvalues of the transformed operator A' are the same as the original operator A. Likewise,

$$\langle a_n'|\psi'\rangle = \langle a_n|U^\dagger U|\psi\rangle = \langle a_n|\psi\rangle \tag{5.34}$$

What we conclude from this is that there are an infinite numbers of ways we can formulate dynamical representations of quantum mechanics based upon unitary transformations.

The Heisenberg picture is based upon the transformation that returns the time-evolved Schrodinger state back to its initial condition,

$$\psi_H(t) = U^\dagger(t, t_o)\psi_S(t) = \psi_s(0) \tag{5.35}$$

Because we never directly observe the state, time evolution is carried by the operators themselves:

$$\mathscr{A}_H(t) = U_S^\dagger(t, t_o)\mathscr{A}_S(t)U_S(t, t_o) \tag{5.36}$$

Upon working through the time derivative of $\mathscr{A}_H(t)$, we find

$$i\hbar \frac{d}{dt}\mathscr{A}_H(t) = [\mathscr{A}_H, H_H] + i\hbar \frac{\partial}{\partial t}\mathscr{A}_H(t) \tag{5.37}$$

where the subscript H denotes an operator in the Heisenberg representation.

The advantage of working in the Heisenberg picture is that there is a very clear connection to the Poisson bracket operation that gives the time evolution of a function in phase space

$$\frac{dA(p,q)}{dt} = \frac{\partial A}{\partial q}\frac{\partial q}{\partial t} + \frac{\partial A}{\partial p}\frac{\partial p}{\partial t} + \frac{\partial A}{\partial t}$$

$$= \frac{\partial A}{\partial q}\frac{\partial H}{\partial p} - \frac{\partial A}{\partial p}\frac{\partial H}{\partial q} + \frac{\partial A}{\partial t} \tag{5.38}$$

$$= \{A, H\} + \frac{\partial A}{\partial t} \tag{5.39}$$

where we have used the canonical relationships

$$\frac{\partial q}{\partial t} = \frac{\partial H}{\partial p} \tag{5.40}$$

$$\frac{\partial p}{\partial t} = -\frac{\partial H}{\partial q} \tag{5.41}$$

Dirac realized this close connection between the classical Poisson bracket and the quantum commutation relation. He proposed that the two are related and that this relation defines an acceptable set of quantum operations:[26]

> The quantum mechanical operators $\hat{f}$ and $\hat{g}$, which in classical theory replace the classically defined functions f and g, must always be such that the commutator of $\hat{f}$ and $\hat{g}$ corresponds to the Poisson bracket of f and g according to
>
> $$i\hbar\{f, g\} = [\hat{f}, \hat{g}] \tag{5.42}$$

where $\hat{f}$ and $\hat{g}$ are operators constructed from the quantum $\hat{x}$ and $\hat{p}$ operators. In other words,

$$[\hat{f}(\hat{x}, \hat{p}), \hat{g}(\hat{x}, \hat{p})] = i\hbar \left(\frac{\partial \hat{f}}{\partial \hat{x}}\frac{\partial \hat{g}}{\partial \hat{p}} - \frac{\partial \hat{g}}{\partial \hat{x}}\frac{\partial \hat{f}}{\partial \hat{p}} \right) \tag{5.43}$$

This is immediately verified if we take $\hat{H}(\hat{x}, \hat{p}) = \hat{p}^2/(2m) + V(\hat{x})$ and write the Heisenberg equations of motion for the momentum and position operators,

$$\frac{d\hat{x}}{dt} = \{\hat{H}, \hat{x}\} = \hat{p}/m \tag{5.44}$$

$$\frac{d\hat{p}}{dt} = \{\hat{H}, \hat{p}\} = -\frac{\partial V(\hat{x})}{\partial \hat{x}} \tag{5.45}$$

Again, we must emphasize that the difference between these equations of motion and their classical counterparts is that here we are dealing with operators rather than with ordinary numbers or functions.

We can extend this idea to any pair of canonical variables. For example, for the case of Boson operators, $[\hat{a}, \hat{a}^\dagger] = 1$, we can write a similar relation for operators composed of products of $\hat{a}$ and $\hat{a}^\dagger$:

$$[\hat{A}(\hat{a}, \hat{a}^\dagger), \hat{B}(\hat{a}, \hat{a}^\dagger)] = \{\hat{A}(\hat{a}, \hat{a}^\dagger), \hat{B}(\hat{a}, \hat{a}^\dagger)\}$$

$$= \frac{\partial \hat{A}}{\partial \hat{a}} \frac{\partial \hat{B}}{\partial \hat{a}^\dagger} - \frac{\partial \hat{B}}{\partial \hat{a}} \frac{\partial \hat{A}}{\partial \hat{a}^\dagger} \qquad (5.46)$$

Again, we can verify this by calculating the Heisenberg equations for a harmonic oscillator,

$$\frac{d\hat{a}}{dt} = \frac{1}{i\hbar}[\hat{H}, \hat{a}] = \frac{1}{i\hbar}\{\hat{H}, \hat{a}\} = i\Omega\hat{a} \qquad (5.47)$$

$$\frac{d\hat{a}^\dagger}{dt} = \frac{1}{i\hbar}[\hat{H}, \hat{a}^\dagger] = \frac{1}{i\hbar}\{\hat{H}, \hat{a}\} = -i\Omega\hat{a}^\dagger \qquad (5.48)$$

The close connection between the Dirac commutation bracket and the Poisson bracket stems from the fact that both are Lie derivatives of one vector field (or operator) with respect to the flow along the other vector field. The connection was first utilized by Dirac to treat constrained systems where the standard Hamiltonian-based mechanics is inadequate. For example, the Pauli exclusion principle is equivalent to imposing an additional constraint on the system such that no two particles can share the same state. Such constraints are handled easily within the context of a Lagrangian formulation.*

Dirac's idea was that we should generalize the Hamiltonian to include any imposed constraint ϕ_j by writing

$$H^* = H + \sum_j c_j \phi_j \qquad (5.49)$$

where the constraints are very small, $\phi_i \approx 0$, and the coefficients are functions of p and q. Typically we arrive at the Hamiltonian equations by taking the variation of H,

$$\delta H = \frac{\partial H}{\partial q}\delta q + \frac{\partial H}{\partial p}\delta p = -\dot{p}\delta q + \dot{q}\delta p \qquad (5.50)$$

so that

$$\left(\frac{\partial H}{\partial q} + \dot{p}\right)\delta q + \left(\frac{\partial H}{\partial p} - \dot{q}\right)\delta p = 0 \qquad (5.51)$$

Since the coefficients c_j are functions of the canonical variables, we cannot separately set δp and δq to zero. The variations must be tangent to the constraints. This can be

---

* One is referred at this point to a more complete discussion of the Poisson bracket formulation of classical mechanics, such as presented in Goldstein's *Classical Mechanics* text.

done by setting

$$\sum_n A_n \delta q_n + \sum_n B_n \delta p_n = 0 \tag{5.52}$$

with

$$A_n = \sum_j u_j \frac{\partial \phi_j}{\partial q_n} \quad \text{and} \quad B_n = \sum_j u_j \frac{\partial \phi_j}{\partial p_n} \tag{5.53}$$

where u_j is an arbitrary function. The constraint is a function of the canonical variables, so we set it to be zero everywhere: $\phi_i(q, p) = 0$ so that our dynamics occurs on the surface defined by constraint. Now, we can write the equations of motion for the canonical variables as

$$\dot{p}_j = -\frac{\partial H}{\partial q_j} - \sum_k u_k \frac{\partial \phi_k}{\partial q_j} \tag{5.54}$$

$$\dot{q}_j = \frac{\partial H}{\partial p_j} - \sum_k u_k \frac{\partial \phi_k}{\partial p_j} \tag{5.55}$$

More generally, the equation of motion for a function of canonical variables becomes

$$\dot{f} = \{f, H^*\} = \{f, H\} + \sum_k u_k \{f, \phi_k\} \tag{5.56}$$

The equations for the u_k's come about by requiring

$$\dot{\phi}_k = 0 = \{\phi_k, H^*\} \tag{5.57}$$

The connection to quantum mechanics is made by requiring the commutator of two operators to be proportional to $i\hbar$ times the modified Poisson bracket (aka Dirac bracket),

$$A \cdot B - B \cdot A = i\hbar \{A, B\}_{DB} \tag{5.58}$$

where

$$\{A, B\}_{DB} = \{A, B\}_{PB} - \sum_{nm} \{A, \phi_n\} M_{nm}^{-1} \{B, \phi_m\} \tag{5.59}$$

defines the Dirac bracket in terms of the Poisson bracket.* The constraint matrix M is formed by taking the Poisson bracket of constraints

$$M_{nm} = \{\phi_n, \phi_m\} \tag{5.60}$$

---

* Here we are taking the constraints to be "second-class" constraints.

5.3 QUANTUM PRINCIPLE OF STATIONARY ACTION

The principle of least action is one of the foundations of classical dynamics. We define the action as the integral of the Lagrangian

$$S = \int_0^t dt' L(\dot{q}(t'), q(t')) = \int_0^t dt' \frac{1}{2} m\dot{q}(t) - V(q) \tag{5.61}$$

where $q(t)$ is a trajectory. Taking the variation $\delta S = 0$ leads to the Euler–Lagrange equation

$$\frac{d}{dt}\frac{\partial L}{\partial \dot{q}} + \frac{\partial L}{\partial q} = 0 \tag{5.62}$$

Substituting $L(\dot{q}, q)$ into the Euler–Lagrange equation yields Newton's equations of motion

$$m\ddot{q} = -\frac{\partial V}{\partial q} \tag{5.63}$$

Consequently, the paths in classical mechanics are the ones by which the action integral is minimized over the entire trajectory.

One wonders, is there a similar principle in quantum mechanics?[27] Do the equations of motion for the time-dependent Schrödinger equation follow from an action principle? The answer is yes. Consider the definition of the transition amplitude between an initial and final state

$$\langle \psi_f | \psi(t_f) \rangle = \langle \psi_f | U(t_f, t_i) | \psi_i \rangle$$
$$= \langle \psi_f | U(t_f, t) U(t, t_i) | \psi_i \rangle$$
$$= \langle \psi_+(t) | \psi_-(t) \rangle \tag{5.64}$$

where $U(t_f, t_i)$ is the Schrödinger time-evolution operator. The two states we have defined are both solutions of the Schrödinger equations

$$i\hbar \partial_t | \psi_-(t) \rangle = H(t) | \psi_-(t) \rangle \quad | \psi_-(t_i) \rangle = | \psi_i \rangle \tag{5.65}$$

$$i\hbar \partial_t \langle \psi_+(t) | = \langle \psi_+(t) | H(t) \quad \langle \psi_+(t_f) | = \langle \psi_f | \tag{5.66}$$

The ket $| \psi_-(t) \rangle$ represents the state at time t knowing that at time t_i the system was in state $| \psi_i \rangle$. Likewise, the bra $\langle \psi_+(t) |$ represents the state of the system at time t knowing that at some future time t_f the system will be in $\langle \psi_f |$. It is important to note that $| \psi_-(t) \rangle$ and $\langle \psi_+(t) |$ are *not* Hermitian conjugates of each other. Also, note that the scalar product

$$\langle \psi_+(t) | \psi_-(t) \rangle = \langle \psi_f | \psi_-(t_f) \rangle = \langle \psi_+(t_i) | \psi_i \rangle \tag{5.67}$$

is invariant of time t.

Now consider the quantity

$$S = \int_{t_i}^{t_f} dt \frac{\langle \phi_+(t) | i\hbar \partial_t - H | \phi_-(t) \rangle}{\langle \phi_+(t) | \phi_-(t) \rangle} - i\hbar \log \langle \phi_f | \phi_-(t_f) \rangle \tag{5.68}$$

which we shall refer to as an action taken as a functional of both the bra and the ket. These we shall take as initial trial vectors for the variation of S. The bra and ket we are using are subject to the boundary conditions

$$|\phi_-(t_i)\rangle = |\phi_i\rangle \tag{5.69}$$

$$\langle\phi_+(t_f)| = \langle\phi_f| \tag{5.70}$$

Taking the variation of S by expanding it in terms of $\langle\delta\psi_+|$ and $|\delta\psi_-|$ yields

$$\delta S = \int_{t_i}^{t_f} dt \left(\frac{\langle\delta\phi_+(t)|i\hbar\partial_t - H - \lambda|\phi_-(t)\rangle}{\langle\phi_+(t)|\phi_-(t)\rangle} \right.$$
$$\left. + \frac{\langle\phi_+(t)|i\hbar\partial_t - H - \lambda|\delta\phi_-(t)\rangle}{\langle\phi_+(t)|\phi_-(t)\rangle} \right) - i\hbar \frac{\langle\phi_f|\delta\phi_-(t_f)\rangle}{\langle\phi_+(t)|\phi_-(t)\rangle} \tag{5.71}$$

where we have defined

$$\lambda(t) = \frac{\langle\phi_+(t)|i\hbar\partial_t - H|\phi_-(t)\rangle}{\langle\phi_+(t)|\phi_-(t)\rangle} \tag{5.72}$$

Integrating by parts and imposing the boundary conditions produces

$$\delta S = \int_{t_i}^{t_f} dt \left(\frac{\langle\delta\phi_+(t)|i\hbar\overrightarrow{\partial}_t - H - \lambda|\phi_-(t)\rangle}{\langle\phi_+(t)|\phi_-(t)\rangle} \right.$$
$$\left. - \frac{\langle\phi_+(t)|i\hbar\overleftarrow{\partial}_t - H - \lambda'|\delta\phi_-(t)\rangle}{\langle\phi_+(t)|\phi_-(t)\rangle} \right) \tag{5.73}$$

where the arrows over the partial derivative operator indicate that the operator acts either to the left or to the right. Clearly, the variation vanishes if both $\langle\phi_+(t)|$ and $|\phi_-(t)\rangle$ obey

$$(i\hbar\partial_t - H(t))|\phi_-(t)\rangle = \lambda(t)|\phi_-(t)\rangle \quad |\phi_-(t_i)\rangle = |\phi_i\rangle \tag{5.74}$$

$$\langle\phi_+(t)|(i\hbar\overleftarrow{\partial}_t + H(t)) = \langle\phi_+(t)|\lambda'(t) \quad \langle\phi_+(t_f)\rangle = \langle\phi_f| \tag{5.75}$$

where we have defined

$$\lambda'(t) = \frac{\langle\phi_+(t)|i\hbar\overleftarrow{\partial}_t + H|\phi_-(t)\rangle}{\langle\phi_+(t)|\phi_-(t)\rangle} \tag{5.76}$$

If we set the boundary conditions $|\phi_i\rangle = |\psi_i\rangle$ and $|\phi_f\rangle = |\psi_i\rangle$, then the trial ket $|\phi_-(t)\rangle$ is related to the Schrödinger ket $|\psi_-(t)\rangle$ by phase factor

$$|\phi_-(t)\rangle = \exp\left[-\frac{i}{\hbar}\int_{t_i}^{t_f} dt\,\lambda'(t)\right]|\psi_-(t)\rangle \tag{5.77}$$

Acting on the left with $\langle\phi_f|$ we obtain the transition amplitude

$$\langle\phi_f|\phi_-(t)\rangle = e^{iS_c/\hbar}\langle\phi_f|\psi_-(t)\rangle \tag{5.78}$$

where the stationary action is

$$S_c = \int_{t_i}^{t_f} dt\,\lambda(t) - i\hbar \log\langle\phi_f|\phi_-(t_f)\rangle = -i\hbar \log\langle\phi_f|\psi(t_f)\rangle \qquad (5.79)$$

Thus, the quantum transition amplitude is given by the stationary value of the action

$$\langle\phi_f|\psi(t_f)\rangle = e^{iS_c/\hbar} \qquad (5.80)$$

Unfortunately, the phases of the two trial vectors are undetermined. For example, if we add on an additional phase so that

$$|\phi'_-(t)\rangle = e^{i\xi(t)/\hbar}|\phi_-(t)\rangle \qquad (5.81)$$

where

$$\xi(t) = \int_{t_i}^{t} dt'\,z(t') \qquad (5.82)$$

this new ket is now a solution of

$$(i\hbar\partial_t - H(t))|\phi'_-(t)\rangle = \lambda'(t)|\phi'_-(t)\rangle \qquad (5.83)$$

where

$$\lambda'(t) = \lambda(t) + z(t) \qquad (5.84)$$

Fortunately, the phase indeterminacy does not change the transition amplitude since the final bra state is modified as well. We can eliminate this indeterminacy by imposing an additional constraint on the system that S is a functional of $|\psi(t)\rangle$ and its Hermitian conjugate $\langle\psi(t)|$:

$$S = \int_{t_1}^{t_2} dt \frac{\langle\psi(t)|i\hbar\partial_t - H|\psi(t)\rangle}{\langle\psi(t)|\psi(t)\rangle} \qquad (5.85)$$

We also assume (as in classical mechanics) that the variations vanish at the boundaries

$$\langle\delta\psi(t_f)| = |\delta\psi(t_i)\rangle = \langle\delta\psi(t_i)| = |\delta\psi(t_i)\rangle \qquad (5.86)$$

What this means is that we are taking S to be a functional of the state vector and its Hermitian conjugate at all times $t_i < t < t_f$ other than the initial and final times. In doing so, our final equations will not depend upon the choice of boundary conditions. We can also set $\langle\psi(t)|\psi(t)\rangle = 1$ for all time to enforce normalization. One can easily verify that $\delta S = 0$ when

$$(i\hbar\partial_t - H)|\psi(t)\rangle = 0 \qquad (5.87)$$

In other words, the action S is stationary with respect to all variations of $|\psi(t)\rangle$ and its Hermitian conjugate provided they are solutions of the Schrödinger equation.

Let us take, for instance, the case where the Hamiltonian is driven by some set of external time-dependent variables (perhaps a set of nuclear coordinates), $q(t)$, such that at time t_i, the state $|\psi(t_i)\rangle$ is an eigenstate of $H(q(t_i))$ and at time t_f, $|\psi(t_f)\rangle$ is an eigenstate of $H(q(t_f))$. What is the stationary action connecting the initial and final states? In other words, can we calculate the transition amplitude $\langle\psi(t_f)|\psi(t_i)\rangle$ such that $\delta\langle\psi(t_f)|\psi(t_i)\rangle = 0$? Let us define the total action as

$$S = S_c + i\hbar \log t_{if} \tag{5.88}$$

where

$$t_{if} = \langle\psi(t_f)|\psi(t_i)\rangle = e^{iS[q]/\hbar} \tag{5.89}$$

is the transition amplitude between the initial and final states, S_c is the classical action

$$S_c = \int_{t_i}^{t_f} \frac{m}{2}\dot{q}^2(t) - V(q(t))dt \tag{5.90}$$

and $S[q]$ is the contribution to the action due to the quantum transition. Again, we use the trick we used above and write this as

$$\langle\psi(t_f)|\psi(t_i)\rangle = \langle\psi_+(t)|\psi_-(t)\rangle \tag{5.91}$$

where $t_f > t > t_i$ is some intermediate time. Setting $V(q) = 0$ and taking the variation of S with respect to $q(t)$ results in the classical equations of motion for $q(t)$:[28,29]

$$m\ddot{q}(t) = -\text{Im}\frac{\langle\psi_+(t)|\frac{\partial H(q(t))}{\partial q}|\psi_-(t)\rangle}{\langle\psi_+(t)|\psi_-(t)\rangle} \tag{5.92}$$

The fact that the transition matrix element also depends upon the path between $q(t_i)$ and $q(t_f)$ means that the resulting force is path dependent. Consequently, in order to determine the path, one must iterate this last equation self-consistently.

The equations of motion (Equation 5.92) were first derived by Phil Pechukas in 1969 starting from a path-integral formulation for the fully quantum mechanical problem of atomic scattering and then making a stationary phase approximation for the nuclear trajectory. Although the approach is very appealing and gives the correct semiclassical path, it is often impossible to converge a unique path if the time interval $t_f - t_i$ is too long and the states are strongly coupled.[30–32]

How do we interpret this last result? Imagine that $q(t)$ represents the scattering trajectory of an atom and the quantum states are internal degrees of freedom (say, the atom's electronic states). At the initial time t_i we prepare the system at $q(t_i)$ in some well-determined quantum state that we will take to be an eigenstate of $H(q(t_i))$. As time evolves, the quantum state $|\psi(t)\rangle$ may no longer be an eigenstate of $H(q(t))$ since the Hamiltonian is changing in time as well. As a result, $|\psi(t)\rangle$ will evolve into a superposition of eigenstates

$$|\psi(t)\rangle = \sum_n |c_n(t)\phi_n(q(t))\rangle \tag{5.93}$$

where the $c_n(t)$ are the transition amplitudes for starting in the initial state and evolving into the nth eigenstate of $H(q(t))$ at some intermediate time $t_i < t < t_f$. Suppose at time t_f we observe the system in state $|\psi_f(t_f)\rangle$, which is now an eigenstate of $H(t_f)$. The path with the least action connecting $|\psi_i(t_i)\rangle$ to $|\psi_f(t_f)\rangle$ satisfies Equation 5.92. In fact, every possible final state has its own unique stationary phase path connecting it to the initial state.

Since we have determined that the quantum state is now $|\psi_f(t_f)\rangle$, we lose all quantum coherence between $|\psi_f(t_f)\rangle$ and any other alternative state at t_f. If we chronicle a series of such events such that at time t_0 we prepare the system in some initial state $|\psi_0(t_0)\rangle$ and at t_1 we determine the quantum state to be $|\psi_1(t_1)\rangle$, at $t_2 > t_1$ we determine the quantum state to be $|\psi_2(t_2)\rangle$ and so on, we can write a history of events:

$$hist_1 = \{\psi_0, \psi_1, \psi_2, \cdots \psi_N\} \tag{5.94}$$

Between each segment, we compute a stationary phase trajectory $q(t)$ according to Equation 5.92. The resulting world-line would depend strongly on the outcome of each quantum transition and the frequency that we measured for the quantum state by choosing the time interval $\delta t = t_2 - t_1$ to be too small,

$$U(t_2, t_1)|\psi_1(t_1)\rangle = \left(1 - i\frac{\delta t}{\hbar}H(t_1) + \cdots\right)|\psi_1(t_1)\rangle$$

$$\approx \left(1 - i\frac{\delta t}{\hbar}\varepsilon_1(q_1) + \cdots\right)|\psi_1(t_1)\rangle \tag{5.95}$$

At time t_2 hardly any quantum evolution will have occurred and we would determine that the atom is still in its original quantum state. In fact, the quantum state will remain in the original eigenstate of H over the course of the trajectory. If our scattering atom was in an electronic excited state and we frequently inquire about its current state, the atom will forever remain in that excited state. An alternative history may be

$$hist_2 = \{\psi_0, \psi_1, \psi_2', \ldots \psi_N'\} \tag{5.96}$$

where $hist_1$ and $hist_2$ are the same until between t_1 and t_2 the system makes a switch from state 1 to state 2'. Up until time t_2, we would have some degree of quantum coherence between the two histories and information can be passed from one world-line to the other.* After t_2 there is no common phase relation between the two paths. This is illustrated in Figure 5.1.

5.4 INTERACTION PICTURE

Very often we are faced with a situation in which the Hamiltonian of the system can be broken into two terms

$$H = H_o + V \tag{5.97}$$

where H_o describes some reference system and V describes some additional interaction. For example, we may choose the reference system to be a harmonic oscillator

---

* This may be starting to sound a bit like the plot line for a *Star Trek* episode.

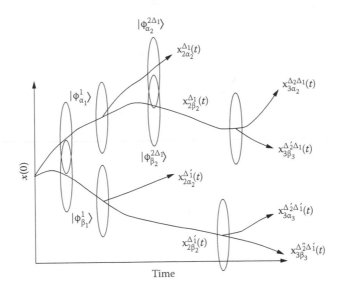

FIGURE 5.1 Illustration of coarse-graining of quantum histories. Starting from an initial point, the system can follow any number of alternative paths as indicated by bifurcations. The ovals surrounding each path indicate quantum fluctuations about the stationary phase paths. Overlapping ovals indicate that two or more paths are quantum mechanically entangled for a short period of time. (From Ref. 7)

with V being some additional coupling that induces some sort of dynamical evolution of the system. With this in mind, we define the *interaction* wave function as

$$\psi_I(t) = e^{+iH_o t/\hbar} \psi_S(t)$$
$$= e^{+iH_o t/\hbar} e^{-iHt/\hbar} \psi_S(0) \tag{5.98}$$

Taking the time derivative, we find

$$i\hbar \frac{d}{dt} \psi_I(t) = V(t)\psi_I(t) \tag{5.99}$$

where

$$V(t) = e^{+iH_o t/\hbar} V e^{-iH_o t/\hbar} \tag{5.100}$$

is the interaction operator written in the Heisenberg representation of the reference system. Since unitary transformations can be visualized as rotations in some N-dimensional space, both the interaction wave function and coupling operator are simultaneously rotated along with the reference system so that any departure from the initial state is entirely due to the interaction. This has the distinct advantage of eliminating the rapidly oscillating terms that appear in the evolution of the Schrödinger state.

Let us now consider the time-evolution operator in the interaction representation. As previously,

$$\psi_I(t) = U_I(t, t_o)\psi_o \tag{5.101}$$

Expanding U_I in time,

$$U_I(t, t_o) = 1 + \frac{1}{i\hbar} \int_{t_o}^{t} dt_1 V(t_1) U(t_1, t_o)$$

$$= 1 + \frac{1}{i\hbar} \int_{t_o}^{t} dt_1 V(t_1) + \frac{1}{(i\hbar)^2} \int_{t_o}^{t} dt_1 \int_{t_o}^{t_1} dt_2 V(t_1) V(t_2) + \cdots \quad (5.102)$$

This is often refered to as the Dyson series and it often serves as the starting point for perturbative theories since each term involves subsequent interactions with the coupling operator. The series can be taken to infinite order

$$U(t, t_o) = \sum_{n=0}^{\infty} U_n(t) \quad (5.103)$$

where each term is given by

$$U_n(t) = \frac{1}{(i\hbar)^n} \int_{t_o}^{t} dt_1 \int_{t_o}^{t_1} dt_2 \cdots \int_{t_o}^{t_{n-1}} dt_n V(t_1) V(t_2) \cdots V(t_n) \quad (5.104)$$

where $t \geq t_1 \geq t_2 \geq \cdots \geq t_n$. The ordering of the operators in the integral is very crucial since we have no *a priori* guarantee that $[V(t), V(t')] = 0$ for $t \neq t'$. To avoid problems associated with time ordering of the operators, we introduce a time-ordering operator

$$P[A(t_1)B(t_2)] = \begin{array}{ll} A(t_1)B(t_2) & \text{for } t_1 \geq t_2 \\ B(t_2)A(t_1) & \text{for } t_2 > t_1 \end{array} \quad (5.105)$$

that has the effect of rearranging a series of operators into chronological order. For example,

$$P[A_i(t_i)A_j(t_j) \cdots A_n(t_n)] = A_i(t_i)A_j(t_j) \cdots A_n(t_n) \quad (5.106)$$

with $t_i \geq t_j \cdots \geq t_n$.

Consider the second term in the expansion of the time-evolution operator:

$$U_2(t, 0) = \frac{1}{(i\hbar)^2} \int_0^t dt_1 \int_0^{t_1} dt_2 V(t_1) V(t_2) \quad (5.107)$$

The implied area of integration on the t_1, t_2 plane is the shaded area above the $t_2 = t_1$ line. On the other hand, in

$$U_2(t, 0) = \frac{1}{(i\hbar)^2} \int_0^t dt_2 \int_0^{t_2} dt_1 V(t_1) V(t_2) \quad (5.108)$$

the implied area of integration is below the $t_1 = t_2$ line. Since both integrals should give the same result, we can write

$$U_2(t, 0) = \frac{1}{2} \frac{1}{(i\hbar)^2} \int_0^t dt_1 \int_0^t dt_2 P[V(t_1) V(t_2)] \quad (5.109)$$

As a result, we can write the time-ordered series as

$$U(t) = 1 + \sum_{n=1}^{\infty} \frac{1}{n!} \frac{1}{(i\hbar)^n} \int_0^t dt_1 \cdots \int_0^t dt_n \, P[V(t_1) \cdots V(t_n)] \qquad (5.110)$$

This is the polynomial expansion for the exponential function, so we can immediately write the interaction evolution operator as

$$U(t) = P \exp\left[-\frac{i}{\hbar} \int_0^t V(t')\,dt' \right] \qquad (5.111)$$

5.5   PROBLEMS AND EXERCISES

Problem 5.1   Demonstrate that each of the above properties of U is true.

Problem 5.2   Using the mixing angle and rotation matrix given in Equation 4.6 show that $THT^\dagger$ is diagonal with eigenvalues $\varepsilon_\pm$.

Problem 5.3   Prove the Kubo identity:

$$[\hat{A}, e^{-\beta \hat{B}}] = e^{-\beta \hat{B}} \int_0^\beta e^{\lambda \hat{B}} [\hat{A}, \hat{B}] e^{-\lambda \hat{B}} d\lambda$$

Problem 5.4   Since the time-dependent Schrödinger equation is a first-order differential equation with respect to time, the state at time t, $\psi(t)$, is uniquely determined by the state at time $t = 0$, $\psi(0)$. In other words, we can write

$$\psi(t) = \hat{S}(t)\psi(0)$$

where $\hat{S}(t)$ is some quantum mechanical operator.

1. Show that $\hat{S}(t)$ satisfies $i\hbar \partial_t \hat{S}(t) = \hat{H}\hat{S}(t)$ where $\hat{H}$ is the Hamiltonian operator. Also, show that $\hat{S}(t)$ is unitary.
2. Show that if $\hat{H}$ is independent of time, $\hat{S}(t)$ takes the form

$$\hat{S}(t) = e^{-i\hat{H}t/\hbar}$$

Problem 5.5   The expectation value of an operator $\hat{B}$ is given by

$$\langle \hat{B}(t) \rangle = \langle \psi(t) | \hat{B} | \psi(t) \rangle$$

1. Show that the time evolution of the Heisenberg operator $\hat{\mathscr{B}} = \hat{S}^{-1}(t)\hat{B}\hat{S}(t)$ satisfies

$$\langle \hat{\mathscr{B}}(t) \rangle = \langle \psi(0) | \hat{S}^{-1}(t)\hat{B}\hat{S}(t) | \psi(0) \rangle$$

2. Show that the time derivative of the Heisenberg operator $\hat{B}(t)$ is given by

$$i\hbar \partial_t \hat{\mathscr{B}}(t) = \hat{\mathscr{B}}\hat{\mathscr{H}} - \hat{\mathscr{H}}\hat{\mathscr{B}}$$

where $\hat{\mathscr{H}} = \hat{S}^{-1}(t)\hat{H}\hat{S}(t)$.

3. Show that if $[\hat{A}, \hat{B}] = \hat{C}$, then the corresponding Heisenberg operators satisfy $[\mathscr{A}, \mathscr{B}] = \mathscr{C}$

Problem 5.6   Prove the following relationship:

$$e^{\hat{L}}\hat{A}e^{-\hat{L}} = \hat{A} + [\hat{L}, \hat{A}] + \frac{1}{2!}[\hat{L}, [\hat{L}, \hat{A}]] + \frac{1}{3!}[\hat{L}, [\hat{L}, [\hat{L}, \hat{A}]]] + \cdots$$

SUGGESTED READING

There are any number of excellent textbooks on quantum mechanics. Listed below are various texts I have found to be particularly useful in preparing this chapter.

1. *Chemical Dynamics in Condensed Phases Relaxation, Transfer and Reactions in Condensed Molecular Systems*, Abraham Nitzan (Oxford Graduate Texts, 2007). This is one of the best interdisciplinary accountings of dynamical processes in the condensed phase.
2. *Quantum Mechanics*, Claude Cohen-Tannoudji, Bernard Diu, and Frank Laloë (Wiley Interscience, 1973)
3. *Quantum Mechanics, A Modern Introduction*, A. Das and A. C. Melissinos (Gordon and Breach, 1986)
4. *Quantum Mechanics*, E. Merzbacher (Wiley, 1961).

6 Quantum Density Matrix

6.1 INTRODUCTION: MIXED VS. PURE STATES

Up until this point, we have concerned ourselves with a description of quantum mechanics centered upon how a state $|\psi\rangle$ evolves in time. Using this we can show that the expectation value of an operator evolves as

$$\langle A(t)\rangle = \langle \psi(t)|\hat{A}|\psi(t)\rangle \tag{6.1}$$

However, for many instances, especially if we are interested in describing relaxation processes, it is useful to introduce the density operator

$$\hat{\rho}(t) = |\psi(t)\rangle\langle\psi(t)| \tag{6.2}$$

taken as the *outer product* of the state vector with itself. From this definition we can write

$$\hat{\rho} = \sum_{mn} c_n^* c_m |m\rangle\langle n| \tag{6.3}$$

$$= \sum_{mn} \rho_{mn} |m\rangle\langle n| \tag{6.4}$$

where the ρ_{mn} are the density matrix elements. Expectation values of operator are then given by the trace

$$\langle A(t)\rangle = \sum_{mn} A_{mn}\rho_{mn} = Tr[\hat{A}\rho(t)] \tag{6.5}$$

where $Tr[\hat{A}\rho(t)]$ denotes the trace operation:

$$Tr[\hat{A}\rho(t)] = \sum_{nn'} A_{nn'}\rho_{n'n}$$

If ρ is diagonal, then

$$Tr[\hat{A}\rho(t)] = \sum_{n} A_{nn}\rho_{nn} = \sum_{n}\langle n|A|n\rangle P_n$$

where $P_n = \rho_{nn}$ is the statistical probability of finding the system in state n. These statistical weights must be such that $P_n \leq 1$ and

$$\sum_{n} P_n = 1$$

Hence, we conclude that *knowing ρ we can compute the statistical average of an operator A*.

161

The diagonal elements of the density matrix, $\rho_{nn} = \langle n|\psi\rangle\langle\psi|n\rangle = p_n$, are the occupation numbers of the nth state while the off-diagonal $\rho_{nm} = \langle n|\psi\rangle\langle\psi|m\rangle$ represent the phase coherences between the n and m states. Since the diagonal elements represent probabilities,

$$1 \geq \rho_{nn} \geq 0$$

so that ρ is a positive semidefinite operator. Moreover, since the trace of a matrix is invariant to representation, the eigenvalues of ρ must be populations as well. Furthermore, the eigenvectors of ρ are the pure states of the ensemble.

We can define a *pure state* as a system that can be described by a single-state vector. Pure states evolve according to the rules of quantum mechanics to form coherent superpositions with well-defined phase relations between the components

$$|\psi(t)\rangle = \sum_n c_n(t)|n\rangle$$

Let us define an initially pure state, $|\psi\rangle = |\lambda\rangle$, with density matrix

$$\rho^\lambda = |\lambda\rangle\langle\lambda|$$

Notice that in this case the density matrix acts as a projection operator such that

$$(\rho^\lambda)^2 = |\lambda\rangle\langle\lambda|\lambda\rangle\langle\lambda| = |\lambda\rangle\langle\lambda| \tag{6.6}$$

Thus we conclude that for a pure state, $Tr[\rho^2] = Tr[\rho] = 1$. This relation holds in *any* representation since the trace operation is invariant to basis transformations. As a result, the evolution of a pure state under the rules of quantum mechanics results in a density matrix,

$$\rho(t) = U|\lambda\rangle\langle\lambda|U^\dagger = U\rho^\lambda U^\dagger$$

that still preserves the invariance under the trace operation so that $Tr[\rho(t)] = 1$. Also,

$$Tr[\rho^2(t)] = Tr[U\rho^\lambda U^\dagger U\rho^\lambda U^\dagger] = 1$$

which allows us to conclude that unitary time evolution transforms an initially pure state into another pure state.

On the other hand, a *mixed state* cannot be described by a single-state vector but consists of a *statistical mixture* of states with a probability p_n of being in any one of them. The members of the mixed state (that is, the ensemble) are independently prepared and there is no phase coherence between them. For example, if our initial system is at thermodynamic equilibrium, we can write it as a mixed state of the form

$$\rho = \frac{1}{Q}\sum_k e^{-\beta H}|n\rangle\langle n| \tag{6.7}$$

where $\beta = 1/kT$. If the $|n\rangle$ are energy eigenstates of H, then

$$\rho = \frac{1}{Q}\sum_k e^{-\beta E_n}|n\rangle\langle n| \tag{6.8}$$

As before, let us define a density matrix for a mixed system as $\rho = p_1|1\rangle\langle 1| + p_2|2\rangle\langle 2|$. Since $p_1 + p_2 = 1$, $Tr[\rho] = 1$. However, $Tr[\rho^2] = p_1^2 + p_2^2 < 1$ unless either $p_1 = 1$ and $p_2 = 0$ or $p_2 = 1$ and $p_1 = 0$.

6.2   TIME EVOLUTION OF THE DENSITY MATRIX

From our definition of the density matrix above,

$$\rho(t) = |\psi(t)\rangle\langle\psi(t)|$$

Taking the time derivative,

$$i\hbar\frac{\partial}{\partial t}\rho = \left(i\hbar\frac{\partial|\psi(t)\rangle}{\partial t}\right)\langle\psi(t)| + |\psi(t)\rangle\left(i\hbar\frac{\partial\langle\psi(t)|}{\partial t}\right) \tag{6.9}$$

Since $i\hbar\partial_t|\psi(t)\rangle = H|\psi(t)\rangle$ and $i\hbar\partial_t\langle\psi(t)| = -\langle\psi|H|$, we have

$$i\hbar\frac{\partial}{\partial t}\rho = H|\psi(t)\rangle\langle\psi(t)| - |\psi(t)\rangle\langle\psi(t)|H \tag{6.10}$$

$$= [H, \rho] \tag{6.11}$$

which is often compactly written as

$$\dot{\rho} = -i\mathcal{L}\rho$$

where $\mathcal{L}$ is the *Liouville superoperator*

$$\mathcal{L}\rho = \frac{1}{\hbar}[H, \rho]$$

This last equation is referred to as the Liouville–von Neumann equation.

Let us take the typical case where the Hamiltonian is composed of a zeroth-order term and an interaction: $H = H_o + V(t)$. The corresponding Liouville–von Neumann equation reads

$$i\frac{\partial\rho}{\partial t} = (\mathcal{L}_o + \mathcal{L}_V(t))\rho \tag{6.12}$$

where the action of $\mathcal{L}_o$ and $\mathcal{L}_V$ on ρ is given by $\hbar\mathcal{L}_o\rho = [H_o, \rho]$ and $\hbar\mathcal{L}_V\rho = [V(t), \rho]$. Formally, the evolution of the density matrix follows from

$$\rho(t) = e^{-i\mathcal{L}t}\rho(0) = \mathcal{U}(t)\rho(0) = U_s(t)\rho(0)U_s^{\dagger}(t)$$

where $U_s(t)$ are the unitary time-evolution operators in the Schrödinger representation and $\rho(0)$ is the initial condition. As noted in Chapter 5, the exponential form is really the result of an expansion in terms of an infinite series. Thus, we can write $\mathcal{U}(t)$ as

$$\mathcal{U} = e^{-i\mathcal{L}_o t} - i\int_0^t dt_1 e^{-i\mathcal{L}_o(t-t_1)}\mathcal{L}_V(t_1)e^{-i\mathcal{L}_o t_1}$$

$$- \int_0^t\int_0^{t_1} dt_2 e^{-i\mathcal{L}_o(t-t_1)}\mathcal{L}_V(t_1)e^{-i\mathcal{L}_o(t_1-t_2)}\mathcal{L}_V(t_2)e^{-i\mathcal{L}_o t_2} + \cdots \tag{6.13}$$

Rearranging this a bit yields

$$\mathcal{U} = e^{-i\mathcal{L}_o t}\left[1 - i\int_0^t dt_1\left(e^{+i\mathcal{L}_o t_1}\mathcal{L}_V(t_1)e^{-i\mathcal{L}_o t_1}\right)\right.$$

$$\left. - \int_0^t\int_0^{t_1} dt_2\left(e^{+i\mathcal{L}_o t_1}\mathcal{L}_V(t_1)e^{-i\mathcal{L}_o t_1}\right)\left(e^{+i\mathcal{L}_o t_2}\mathcal{L}_V(t_2)e^{-i\mathcal{L}_o t_2}\right) + \cdots\right] \tag{6.14}$$

Taking each term in the () as a Heisenberg operator evolving under $\mathscr{L}_o$, we can write this as

$$\mathscr{U} = e^{-i\mathscr{L}_o t}\left[1 - i\int_0^t dt_1 \mathscr{L}_{VI}(t_1) - \int_0^t\int_0^{t_1} dt_2 \mathscr{L}_{VI}(t_1)\mathscr{L}_{VI}(t_2) + \cdots\right] \quad (6.15)$$

This allows us to define the propagator for the *interaction* representation as

$$\mathscr{U}_I(t) = e^{-i\mathscr{L}_o t}\mathscr{U}(t) \quad (6.16)$$

6.3   REDUCED DENSITY MATRIX

Up until now, we have labeled state space using some generic index n. In general, however, n can be a collection of subindices labeling, for example, the spin, angular momentum, and so forth of the state. It could also include continuous variables such as the position or momentum of the state. In general we really should write

$$\rho_{nn'} = \rho_{n_1 n_2 \ldots, n_1', n_2' \ldots}$$

Say, for example, we are interested in only one aspect of the system or in some particular property, such as the spin or energy. We can then retain only the relevant indices and define in such a way a *reduced* density matrix. Most commonly, the reduced density matrix is used in cases where the total state space is partitioned into interacting (or noninteracting) subsystems, such as the internal states of a molecule coupled to the normal modes of a surrounding environment. In such cases we can factor the total density matrix into a tensor product

$$\rho_{AB} = \rho_A \otimes \rho_B$$

If we take states $|i\rangle$ as spanning space A and states $|j\rangle$ as spanning B so that $|ij\rangle$ spans the entire composite space, the tensor product is written as

$$\rho = \sum_{ij}\sum_{kl} \rho_{ij,kl}|ij\rangle\langle kl|$$

In the case of a molecule embedded in a matrix, we may only be explicitly interested in the part of ρ_{AB} corresponding to the internal states of the molecule. Thus, we define the reduced density matrix for this subspace as

$$\rho_A = \sum_{ik} \sigma_{ik}|i\rangle\langle k|$$

where the coefficients are obtained by

$$\sigma_{ik} = \sum_j \rho_{ij,kj}$$

This is a "partial" trace since it involves summing only over states belonging to space B.

The expectation value of any operator acting in space A can be computed using ρ_A

$$\langle A \rangle = Tr[\rho_A A]$$

However, if we are interested in quantities involving space B or correlations between subspaces A and B, we need the full density matrix.

For example, consider a pair of particles that are entangled in a superposition state

$$|\Phi^-\rangle = \frac{1}{\sqrt{2}}(|00\rangle - |11\rangle)$$

where $|0\rangle$ and $|1\rangle$ label, say, the ground and excited states of each particle. The density matrix for this system is then

$$\hat{\rho} = |\Phi^-\rangle\langle\Phi^-|$$

$$= \frac{1}{\sqrt{2}}[|00\rangle - |11\rangle)\frac{1}{\sqrt{2}}(\langle00| - \langle11|)$$

$$= \frac{1}{2}(|00\rangle\langle00| - |00\rangle\langle11| - |11\rangle\langle00| + |11\rangle\langle11|] \qquad (6.17)$$

Say, for example, you want to find the reduced density matrix for the second particle. For this you would take the partial trace over the first

$$\hat{\rho}_2 = tr_1(\hat{\rho})$$

$$= \frac{1}{2}(tr\,(|0\rangle\langle0|)|0\rangle\langle0| - tr\,(|0\rangle\langle1|)|0\rangle\langle1| - tr\,(|1\rangle\langle0|)|1\rangle\langle0| + tr\,(|1\rangle\langle1|)|1\rangle\langle1|)$$

$$= \frac{1}{2}[\langle0|0\rangle|0\rangle\langle0| - \langle0|1\rangle|0\rangle\langle1| - \langle1|0\rangle|1\rangle\langle0| + \langle1|1\rangle|1\rangle\langle1|]$$

$$= \frac{1}{2}(|0\rangle\langle0| + |1\rangle\langle1|)$$

$$= \frac{1}{2}\hat{\mathbb{1}} \qquad (6.18)$$

Notice the reduced density matrix for the second particle tells us the probabilities of that particular particle being in either state $|0\rangle$ or $|1\rangle$ with no mention or regards to the state of the other particle. Notice also that ρ_2 represents a *mixed* state for particle #2. We cannot determine with complete certainty which state particle #2 is in because it is not in one. Because of its entanglement with particle #1, particle #2 is not in a single definable state and likewise for particle #1.

If one were to make a measurement of the state of #1 (for example, $|1\rangle$ emits a photon to decay to $|0\rangle$)), the total system would be forced to be in state $|00\rangle$. If we were to then calculate the reduced density matrix for #2, we would find $\rho_2 = |0\rangle\langle0|$ indicating that #2 has a 100% chance of being in state $|0\rangle$ and a 0% chance of being in state $|1\rangle$. This is the "spooky" nature of quantum mechanics. Measuring the state of one part of an entangled pair determines the state of the other.

6.3.1 VON NEUMANN ENTROPY

Recognizing the connection between statistical mechanics and the quantum density matrix, John von Neumann formulated the concept of entropy as an extension of the Gibbs and Shannon entropy to a quantum mechanical system. The von Neumann entropy is defined as

$$S[\rho] = -Tr[\rho \log \rho]$$

Using the definition of the density matrix for a thermal mixture, we can immediately see that $k_B S[\rho]$ is the correct thermodynamic entropy. Just as in statistical mechanics, S gives a measure of the number of thermodynamically accessible states under a given set of thermodynamic conditions, the von Neumann entropy gives the degree departure of the system from a pure state. In essence, it gives a measure of the degree of mixture.

Some properties of the von Neumann entropy functional $S[\rho]$ include

- $S[\rho] = 0$ only for a pure state.
- The maximum value of $S[\rho] = \log N$ for maximally mixed states where N is the number of states (that is, the dimension of the Hilbert space).
- $S[\rho]$ is invariant under basis transformation of ρ.
- $S[\rho]$ is *concave*. In other words, given a set of positive numbers $\lambda_i > 0$ and density matrices ρ_i,

$$S\left[\sum_i \lambda_i \rho_i\right] \geq \sum_i \lambda_i S[\rho_i]$$

- $S[\rho]$ is additive. That is, given two independent systems each with density matrix ρ_A and ρ_B, $S[\rho_A \otimes \rho_B] = S[\rho_A] + S[\rho_B]$. If instead ρ_A and ρ_B are the reduced density matrices of some general system ρ_{AB}, then

$$|S[\rho_A] - S[\rho_B]| \leq S[\rho_{AB}] \leq S[\rho_A] + S[\rho_B]$$

This last property is termed *subadditivity*. In a quantum mechanical system, the entropy of the composite system can in fact be lower than the sum of the entropies of its parts. This is the case when there is any degree of coherence between parts A and B.

6.4 THE DENSITY MATRIX FOR A TWO-STATE SYSTEM

To illustrate these properties, let us consider the case of a degenerate two-state system with some coupling β with Hamiltonian

$$H = E \begin{pmatrix} 1 & 0 \\ 0 & 1 \end{pmatrix} + \hbar\beta \begin{pmatrix} 0 & 1 \\ 1 & 0 \end{pmatrix}$$

The equations of motion for the density matrix elements in this basis are thus

$$\dot{\rho}_{11} = i\beta(\rho_{12} - \rho_{21})$$
$$\dot{\rho}_{22} = -i\beta(\rho_{12} - \rho_{21})$$

$$\dot{\rho}_{12} = i\beta(\rho_{11} - \rho_{22})$$
$$\dot{\rho}_{21} = -i\beta(\rho_{11} - \rho_{22}) \tag{6.19}$$

We can solve these equations in a number of ways, the easiest being to take the time derivative of $\dot{\rho}_{11}$

$$\ddot{\rho}_{11} = i\beta(\dot{\rho}_{12} - \dot{\rho}_{21})$$
$$= -2\beta^2(\rho_{11} - \rho_{22}) \tag{6.20}$$

Since $\rho_{11} + \rho_{22} = 1$,

$$\ddot{\rho}_{11} = 2\beta^2 - 4\beta^2\rho_{11}$$

Solving this last differential equation produces

$$\rho_{11}(t) = \frac{1}{2} + C_1 \cos(2\beta t) + C_2 \sin(2\beta t)$$

with C_1 and C_2 being constants of integration. Setting $\rho_{11}(0) = 1$ yields

$$\rho_{11}(t) = \cos^2(\beta t)$$

and

$$\rho_{22}(t) = \sin^2(\beta t)$$

for the populations. Obtaining the coherences

$$\dot{\rho}_{12} = i\beta(\rho_{11} - \rho_{22})$$

results in the integral

$$\rho_{12}(t) = i\beta \int_0^t dt'(\cos^2(\beta t') - \sin^2(\beta t'))$$

$$= \frac{i}{2} \sin(2\beta t) \tag{6.21}$$

The time evolution of the populations and coherence are shown in Figure 6.1. Notice that at $t\beta = \pi/2$, the original population in state 1 has been entirely transferred to state 2 and entirely returns to the original state every $\beta t = \pi$. Notice that in order to transfer population from state 1 to state 2, one must first establish a coherence ρ_{12} between the two states. Only once this coherence has been established can a population be effectively transferred. If this coherence is destroyed—for example, though interaction with an external environment—the ability to transfer a population from 1 to 2 and back is effectively diminished.

6.4.1 Two-Level System under Resonance Coupling—Revisited

Let us now revisit our discussion of what happens when a system is coupled to a time-dependent driving field. For example, consider the case of a two-state system with energies $E_{1,2} = \pm\hbar\omega_o$ coupled to a periodic driving field, such as a laser or a

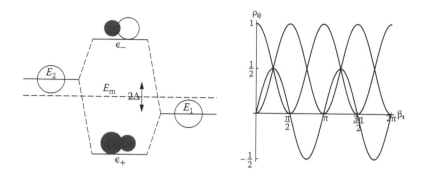

FIGURE 6.1 Time evolution of various components of the density matrix for a degenerate two-state system with coupling $\hbar\beta$.

single phonon, with frequency Ω. This system is shown schematically in Figure 6.2a. Whereas in our previous discussion of this system we assumed that the driving field remained on essentially forever, here we shall consider the case where the field is switched on at time $t = 0$ and then switched off at some later time. For the sake of discussion, let the two states in question be two electronic states of some system and the external field be the electric field. The Hamiltonian we consider is given by

$$H = \hbar\omega\sigma_z + \hbar(\mathscr{E}e\hat{x}\cos(\Omega t)) \tag{6.22}$$

where $\hat{\mu} = e\hat{x}$ is the x component of the electric dipole operator and Ω is the laser frequency with intensity $\mathscr{E}$. Let us assume that $\hat{x}$ only couples 1 and 2 so that its matrix elements are

$$\mu = \langle 1|e\hat{x}|2\rangle$$

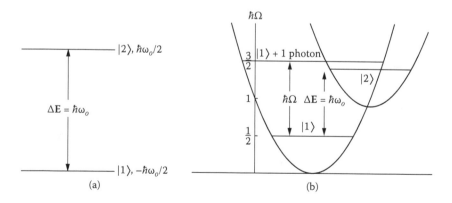

FIGURE 6.2 (a) States 1 and 2 with energies $\pm\hbar\omega_o$ are coupled to an external driving field with frequency Ω. (b) Energy level diagram showing how state $|1\rangle$ is *dressed* by its interaction with the photon field so that it becomes nearly degenerate with state $|2\rangle$. Note: The horizontal offset between the wells is strictly for clarity.

Thus, our Hamiltonian in the $\{|1\rangle, |2\rangle\}$ basis becomes

$$H = \hbar\omega_o\sigma_z + \hbar\mu\mathcal{E}\cos(\Omega t)\sigma_x$$

From the discussion in Chapter 2, we identify the Rabi frequency as

$$\omega_1 = \mu\mathcal{E}$$

and write the equations of motion for the density matrix elements as

$$\dot{\rho}_{11} = i\omega_1\cos(\Omega t)(e^{i\omega_o t}\rho_{12} - e^{-i\omega_o t}\rho_{21}) \tag{6.23}$$

$$\dot{\rho}_{22} = -i\omega_1\cos(\Omega t)(e^{i\omega_o t}\rho_{12} - e^{-i\omega_o t}\rho_{21}) \tag{6.24}$$

$$\dot{\rho}_{12} = i\omega_i\cos(\Omega t)e^{-i\omega_o t}(\rho_{11} - \rho_{22}) \tag{6.25}$$

$$\dot{\rho}_{21} = \dot{\rho}_{12}^* \tag{6.26}$$

Thus far, our analysis is exact. However, to proceed we need to make a judicious approximation in order to simplify our analysis. First, note that when we combine the cosine and exponential, we arrive at terms that are of the form

$$e^{\pm i(\omega_o + \Omega)t} \quad \text{and} \quad e^{\pm i(\omega_o - \Omega)t}$$

If the laser frequency is such that we are nearly resonant with $\hbar\omega_o$, then only the terms with $\omega_o - \Omega \approx 0$ will give a contribution to the transition rate between the two states. The other terms with $\omega_o + \Omega \approx 2\omega_o$ are off resonance and will give vanishingly little to the transition rate. Thus we define the *rotating wave approximation* or RWA by neglecting all off-resonant terms. This allows us to simplify the above equations as

$$\dot{\rho}_{11} = i\frac{\omega_1}{2}\left[e^{i(\omega_o - \Omega)t}\rho_{12} - e^{-i(\omega_o - \Omega)t}\rho_{21}\right] \tag{6.27}$$

$$\dot{\rho}_{22} = -i\frac{\omega_1}{2}\left[e^{i(\omega_o - \Omega)t}\rho_{12} - e^{-i(\omega_o - \Omega)t}\rho_{21}\right] \tag{6.28}$$

$$\dot{\rho}_{12} = i\frac{\omega_1}{2}e^{i(\omega_o - \Omega)t}(\rho_{11} - \rho_{22}) \tag{6.29}$$

$$\dot{\rho}_{21} = \dot{\rho}_{12}^* \tag{6.30}$$

If we are exactly on resonance, then all the exponential terms become unity and the equations reduce to

$$\dot{\rho}_{11} = i\frac{\omega_1}{2}(\rho_{12} - \rho_{21}) \tag{6.31}$$

$$\dot{\rho}_{22} = -i\frac{\omega_1}{2}(\rho_{12} - \rho_{21}) \tag{6.32}$$

$$\dot{\rho}_{12} = i\frac{\omega_1}{2}(\rho_{11} - \rho_{22}) \tag{6.33}$$

$$\dot{\rho}_{21} = \dot{\rho}_{12}^* \tag{6.34}$$

The solutions read

$$\rho_{11} = \cos^2(\omega_1 t/2) \tag{6.35}$$

$$\rho_{22} = \sin^2(\omega_1 t/2) \tag{6.36}$$

$$\rho_{12} = \frac{i}{2}\sin(\omega_1 t) \tag{6.37}$$

$$\rho_{21} = -\frac{i}{2}\sin(\omega_1 t) \tag{6.38}$$

These are identical to the equations we had for the degenerate two-state system. Thus, on resonance, the dynamics of a nondegenerate two-state system becomes identical to that of a degenerate system.

This can be understood by considering Figure 6.2b. We can imagine that the effect of coupling a two-state system to a driving field is identical to the case where we have two wells representing the energy levels of the system plus the quantized energy levels of a harmonic field. When the interaction is switched off, the harmonic field is in its ground state, that is, there are no photons (or phonons) present. When the field is switched on, we have at least one phonon present and the system is now in the second-energy level of the left-hand well in Figure 6.2b. Depending upon the frequency of the laser, this state ($|1\rangle$ + 1 photon) is now energetically closer to $|2\rangle$. When the resonance condition is met, the system oscillates between the $|1\rangle$ + 1 photon state and the $|2\rangle$ + 0 photons state at the Rabi frequency ω_1. The coherent transfer of population from one state to the other is termed transient nutation.

For a laser field, we can easily control the duration of the interaction, and when the field is switched off,

$$\dot{\rho}_{11} = \dot{\rho}_{22} = 0$$

but

$$\dot{\rho}_{12} = +i\omega_o\rho_{12} = \dot{\rho}_{21}^*$$

If the field is switched off at some later time $t_1 > 0$, then for all time after t_1 the populations will remain constant with $\rho_{11} = \rho_{11}(t_1)$, $\rho_{22} = \rho_{22}(t_1)$. However, the coherences will continue to evolve as

$$\rho_{12} = e^{i\omega_o t}\rho_{12}(t_1) \ \& \ \rho_{21} = e^{-i\omega_o t}\rho_{21}(t_1)$$

This is free precession in that the system continues to evolve even though no further population is being transferred. State 1 is effectively locked into a superposition with state 2.

If the duration of the pulse is such that $\omega_1 t_1 = \pi$, then for times $t > \pi/\omega_1$, $\rho_{11} = 0$ and $\rho_{22} = 1$. In other words, we have achieved a perfect population inversion. If we are discussing a spin 1/2 system, we can imagine that all the spins in the system have been flipped from up to down (or vice versa). Such pulses are termed "pi" pulses. We can also define a "pi-over-two" pulse by setting the pulse duration to be such that $\omega_1 t_1 = \pi/2$. In that case, the magnitudes of the two coherences ρ_{12} and ρ_{21} are at their maximal values. In general, we can define a "flip-angle" $\theta = \omega_1 t_1$ such that following a pulse of duration $t_1 = \theta/\omega_1$, the density matrix elements are exactly

$$\rho_{11}(t_1) = \cos^2(\theta/2) \tag{6.39}$$

$$\rho_{22}(t_1) = \sin^2(\theta/2) \tag{6.40}$$

$$\rho_{12}(t_1) = i\,\sin(\theta)/2 \tag{6.41}$$

$$\rho_{21}(t_1) = -i\,\sin(\theta)/2 \tag{6.42}$$

To appreciate the effect of all this on the system, consider what happens to the evolution of an observable, such as the average polarization $\langle\mu\rangle$ following preparation by a pulse of duration $\theta = t_1\omega_1$. The x component of the polarization is given by

$$\langle\mu_x(t)\rangle = Tr[\rho(t)\hat{\mu}] = e\,Tr[\rho(t)\hat{x}]$$

For the two-state system at hand, the polarization operator in the $\{|1\rangle, |2\rangle\}$ basis is given by

$$\hat{\mu}_x = \mu \begin{pmatrix} 0 & 1 \\ 1 & 0 \end{pmatrix}$$

Thus, for $t > t_1$

$$\langle\mu_x\rangle = -\mu\,\sin(\theta)\sin(\omega_o t)$$

In other words, the x component of the polarization vector of the sample oscillates between $\pm\mu\sin(\theta)$ at the Bohr transition frequency. According to classical electrodynamics, an oscillating electric field must radiate at its oscillation frequency. Consequently, this polarization will eventually lead to radiative decay in any realistic physical situation. We can introduce a phenomenological radiative decay as an afterthought to the dynamics by requiring the population of state 2 to be exponentially damped (corresponding to radiative decay to state 1). As a result, we expect

$$\langle\mu_x(t)\rangle_{rad} = -\mu\,\sin(\theta)\sin(\omega_o t) \times e^{-\gamma t}$$

Since the coupling of a molecular or atomic state to the electromagnetic field is an intrinsic property of that state, this radiative decay time $1/\gamma$ is referred to as the *natural lifetime* and the exponential decay of the time signal results in a Lorentzian spectral line shape centered at the transition frequency with full-width at half maximum equal to γ. This spectral shape assumes that the entire sample is composed of identical systems each identically prepared. Hence, the resulting spectral line shape is termed the "homogeneous line shape."

However, in reality no sample is composed of an ensemble of identical components. In a realistic sample, each component may sit in a slightly different environment and have a slightly different energy gap $\hbar\omega_{ok}$. If we let P_k be the probability that any given component of the system has energy gap ω_{ok}, then the total density matrix for the ensemble is the direct sum of the density matrices for the components weighted by the probability P_k

$$\rho = \sum_k P_k \rho_k$$

with

$$\sum_k P_k = 1$$

Note that if we were to write out the full density matrix in matrix form, it would be in block-diagonal form such as:

$$\rho = \begin{pmatrix} \rho_1 & 0 & \cdots & 0 \\ 0 & \rho_2 & 0 & \cdots \\ \vdots & 0 & \ddots & 0 \\ 0 & 0 & 0 & \rho_n \end{pmatrix} \tag{6.43}$$

and P_k would be the fractional number of subsystems with energy gap $\hbar\omega_{ok}$. Since there is no coupling between subsystems, the off-diagonal blocks will remain 0 for all times. In such a case, the density matrix of the entire system is a statistical mixture of its component subsystems.

To calculate the polarization of the ensemble as a function of time, we need to evaluate

$$\langle \mu_x(t) \rangle = Tr[\rho\hat{\mu}_x]$$
$$= \sum_k P_k Tr[\rho_k\hat{\mu}]$$
$$= -\mu \sin\theta \sum_k P_k \sin(\omega_{ok}t) \tag{6.44}$$

To make things concrete, let us assume that the energy gaps are normally distributed about some average energy gap such that the probability of a subsystem having energy gap $\hbar\omega_{ok}$ is given by

$$P(\omega_{ok}) = \frac{1}{\sqrt{2\pi\sigma^2}}e^{-(\omega_{ok}-\overline{\omega})^2/2\sigma^2}$$

Taking the sum over subsystems to a continuous integral and integrating over all frequencies results in

$$\langle \mu_x(t) \rangle = -\mu \sin\theta \sin(\overline{\omega}t)e^{-\sigma^2 t^2/2} \tag{6.45}$$

where $t = 0$ is taken to be at the end of the initial polarizing pulse.

The decay of the polarization of the sample in this case is due to the fact that each member of the ensemble is contributing a slightly different Fourier component to the entire signal. Initially all the polarization components are in lockstep with each other and contribute constructively to the total polarization. However, since each member is oscillating at its own frequency, very soon we will have an equal likelihood of finding any member of the ensemble with any possible orientation of its polarization. This decay of the signal due to variations in the environment gives rise to an *inhomogeneous* broadening of the spectral line shape associated with this signal.

Again, in a real physical system there will be a natural lifetime associated with the excitation giving rise to a homogeneous contribution to the spectral line shape.

Whereas the inhomogeneous contribution is entirely due to the distribution of environments, the homogeneous contribution is an intrinsic atomic or molecular property and will at best be weakly dependent upon the environment surrounding the system. In the parlance of the field, the natural lifetime giving rise to the homogeneous linewidth is called the "longitudinal" relaxation time or T_1 and the various contributions to the total inhomogeneous line shape are lumped into a single timescale T_2^*, which is termed the "transverse" relaxation time.

6.4.2 PHOTON ECHO EXPERIMENT

Consider the following experiment. At time $t = 0$ we excite the system with a resonant $\pi/2$ pulse ($\omega_o t_1 - \pi/2$). As just discussed above, this results in a polarization of the sample, which for $t > t_1$ decays according to Equation 6.45. At some time $t_d = t_2 - t_1$ later, we apply a second resonant pulse again of twice this duration (that is, a π pulse). This has the effect of reversing the precession of the dipoles in the sample such that after again some delay time $t_d' = t_4 - t_3 = t_d$, the dipoles are once again perfectly aligned, giving rise to a spontaneous repolarization of the sample, or an "echo." This time series of pulses is shown in Figure 6.3. This aligned state will inevitably decay due to the same dephasing that occurred following the original $\pi/2$ pulse.[33] The "spin-echo" was first demonstrated for nuclear spins by Hahn in 1950.[34,35] Hahn's original paper on this is one of the most cited papers in experimental physics with well close to 1000 citations according to the *Physical Review* online archive. Later, an analogous "photon-echo" effect was observed in 1964 by Kurnit, Abella, and Hartmann.[36,37] By varying the various pulse widths and delays, a whole host of "two-dimensional" spectroscopy experiments can be devised. While such spectroscopies are now commonplace in the field of magnetic resonance spectroscopy, multidimensional spectroscopies are currently under active development for UV/visible and infrared ranges.

Aside from being an interesting physical phenomenon, the photon echo experiment can resolve the homogeneous from inhomogeneous contributions to the spectral line shape. If $T_1 \gg T_2^*$ and we can space the two pulses by some delay time $t_d < T_1$, we can determine the T_1 time by monitoring the attenuation of the intensity of the echo pulse as a function of the t_d. In doing so, we can unambiguously separate the two components of the total spectral line shape.

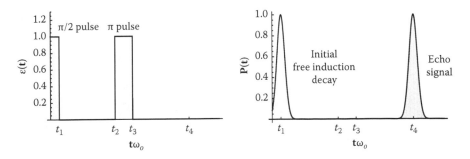

FIGURE 6.3 Photon echo time sequence. Left: The two input pulses with areas $\pi/2$ and π. Right: On same time scale are the resulting output signals for the macroscopic polarization $P(t)$ in the form of two free induction decay signals, each with nearly identical lifetimes.

6.4.3 RELAXATION PROCESSES

We conclude our discussion of the density matrix with a phenomenological description of the relaxation of the density matrix. In a later chapter we will be more specific as to how quantum relaxation occurs when a system is in thermal contact with a bath environment. For the time being, however, we simply assume that the net contribution of the coupling between the chromophore and its surroundings and the radiative decay of the chromophore's excited state can be described with a few parameters that can be related to spectroscopic observables such as T_1 and T_2^*.

As a starting point for discussion, consider the transfer of an excitation (exciton) from one molecular species (A) to a nearby neighbor (B). The reaction we consider can be written as

$$A^* + B \leftrightharpoons A + B^*$$

I have written this as a reversible reaction since the interaction that facilitates the forward transfer also facilitates the reverse transfer. We can write the Schrödinger equation describing this reaction in matrix form as

$$\begin{pmatrix} E_A & V \\ V & E_B \end{pmatrix} \begin{pmatrix} a(t) \\ b(t) \end{pmatrix} = i\hbar \begin{pmatrix} \dot{a}(t) \\ \dot{b}(t) \end{pmatrix} \tag{6.46}$$

where V is the matrix element coupling the two states and E_A and E_B are their energies. The coefficients, $a(t)$ and $b(t)$, are the probability amplitudes for observing the excitation on A and B, respectively, at time t. Take, for example, the case where the two species are identical so that $E_1 = E_2$ and that at time $t = 0$, A is photoexcited so that $a(0) = 1$ and $b(0) = 0$. We can immediately write the solution as

$$a(t) = \cos(Vt/\hbar) \ \& \ b(t) = \sin(Vt/\hbar) \tag{6.47}$$

where we see that the exciton is passed back and forth between A and B at the Rabi frequency, $\Omega = V/\hbar$. This oscillation is due to the fact that the initial state is not a stationary state.

We know, however, that in a thermal system, energy transfer can be irreversible due to contact and mixing between the donor and acceptor species with the solvent media or matrix in which the two molecules are embedded. To consider the case of irreversible transfer, we need to use a density matrix approach $\rho = |\psi\rangle\langle\psi|$ where the diagonal elements of the density matrix, $\rho_{11} = |a(t)|^2$ and $\rho_{22} = |b(t)|^2$, are the populations of each state and the off-diagonal elements, $\rho_{12} = a^*(t)b(t)$ and $\rho_{21} = b^*(t)a(t)$, are the coherences between the two states. For an isolated system, the time evolution of the density matrix is given by the Liouville–von Neumann equation,

$$i\hbar\dot{\rho} = [H, \rho] \tag{6.48}$$

However, for a system imbedded in an environment, we need to consider the relaxation populations and coherences to their equilibrium values. For the moment, we take a phenomenological approach and define a longitudinal relaxation time T_1 as the time scale for the populations of each state to relax to its equilibrium value and a transverse relaxation time T_2 as the time scale for coherence relaxation. For a statistical mixture, $\rho_{12} = \rho_{21} = 0$ and we can write the equations of motion for ρ as

$$\frac{\partial \rho_{ii}}{\partial t} = \frac{1}{i\hbar}[H, \rho]_{ii} + \frac{1}{T_1}\left(\rho_{ii}^{(eq)} - \rho_{ii}\right) \tag{6.49}$$

for the diagonal terms and

$$\frac{\partial \rho_{ij}}{\partial t} = \frac{1}{i\hbar}[H, \rho]_{ij} - \frac{1}{T_2}\rho_{ij} \tag{6.50}$$

for the coherences. Since we are dealing with the transfer of an electronic excitation from one species to the next, we can assume that at thermal equilibrium, all of the population is in the ground electronic state $|0\rangle$, which we have not explicitly included. If we take E_A and $E_B \gg kT$, the thermal populations of the exciton states is vanishingly small and we can include the ground state only as a "sink" such that the total population $\rho_{11} + \rho_{22} + \rho_{00} = 1$. Thus, our equations of motion for the two excited states read

$$\dot{\rho}_{11} = -\frac{i}{\hbar}V(\rho_{21} - \rho_{12}) - \frac{1}{\tau_A}\rho_{11} \tag{6.51}$$

$$\dot{\rho}_{22} = +\frac{i}{\hbar}V(\rho_{21} - \rho_{12}) - \frac{1}{\tau_B}\rho_{22} \tag{6.52}$$

$$\dot{\rho}_{12} = -\frac{i}{\hbar}V(\rho_{22} - \rho_{11}) - \frac{1}{T_2}\rho_{12} - \frac{\Delta E}{i\hbar}\rho_{12} \tag{6.53}$$

$$\dot{\rho}_{21} = +\frac{i}{\hbar}V(\rho_{22} - \rho_{11}) - \frac{1}{T_2}\rho_{21} + \frac{\Delta E}{i\hbar}\rho_{21} \tag{6.54}$$

Here, τ_A and τ_B are the radiative lifetimes of the A and B excitons and ΔE is the energy difference. We now consider various limits to understanding the various physical regimes described by these equations.

In the case of strong coupling, we take the Rabi frequency to be much greater than the radiative rates, $\Omega = 2V/\hbar \gg 1/\tau_A \& 1/\tau_B$, as well as the dephasing rate, $1/T_2$. In this case, the exchange between A and B is far more rapid than any other process in the system. Our equations of motion reduce to

$$\ddot{\rho}_{11} + \Omega^2 \rho_{11} = 1/2 \tag{6.55}$$

and we have the same oscillatory behavior as previously. A comparison between the numerically exact solution and the approximate solution is shown in Figure 6.4(a) for the case of $\Delta E = 0.1$, $V = 1/2$, and $\tau = T_2 = 10^4$.

Identical systems: For the case of exciton exchange between two identical systems, we have $\tau_A = \tau_B = \tau$ and $\Delta E = 0$. In this limit, we can arrive at an equation for the population in state 1:

$$\ddot{\rho}_{11} + \left(\frac{1}{\tau} + \frac{1}{T_2}\right)\dot{\rho}_{11} + \frac{1}{T_2}\left(\frac{1}{\tau} + T_2\Omega^2\right)\rho_{11} = \Omega^2/2 \tag{6.56}$$

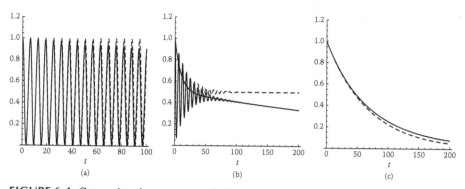

FIGURE 6.4 Comparison between numerically exact evaluation and various approximations. (a) Strong coupling with no decay of population or dephasing $1/T_2, 1/\tau \to 0$. (b) Identical systems ($\Delta E = 0$) with $1/T_2 \gg 1/T_1$. (c) Rapid dephasing with $1/T_2 \gg \Omega$.

These are the equations of motion for a damped oscillating system. Taking the initial condition to be $\rho_{11}(0) = 1$ and ignoring the oscillatory part, we obtain the solution for the overdamped decay of the initial population

$$\rho_{11, decay}(t) = \frac{1}{2} e^{-t/\tau}(1 - \exp(-t/T_2)) \tag{6.57}$$

The results of this case are shown in Figure 6.4(b) for the case of $\Delta E = 0$, $V = 1/2$, $\tau = 500$, and $T_2 = 10$. It is interesting to note that the approximate solution given by Equation 6.57 does not decay at long times.

Rapid dephasing: As a final limit, we take the case where the dephasing time is short compared with the radiative lifetime. This gives us the case of a damped oscillator:

$$\dot{\rho}_{11} + \left(\frac{1}{\tau_A} + \frac{\Omega^2 T_2/2}{1 + (\Delta E T_2/\hbar)^2} \right) \rho_{11} = 0 \tag{6.58}$$

The solution we immediately obtain is

$$\rho_{11}(t) = \exp \left[-\left(\frac{1}{\tau_A} - W \right) t \right] \tag{6.59}$$

with

$$W = \frac{2|V|^2}{\hbar} \frac{T_2}{1 + (T_2 \Delta E/\hbar)^2} \tag{6.60}$$

In this limit, we see energy transfer as a truly irreversible processes with rate constant W, which gives the probability of transfer from A to B per unit time. This estimate works well if the dephasing time is the shortest time scale in the system. In Figure 6.4(c) we show the case for $T_2 = 0.05$, $V = 0.1$, $\Delta E = 1$, and $\tau = 100$. The dashed curve is the approximate solution with the solid curve being the numerically exact solution.

Note that if we take the dephasing time to be extremely short, $T_2 \to 0$, then the transfer rate vanishes. This underscores the importance of the buildup of quantum coherence between the two coupled states. In many regards, this is much like the old phrase that a watched pot never boils. If we take T_2 as a time scale by which the environment queries the system as to which state it happens to be in at a particular instant in time, the system must be found in one state or the other and hence immediately after that instant in time one or the other of the populations must be *exactly* 1 and the other *exactly* 0 and all the coherences between the two must *exactly* vanish. As a result, even if the states are strongly coupled, if $T_2 \ll \hbar / V$, the exciton is effectively localized on the initial state forever.

General solution: We now seek a general solution to the equations of motion in the form of a damped oscillator. Notice that if we integrate Equation 6.59 over all time, we obtain an equation of the form

$$\int_0^\infty dt \rho_{11}(t) = \left(\frac{1}{\tau_A} + \overline{W} \right)^{-1} \tag{6.61}$$

For now, let us differentiate the approximate rate W from Eq. (6.60) from $\overline{W}$, which we will obtain from the exact solution, and look for the case in which the two are identical.

At this point it is best to work with the Laplace transformed versions of the equations of motion,

$$\mathscr{L}\rho = \tilde{\rho}(s) = \int_0^\infty e^{-st} \rho(t) dt \tag{6.62}$$

and

$$\mathscr{L}\dot{\rho} = s\tilde{\rho} - \rho(0) \tag{6.63}$$

First, the transformed equations must be true for all values of s, so we take the case of $s = 0$. Secondly, the initial conditions are such that only $\rho_{11}(0) = 1$ with all other elements equal to zero. Thus, our Laplace transformed equations of motion reduce to a set of algebraic equations [where we take $\tilde{\rho} = \tilde{\rho}(0)$]:

$$-1 = \frac{V}{i\hbar}(\tilde{\rho}_{21} - \tilde{\rho}_{12}) - \frac{1}{\tau_A}\tilde{\rho}_{11} \tag{6.64}$$

$$0 = -\frac{V}{i\hbar}(\tilde{\rho}_{21} - \tilde{\rho}_{12}) - \frac{1}{\tau_B}\tilde{\rho}_{22} \tag{6.65}$$

$$0 = \frac{V}{i\hbar}(\tilde{\rho}_{22} - \tilde{\rho}_{12}) - \frac{1}{T_2}\tilde{\rho}_{12} + \frac{\Delta E}{i\hbar}\tilde{\rho}_{12} \tag{6.66}$$

$$0 = -\frac{V}{i\hbar}(\tilde{\rho}_{22} - \tilde{\rho}_{12}) - \frac{1}{T_2}\tilde{\rho}_{21} + \frac{\Delta E}{i\hbar}\tilde{\rho}_{21} \tag{6.67}$$

After quite a bit of tedious algebra, we obtain the final result in the desired form

$$\tilde{\rho}_{11}^{-1} = \frac{1}{\tau_A} + \overline{W} \tag{6.68}$$

where the exact rate is given by

$$\overline{W} = \frac{2|V|^2 T_2/\hbar}{1 + (T_2 \Delta E/\hbar)^2 + (2|V|^2/\hbar)T_2\tau_B} \tag{6.69}$$

In the case of weak interaction $2T_2\tau_b V^2/\hbar \ll 1$, we recover $\overline{W} = W$ from above. In the case of very strong interaction, $\overline{W} \rightarrow 1/\tau_B$ and the transfer is dominated by the radiative lifetime of B. It is important to note that the transfer is completely independent of the coupling V.

Typically in molecular systems, τ_B is not determined by the radiative lifetime from the excited state to the ground state but by the vibrational relaxation time, which is typically on the order of a ps rather than ns. For the moment, we shall ignore the internal dynamical contributions to energy transfer and focus our attention on the electronic contributions.

6.5 DECOHERENCE

Certainly one of the hallmarks of quantum dynamics is the fact that over the course of the evolution of a system, it evolves from some initially prepared state to a superposition of states. This leads to one of the effects that make quantum mechanics interesting, namely, interference. The most common thought experiment (and very colorfully described in Feynman's book) is where electrons are shot from some source toward a blocking screen with two small parallel slits. If the slits are close enough together, then there is equal likelihood for an electron to go through either slit. For classical electrons, this would result in the accumulation of two distributions of electrons behind the screen: those that went though slit #1 and those that went through slit #2. However, what is observed is an interference pattern consistent with a plane wave passing through both slits and interfering constructively and destructively on the other side—much like water waves. Since electrons are not divisible into chunks of partial electrons, we have to assume that each electron passing through the observing screen went through one or the other slits and not through both.

By its very nature, quantum mechanics likes to explore all equivalent (or nearly equivalent) alternatives. Yogi Berra put it best in saying, "When you come to a fork in the road, you take it." Perhaps the "Yogi Berra" rule of quantum mechanics is "When you come to a fork in the road, you take both paths." The story behind this quote is that Yogi lived at the end of a cul-de-sac. So, if you were going to Yogi's house, you would eventually have to take the left or right turn . . . both of which would land you at Yogi's house. I wonder how many Brooklyn Dodgers were lost through destructive interference this way.

Returning to the double-slit thought experiment, if we try to monitor the flux of electrons through either hole, then we force the electron's wave function to localize every time we observe an electron passing by. Say we observe the flux using a laser beam so that the scatter of a photon by the electron indicates the electron's passage. If we turn the light intensity down low so that some electrons go by undetected, we partially recover the interference pattern such that the resulting distribution of electrons on the final detector represents the weighted sum of electrons that got caught (with no interferrence) and those that slipped by undetected.

6.5.1 DECOHERENCE BY SCATTERING

From an operational standpoint, dephasing and decoherence are identical in that they both cause the off-diagonal elements of the density matrix to relax to zero so as to produce a statistical mixture. However, this "relaxation of the off-diagonal elements" statement is not entirely rigorous, although one finds this terminology throughout the literature. A more precise statement is that through the interaction with an environment, the density matrix becomes diagonal and is a *basis of eigenstate of the system-bath interaction operator.* In this representation, $[V_{sb}, \sigma] = 0$. Zurek refers to such states as "pointer states" since it is into these states that the system is directed through its interaction with the bath.

Dephasing refers to the fact that if we have an ensemble of identically prepared systems, each evolves in a slightly different (static) environment. When we average over the ensemble, the accumulated phase differences interfere destructively and sum to zero, leading to a loss of polarization of the sample. The fact that the ensemble can be repolarized as in the spin-echo experiment indicates that dissipative and irreversible effects are not important on the time scale of the spin-echo experiment. Decoherence, on the other hand, is the irreversible loss of phase coherence between states for a *single* subsystem embedded in an environment. It results from the fact that the eigenstates of the subsystem become intertwined with the eigenstates of the environment. In a somewhat prosaic sense, the environment makes a series of weak measurements on the system, asking it, "What state are you in now?" If the system answers, "OK, now I'm in state #2!" then the potential force coupling the environment to the system will be the force for state #2. Furthermore, since the environment has determined that the system is in fact in state #2, the reduced density matrix for the system must at that instance correspond to a pure state with the system's state vector in state #2. As the coupled systems continue to evolve, each will ascertain from the other information concerning how it is to evolve. Consequently, one can easily imagine that this dialog could be recorded in the form of a series of events indicating the state of the system at various time steps.

We can formalize this somewhat by letting $|x\rangle$ be the position state of a particle and $|X\rangle$ be the macroscopic environment. We shall assume for the time being that the interaction between the particle and its surroundings is purely elastic with no recoil. If we prepare the system in a given state at time $t = 0$ and let it interact with the surroundings, then we have the following:

$$|x\rangle|X\rangle \mapsto |x\rangle|X_x\rangle = |x\rangle S_x|X\rangle$$

where the x subscript denotes that the surroundings have interacted with the quantum particle and are thus entangled. S_x is the scattering matrix describing the process. If instead we start with a superposition state

$$\int dx \phi(x)|x\rangle|X\rangle \mapsto \int dx \phi(x)|x\rangle S_x|X\rangle$$

the reduced density matrix for the system can be written as

$$\rho(x, x') = \phi(x)\phi^*(x') \mapsto \phi(x)\phi^*(x')\langle X|S_x^\dagger S_x|X\rangle$$

In order to estimate $\langle X | S_x^\dagger S_x | X \rangle$, let us consider the scattering light in terms of incoming and outgoing plane waves and that the wavelength of the light is long enough that it cannot resolve distances smaller than $|x - x'| \ll \lambda$ where $\lambda = c/v$ is the wavelength of the scattering light. If this is the case, then working in the wave-vector representation,

$$S_x(k, k') = S(k, k')e^{-i(k-k')x}$$

where $S(k, k')$ is the scattering matrix, which we can relate to the form factor of the interaction. As the result of this impulsive approximation, we can write

$$\rho(x, x'; t) = \rho(x, x'; 0) \exp[-\Lambda(x - x')^2 t]$$

where

$$\Lambda = \frac{k^2 N v \sigma_{eff}}{V}$$

with k as the wave vector, Nv/V the incoming flux that we can relate to the collision frequency, and σ_{eff} the effective cross-section. Λ is the decoherence rate given as the number of scattering events per unit time per unit area. It is also (formally) equivalent to the dephasing rate $1/T_2$ introduced in the last chapter. Here, at least, we have some inkling of how the dephasing process may actually occur.

If we extend this analogy to a quantum state in a solvent environment, such as a solvated electron or an excited state of a molecule, then we can relate the effective cross-section to the molecular radius $\sigma_{eff} = \pi r^2$; k is given by the de Broglie wavelength of the scattering particle, $k = 2\pi/\lambda_{th}$, and v is replaced by the mean thermal velocity $v = \sqrt{kT/m}$. Pulling this together, we arrive at a simple estimate for the decoherence rate for condensed phase states:

$$\Lambda = \frac{2\pi^2 \sqrt{m}(kT)^{3/2} r^2}{h^2} \frac{N}{V}$$

Still working within the impulsive approximation, the resulting equation of motion for the reduced density matrix of quantum particle is

$$i\hbar \frac{\partial \rho}{\partial t} = [H, \rho] - \frac{i}{\hbar} \Lambda[x, [x, \rho]]$$

It is interesting to note that this model also results if we consider the equations of motion for a particle randomly kicked by a Gaussian noise term. The equivalent Hamiltonian for this reads

$$H = H_0 + \sum_i \lambda_i(t) V_i \tag{6.70}$$

where H_0 and V_i are arbitrary Hermitian operators and Gaussian stochastic coefficients $\lambda_i(t)$ with average mean $\overline{\lambda_i(t)} = \overline{\lambda_i} = 0$ and second moments given by

$$\overline{\lambda_i(t)\lambda_j(t')} = g_{ij}\delta(t - t') \tag{6.71}$$

Hamiltonians such as this can model a wide variety of physical situations where the motion or transport is driven by an external field. One such example is the case of Förster resonant excitation transfer (FRET) in biomolecules where the migration and diffusion of an initial electronic excitation within the system is dependent upon the local environmental and conformational fluctuations or the transport of a proton in a channel in which the tunneling barriers between sites are modulated by the environment in some stochastic way. A useful property of such Hamiltonians is that one can explicitly average over the noise when calculating the time evolution of the density matrix or expectation values of various observables. Noise-averaged time evolution for various specific forms of the Hamiltonian in Equation 6.70 have been considered previously.[38–42]

Let us take a brief technical departure and consider the ramifications of this model. In general, the quantum density matrix satisfies the Liouville–von Neumann equation,

$$i\frac{\partial \rho}{\partial t} = (\mathscr{L}_0 + \mathscr{L}_V(t))\rho \qquad (6.72)$$

Action of the superoperators $\mathscr{L}_0$ and $\mathscr{L}_V$ on ρ is given, respectively, by $\hbar\mathscr{L}_0\rho = [H_0, \rho]$ and $\hbar\mathscr{L}_V(t)\rho = \sum_i \lambda_i(t)[V_i, \rho]$. The density matrix at time t is given in terms of the density matrix $\rho(0)$ at time $t = 0$ by

$$\rho(t) = \mathscr{U}(t)\rho(0) \qquad (6.73)$$

Here the time-evolution superoperator $\mathscr{U}(t)$ is given by the following infinite series:

$$\mathscr{U}(t) = e^{-i\mathscr{L}_0 t} - i\int_0^t d\tau\, e^{-i\mathscr{L}_0(t-\tau)}\mathscr{L}_V(\tau)e^{-i\mathscr{L}_0 \tau}$$

$$- \int_0^t d\tau \int_0^\tau d\tau'\, e^{-i\mathscr{L}_0(t-\tau)}\mathscr{L}_V(\tau)e^{-i\mathscr{L}_0(\tau-\tau')}\mathscr{L}_V(\tau')e^{-i\mathscr{L}_0 \tau'} + \dots \qquad (6.74)$$

Noise-averaged expectation values of an operator O are computed using

$$\overline{\langle O(t)\rangle} = Tr\,(\overline{O(t)\mathscr{U}(t)\rho(0)}) \qquad (6.75)$$

It is assumed here that operator O can have an explicit time dependence. When performing averages as in Equation 6.75, we need to distinguish two types of operators: those with and those without stochastic coefficients $\lambda_i(t)$. In the latter case, averaging over noise as in Equation 6.75 reduces to an averaging of the evolution superoperator $\mathscr{U}(t)$ and such expectation values can be calculated with the noise-averaged density matrix.

Noise averaging of $U(t)$ can be performed by taking averages for each term in the series and then resumming the series. This involves averaging products of the stochastic coefficients. Since $\lambda_i(t)$ is sampled from a Gaussian deviate, all terms involving an odd number of coefficients necessarily vanish. Furthermore, any term with an even number of coefficients can be written as a sum of all possible products of second moments. However, due to the order of integrations over time in Equation 6.74 and the fact that second moments in Equation 6.71 involve delta functions in time, only one product from the sum contributes to the average after all the time integrations are performed. For example, consider the fourth-order average:

$$\overline{\lambda_i(t)\lambda_j(\tau)\lambda_k(\tau')\lambda_l(\tau'')}$$

Using Equation 6.71 for the second moments, we have

$$\overline{\lambda_i(t)\lambda_j(\tau)\lambda_k(\tau')\lambda_l(\tau'')} = g_{ij}g_{kl}\delta(t - \tau)\delta(\tau' - \tau'')$$
$$+ g_{ik}g_{jl}\delta(t - \tau')\delta(\tau - \tau'')$$
$$+ g_{il}g_{jk}\delta(t - \tau'')\delta(\tau - \tau') \qquad (6.76)$$

Since the region of integration for the fourth-order term in the series (Equation 6.74) is $t \geq \tau \geq \tau' \geq \tau''$, we can see that only the $g_{ij}g_{kl}\delta(t - \tau)\delta(\tau' - \tau'')$ term will contribute to the series. Similar analysis can be applied to all even-order terms.

Averaging over noise and resumming the series (Equation 6.74) produces

$$\overline{\mathscr{U}(t)} = e^{-i\mathscr{L}_0 t - \mathscr{M}t} \qquad (6.77)$$

where the action of superoperator $\mathscr{M}$ on the density matrix ρ is given by

$$\mathscr{M}\rho = \frac{1}{2\hbar^2}\sum_{ij} g_{ij}[V_i, [V_j, \rho]] \qquad (6.78)$$

It follows from Equation 6.77 that the noise-averaged density matrix satisfies the following equation:

$$i\frac{\partial\overline{\rho}}{\partial t} = (\mathscr{L}_0 - i\mathscr{M})\overline{\rho} \qquad (6.79)$$

This allows one to see an interesting connection between the noise-averaged time evolution of the density matrix for the noisy system and the time evolution of the reduced density matrix for an open quantum system.

Consider the case when the correlation matrix g_{ij} in Equation 6.78 is diagonal, that is, $g_{ij} = g_{ii}\delta_{ij}$. In this case, Equation 6.79 can be rewritten as

$$i\frac{\partial\overline{\rho}}{\partial t} = \frac{1}{\hbar}[H_0, \overline{\rho}] - \frac{i}{2\hbar^2}\sum_i g_{ii}[V_i, [V_i, \overline{\rho}]] \qquad (6.80)$$

Replacing $g_{ii} = \Lambda$ and $V_i = x$, we arrive at the equation we had above for the implusively monitored particle.

Even though we have stated that there is no recoil between the quantum particle and the scattering particle, the energy of the quantum particle must increase with time. In general, the Ehrenfest equations of motion for the expectation values of operators are given by

$$\frac{d\langle O\rangle}{dt} = \frac{d}{dt}Tr(O\rho) = Tr\left(O\frac{d\rho}{dt}\right)$$

For example, if we consider the Ehrenfest equations of motion for position, momentum, and energy, we find

$$\frac{d\langle x\rangle}{dt} = \frac{\langle p\rangle}{m} \qquad (6.81)$$

$$\frac{d\langle p\rangle}{dt} = -\left\langle\frac{dV}{dx}\right\rangle \qquad (6.82)$$

$$\frac{d\langle H\rangle}{dt} = 0 \qquad (6.83)$$

In other words, when we eliminate the interaction with the scattering particles, energy is conserved as we expect. However, including the recoilless interaction,

$$i\hbar \frac{\partial \rho}{\partial t} = [H_o, \rho] - \frac{i}{\hbar} \Lambda [x, [x, \rho]] \tag{6.84}$$

$$\frac{d\langle x \rangle}{dt} = \frac{\langle p \rangle}{m} \tag{6.85}$$

$$\frac{d\langle p \rangle}{dt} = -\left\langle \frac{dV}{dx} \right\rangle \tag{6.86}$$

$$\frac{d\langle H \rangle}{dt} = +\frac{\Lambda}{m} \tag{6.87}$$

where m is the mass of the quantum particle. Thus, the average energy must *increase* due to the noisy interaction. One can also conclude that the necessary condition for energy conservation even in this case is that $[H_o, V_i] = 0$ for all operators describing the coupling between the quantum particle and the scatterer.

We can construct equations of motion that do lead to $d\langle H \rangle/dt = 0$ at long times by including a frictional term. For example, the equations of motion for a classical Brownian particle can be written as

$$m\ddot{x} + \eta \dot{x} + V' = f(t)$$

where η is the relaxation rate and $f(t)$ is a noise source with the properties $\overline{f(t)} = 0$ and $\overline{f(t)f(t')} = 2\eta kT\delta(t - t') = 2\Lambda\delta(t - t')$. In other words, the relaxation rate η is directly proportional to the frequency (and strength) of the interaction with the noisy field. Analyzing the classical equations results in the following noise-averaged quantities:

$$\frac{d}{dt}\bar{x} = \bar{p}/m \tag{6.88}$$

$$\frac{d}{dt}\bar{p} = -\overline{V'} - \eta\bar{p}/m \tag{6.89}$$

$$\frac{d}{dt}\bar{E} = \frac{2}{\eta m}\left[\frac{kT}{2} - \frac{\overline{p^2}}{2m}\right] \tag{6.90}$$

As the system relaxes, the average kinetic energy becomes equal to $kT/2$ even though the average momentum $\bar{p}$ relaxes to zero. Also, it is important to notice that $\overline{p^2} \neq \bar{p}^2$. To actually solve these equations, we also need to also work out the equations for the higher-order averages, $\overline{p^2}, \overline{x^2}$, and so on.

Similarly, for quantum systems, we can insert a frictional term into the equations of motion for the density matrix:

$$i\frac{\partial}{\partial t}\rho = \frac{1}{\hbar}[H_o, \rho] + \frac{\eta}{2m\hbar^2}[x, \{p, \rho\}] - i\frac{\eta kT}{\hbar^2}[x, [x, \rho]] \tag{6.91}$$

where $\{\cdots\}$ is the anticommutation bracket $\{A, B\} = AB - BA$. Again, working out the Ehrenfest equations of motion for the averaged quantities,

$$\frac{d}{dt}\langle x \rangle = \langle p \rangle \tag{6.92}$$

$$\frac{d}{dt}\langle p \rangle = -\left\langle \frac{dV}{dx} \right\rangle - \frac{\eta}{m}\langle p \rangle \tag{6.93}$$

$$\frac{d}{dt}\langle H \rangle = \frac{2\eta}{m}\left[\frac{kT}{2} - \left\langle \frac{p^2}{2m} \right\rangle \right] \tag{6.94}$$

As in the classical case, the system relaxes to some thermal distribution such that its final kinetic energy is identical to the thermal energy. Also, one needs to be aware that these equations of motion are not closed. In fact, for the general case, this is the beginning of a hierarchy of equations.[43–48]

6.5.2 THE QUANTUM ZENO EFFECT

The quantum Zeno effect is the prediction that an unstable particle (such as an excited state of an atom) will never decay if it is continuously observed. The effect was first predicted by Leonid Khalfin in 1959[49] and the term "quantum Zeno effect" was later coined by Sudarshan and Misra almost 20 years later.[50] In the analysis above, we have shown that whenever the environment interacts with a quantum subsystem, the coherence terms in the reduced density matrix decay to zero. The rate of decay depends upon how often the environment queries the quantum state of the subsystem.

A simple experiment to test this idea was proposed by Cook;[51] it consisted of three levels in which two of the levels, (2) and (3), were resonantly coupled to a common ground state (1). Spontaneous decay between (2) and (1) is assumed to be negligible (Figure 6.5a). If the system is (1) at $t = 0$ and a resonant interaction is applied, a superposition, is created between (1) and (2) that transfers population between (1) and (2) at the Rabi frequency Ω. As we have seen previously, the transient populations in (1) and (2) are then given by

$$P_1(t) = \cos^2(\Omega t/2) \tag{6.95}$$

$$P_2(2) = \sin^2(\Omega t/2) \tag{6.96}$$

Suppose at some time such that $\Omega t \ll 1$ we measure the state of the system and $P_1 \approx 1$ and $P_2 \approx \Omega^2 t^2/4 \ll 1$. Likewise, if we had prepared the system in (2), we would have the reverse situation where $P_2 \approx= 1$ and $P_1 \approx= 0$. If level (3) can only decay to (1), say, due to a selection rule, then we can perform a measurement on the system by driving the (1) $\rightarrow$ (3) transition with an optical pulse.

The proposed experiment goes as follows. First, we prepare the system in state (2) by driving the (1) $\rightarrow$ (2) transition with a π-pulse of duration $T = \pi/\Omega$ while simultaneously applying a series of short measurement pulses. The duration of the measurement pulse is assumed to be much less than T. Suppose the system is in (1) at $t = 0$ and a π pulse is applied. In the absence of the probe pulse, $P_2(T) = 1$. For a

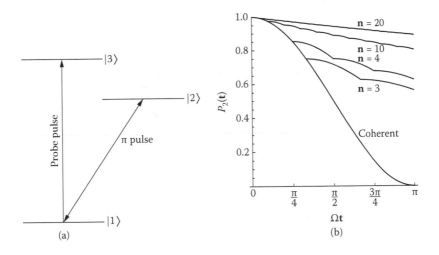

FIGURE 6.5 (a) Three-level scheme used in Cook's proposed experiment for testing the quantum Zeno effect. (b) Predicted transient populations of state (2) following n impulsive probe pulses.

two-state system, we can represent the density matrix as

$$\rho(t) = R_1(t)\sigma_x + R_2(t)\sigma_y + R_3(t)\sigma_z$$

The equations of motion for the polarization vector $\vec{R} = (R_1, R_2, R_3)$ are given by

$$d\vec{R}/dt = \vec{\omega} \times \vec{R}$$

with $R(0) = (0, 0, -1)$ and $\vec{\omega} = (1, 0, 0)$. Following preparation and in the absence of any further interactions, the polarization vector precesses about $\vec{\omega}$ at the Rabi frequency. (Note that we have ignored the counter-rotating terms.)

Now assume that n probe pulses are applied at times $\tau_k = k\pi/(n\Omega)$ where $k = 1, \ldots, n$. Just before the first probe pulse at $t = \pi/(n\Omega)$, the polarization vector is

$$\vec{R} = (0, \sin(\pi/n), -\cos(\pi/n))$$

The probe pulse collapses the wave function (that is, eliminates the coherences) while leaving the populations unchanged. In other words, after the pulse, $R_1(t_+) = R_2(t_+) = 0$ and $R_3(t_+) = (0, 0, -\cos(\pi/n))$. For all intents and purposes, $\vec{R}(t_+)$ is identical to $\vec{R}(0)$ except that its magnitude is now $|R| = |\cos(\pi/n)|$. Consequently, after a sequence of n pulses, $\vec{R}(T) = (0, 0, -\cos^n(\pi/n))$. Since R_3 is the difference between the two populations, $R_3 = P_2 - P_1$ and $P_1 + P_2 = 1$, it is easy to see that

$$P_2(T) = (1 + R_3(T))/2$$

$$= (1 - \cos^n(\pi/n))/2 \tag{6.97}$$

Expanding $\cos(\pi/n)$ as a power series and using

$$\lim_{n \to \infty} (1 - x/n)^n = e^{-x}$$

we find that in the limit of rapid probe pulses

$$P_2(T) = (1 - \exp(-\pi^2/(2n)))/2 \tag{6.98}$$

The predicted transient populations of state (2) are shown in Figure 6.5b where we have assumed the probe pulse to be impulsive. As the frequency probe pulses increase, the population in (2) does not decay to state (1).

This scheme was used by Itano et al.[50] in 1990 to examine the effect of measurement on a quantum superposition states. In this experiment, approximately 5000 $^9Be^+$ ions were held in a Penning trap and laser cooled to below 250 mK. In a magnetic field, the $2s^2 S_{1/2}$ ground state of Be^+ is split into hyperfine levels similar to what is shown in Figure 6.5(a). Radio-frequency (rf) transitions can occur between the $(m_I, m_J) = (3/2, 1/2)$ and $(1/2, 1/2)$ sublevels. A resonant rf pulse would place nearly all the atoms in the upper $(3/2, 1/2)$ state (2), depopulating the lower state. A second UV pulse resonant with the transition between the lower $2s^2 S_{1/2}$ state (1) and one of the $2p^2 P_{3/2}$ states (3) with quantum numbers $(m_I, m_J) = (3/2, 1/2)$, which only decays to state (1). The results of this experiment are shown in Figure 6.6 where we have plotted both the predicted and experimental transition probabilities between states (1) and (2). The agreement is well within the 0.02% statistical uncertainty of the experiment due to photon counting.

6.6   SUMMARY

In this chapter we have described the relaxation of a quantum system through a rather phenomenological approach. We have not thus far described the process for connecting the various relaxation time scales to a molecular-level description of the interaction between an individual molecule and an environment. This we reserve for later discussion and refer the interested reader to other texts and sources:

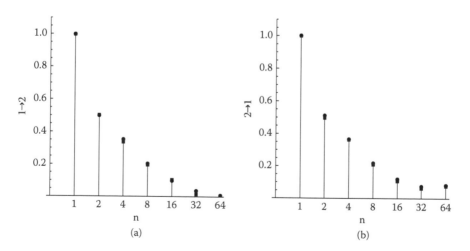

FIGURE 6.6 Comparison between predicted and experimental transition probabilities following n probe pulses for the (a) $1 \rightarrow 2$ (b) $2 \rightarrow 1$ transition. Experimental data from Ref. 50.

1. *Chemical Dynamics in the Condensed Phase*, Abraham Nitzan (Oxford, UK: Oxford University Press, 2007),
2. *Quantum Mechanics in Chemistry*, George C. Schatz and Mark A. Ratner (Mineola, NY: Dover, 2002).

Both of these texts provide excellent pedagogic descriptions and technical details for deriving the reduced equations of motion for a system interacting with an environment.

6.7 APPENDIX: WIGNER QUASI-PROBABILITY DISTRIBUTION

Not long after the introduction of the Schrödinger equation with its wave-function solutions, Eugene Wigner proposed a means to study quantum dynamics in phase space and thereby systematically include quantum mechanical corrections to classical dynamics.[52] The resulting function of interest is a "quasi-probability" distribution function over the p, x classical phase-space variables. The function is defined in terms of a Fourier transform of the density matrix (or wave function)

$$W(x, p) = \frac{1}{\pi \hbar} \int dy \psi(x + y/2) \psi^*(x - y/2) e^{ipy/\hbar} \qquad (6.99)$$

$$= \frac{1}{\pi \hbar} \int dy \rho(x - y/2, x + y/2) e^{ipy/\hbar} \qquad (6.100)$$

Here, x and p denote the position and momentum. However, in principle, one can use any pair of conjugate variables—for example, the real and imaginary parts of a field or the time and frequency components of a signal. We can also take the inverse transform and recover the original density matrix

$$\rho(x + y/2, x - y/2) = \int W(x, p) e^{ipy/\hbar} dp \qquad (6.101)$$

$W(x, p)$ is a generating function for all spatial autocorrelation functions of a given quantum mechanical wave function $\psi(x)$. For example, taking the derivative of W

$$\frac{1}{i} \frac{\partial W}{\partial p} \bigg|_{p=0} = \frac{1}{\pi \hbar} \int dy \rho(x - y, x + y) y = \langle x \rangle \qquad (6.102)$$

gives the expectation value of a conjugate variable. Thus, W corresponds to the quantum density matrix in the map between real phase-space functions and Hermitian operators introduced by Hermann Weyl in 1927, in a context related to representation theory in mathematics (cf. Weyl quantization in physics).[53-55] In effect, it is the Weyl transform of the density matrix. Similar transforms were later rederived by J. Ville in 1948 as a quadratic (in signal) representation of the local time-frequency energy of a signal.[56] Furthermore in 1949, José Enrique Moyal, who had also rederived it independently, recognized it as the quantum moment-generating functional and, thus, as the basis of an elegant encoding of all quantum expectation values and, hence, quantum mechanics, in phase space (cf. Weyl quantization).[57] It has applications in statistical mechanics, quantum chemistry, quantum optics, classical optics, and signal analysis in such diverse fields as electrical engineering, seismology, biology, speech processing, and engine design.

A classical particle has a definite position and momentum, and hence it is represented by a point in phase space, x, p. Given a collection or ensemble of particles, the probability of finding a particle in an infinitesimal phase-space volume $dxdp$ about the point x, p at time t is given by $P(x, p;t) = \rho(x, p)dx\, dp$, which has the property that $P \geq 0$,

$$\int dxdp\rho(x, p) = N \qquad (6.103)$$

where N is the number of particles in the ensemble. The time development for the phase-space distribution $\rho(x, p)$ is given by the Liouville equation:

$$\frac{d\rho}{dt} = \frac{\partial \rho}{\partial t} + \frac{\partial \rho}{\partial q}\frac{\partial H}{\partial p} - \frac{\partial \rho}{\partial p}\frac{\partial H}{\partial q} = 0 \qquad (6.104)$$

where H is the Hamiltonian governing the motion of the particles. Thus, the equation of motion governing the classical phase-space density is

$$\frac{\partial \rho}{\partial t} = -\frac{\partial \rho}{\partial q}\frac{\partial H}{\partial p} + \frac{\partial \rho}{\partial p}\frac{\partial H}{\partial q} = -\{\rho, H\} \qquad (6.105)$$

or, in terms of the Liouville operator, $\hat{L}$,

$$\frac{\partial \rho}{\partial t} + \hat{L}\rho = 0 \qquad (6.106)$$

One can also recast this equation as

$$\frac{\partial \rho}{\partial t} + \frac{p}{m}\cdot \nabla_x\rho + F\cdot \nabla_p\rho = 0 \qquad (6.107)$$

In astrophysics this is called the Vlasov equation, or sometimes the collisionless Boltzmann equation, and is used to describe the evolution of a large number of collisionless particles moving in a potential. In classical statistical mechanics, N can become very large. Consequently, setting $\partial \rho/\partial t = 0$ gives the stationary or equilibrium density for an ensemble of microstates. In particular, this is satisfied by the Boltzmann distribution, $\rho \propto \exp(-H\beta)$ where $\beta = 1/k_B T$.

The physical interpretation of the classical Liouville equation is that we imagine the density enclosed in small-volume elements as seen by following along a trajectory $x(t)$, $p(t)$. Since $d\rho/dt = 0$, there is no net flux of probability in or out of the small-volume element. Alternatively, one can imagine a cloud of points in phase space. It is straightforward to show that as the cloud stretches in one dimension, say, in x, it shrinks in the other dimension p so that the volume $\Delta x \Delta p$ remains a constant. This classical interpretation fails for a quantum particle due to the uncertainty principle, which forbids the precise simultaneous determination of both x and p. Instead, the above quasi-probability Wigner distribution plays an analogous role but does not satisfy all the properties of a conventional probability distribution; and, conversely, it satisfies boundedness properties unavailable to classical distributions. Moreover, the Wigner distribution can and normally does go negative for states that have no classical analog—and is a convenient indicator of quantum mechanical interference. This can be seen in Figure 6.7 where we have plotted the Wigner function for the first three eigenstates of the harmonic oscillator. For every state other than the $n = 0$ ground state, one can clearly see regions where W is positive and regions where W is negative. Hence, $W(x, p; t)dxdp$ cannot be interpreted as the probability of finding a particle in an infinitesimal phase-space volume $dxdp$ about the point x, p.

The function itself has a number of useful properties. First, $W(p, x)$ is real. Secondly, the x and p distributions are given by the marginals

$$\int dp W(x, p) = \rho(x, x) \tag{6.108}$$

If ψ is a pure state, then $\rho(x, x) = |\psi(x)|^2$. Likewise,

$$\int dx W(x, p) = \rho(p, p) \tag{6.109}$$

yields the momentum distribution. Again for a pure state, $\rho(p, p) = |\tilde{\psi}(p)|^2$. Finally

$$\int dxdp W = Tr(\rho) = 1 \tag{6.110}$$

That W is real and it can give both the momentum and position distributions implies that W can be negative somewhere.

In order to compute physical quantities, we need to first transform the quantum mechanical operators into the Wigner representation

$$B_W(x, p) = \int dy \langle x + y/2|\hat{B}|x - y/2\rangle e^{ipy/\hbar} \tag{6.111}$$

where the W subscript denotes the "Wigner-ized" operator. This may sound grand, however, in practice, it is actually quite simple for operators involving only position and momentum variables. For example, if operator A is a function of the position operator, $\hat{q}$, then

$$A_W(x) = \int \langle q + y/2|A(q)|q - y/2\rangle e^{ipy/\hbar} = A(\hat{q}) \tag{6.112}$$

Likewise for the momentum operator. Where we do have to be careful, however, is in taking the Wigner transform of operator products

$$(A \cdot B)_W = A_W \exp\left[\frac{i\hbar}{2}\hat{\Lambda}\right] B_W \tag{6.113}$$

where $\hat{\Lambda}$ is the Poisson bracket operator defined as

$$\hat{\Lambda} = \left[\frac{\overleftarrow{\partial}}{\partial x}\frac{\overrightarrow{\partial}}{\partial p} - \frac{\overleftarrow{\partial}}{\partial p}\frac{\overrightarrow{\partial}}{\partial x}\right] \tag{6.114}$$

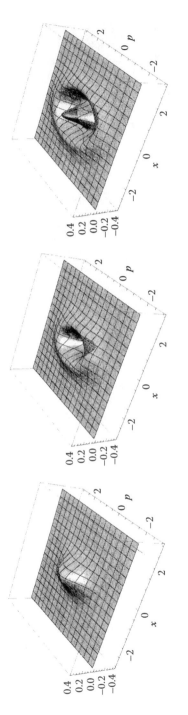

FIGURE 6.7 Wigner distribution for the first three states of the harmonic oscillator.

We need to pay attention to the direction of the arrows in this last expression since they indicate the *direction of operation for the partial derivative*. For example,

$$A_W \hat{\Lambda} B_W = \frac{\partial A}{\partial x} \frac{\partial B}{\partial p} - \frac{\partial A}{\partial p} \frac{\partial B}{\partial x} \tag{6.115}$$

We can also easily see that

$$(A \cdot B)_W = A_W \exp\left[\frac{i\hbar}{2} \hat{\Lambda}\right] B_W = B_W \exp\left[-\frac{i\hbar}{2} \hat{\Lambda}\right] A_W \tag{6.116}$$

This allows us to construct the Wigner transform of a commutator as

$$[A, B]_W = (AB)_W - (BA)_W \tag{6.117}$$

$$= 2i A_W \sin(\hbar \hat{\Lambda}/2) B_W \tag{6.118}$$

From this we can construct an expansion in powers of $\hbar$ using

$$\frac{2}{\hbar} \sin(\hbar \hat{T}/2) = \hat{T} - \frac{\hbar^2}{24} \hat{T}^3 + \cdots \tag{6.119}$$

where we notice that the lowest-order term involving $\hbar$ enters in only the second term and beyond and the leading-order term is simply the classical Poisson bracket operator. Such expansions are extremely useful in evaluating the time evolution of the Wigner distribution

$$\frac{\partial W}{\partial t} = -\frac{i}{\hbar}([H, \rho])_W$$

$$= \frac{2}{\hbar} H_c \sin(\hbar \hat{\Lambda}/2) W \tag{6.120}$$

$$= -i \mathscr{L}_W W \tag{6.121}$$

where H_c is the classical Hamiltonian. Applying the Poisson operator yields

$$\frac{\partial W}{\partial t} = -\frac{p}{m} \frac{\partial W}{\partial x} - \frac{2}{\hbar} \sin\left(\frac{\hbar}{2} \frac{\partial}{\partial x} \frac{\partial}{\partial p}\right) V(x) W(x, p) \tag{6.122}$$

$$= -\frac{p}{m} \frac{\partial W}{\partial x} - \frac{2}{\hbar} \sum_{n=0}^{\infty} (-1)^n \left(\frac{\hbar}{2}\right)^{2n+1} \frac{1}{(2n+1)!} \frac{\partial^{2n+1} V(x)}{\partial x^{2n+1}} \frac{\partial^{2n+1} W(x, p)}{\partial p^{2n+1}}$$

where in the first equation it is assumed that the $\partial/\partial x$ operates only on the $V(x)$ term and the $\partial/\partial p$ acts only on the $W(x, p)$ term. Finally, there is an equivalent form given by Groenewold that reads[58]

$$\frac{\partial W}{\partial t} = -\frac{p}{m} \frac{\partial W}{\partial x} + \frac{1}{i\hbar} \left(V\left(x + \frac{i\hbar}{2} \frac{\partial}{\partial p}\right) - V\left(x - \frac{i\hbar}{2} \frac{\partial}{\partial p}\right)\right) W(x, p) \tag{6.123}$$

This last term is especially useful when $V(x)$ can be expressed as a polynomial in x. Operationally, we perform the Taylor series expansion, then replace the x operator with $x \pm (i\hbar/2) d/dp$. For example, for the harmonic potential, the potential term yields

$m\omega^2 x \, dW/dp$, which is precisely what we get for the classical Liouville equation. Only at order of $V(x) \propto a_3 x^3/3$ do we begin to see quantum terms appearing in the equations of motion. For example, for the cubic potential,

$$\frac{\partial W}{\partial t} = -\frac{p}{m}\frac{\partial W}{\partial x} + a_3 x^2 \frac{\partial W}{\partial p} - \frac{\hbar^2}{12} a_3 \frac{\partial^3 W}{\partial p^3}$$

Let us examine the various terms in Equations 6.122 and 6.123. The first term on the right-hand side of the two equations comes from the kinetic energy operator and, depending upon the context, is termed the "drift," "streaming," or "advection" term. This term is also present in the classical Liouville equation (Equation 6.107). In fact, simply expanding the potential term in powers of $\hbar$ and then setting $\hbar \to 0$ produces the classical Liouville equation. Thus, all quantum effects enter in through the Wignerized potential, $V_W(x, p-p')$, which is nonlocal in momentum and basically redistributes the Wigner function along all possible momenta p for a given position x. The rough picture here is that particles that have been scattered by the potential at point $x \pm y/2$ interfere with particles scattering at different points. In this way, we sample over all possible pathways the particle can take.[59]

6.7.1 Wigner Representation on a Lattice: Exciton Diffusion

So far we have discussed the Wigner function for a continuous system. One can also derive a Wigner representation for a discretized system of a particle (or quasi particle such as an excitation) hopping from one site to the next. Consider, for example, the simple case of an exciton hopping between different sites as in the Frenkel exciton model:[60–62]

$$H = \hbar \sum_n \Omega_n B_n^\dagger B_n + \sum_{n \neq m} J_{nm} \left(B_n^\dagger B_m + B_m^\dagger B_n \right) \tag{6.124}$$

where the B_n and $B_n^\dagger$ destroy and create an exciton on site n. We can write this in a basis as

$$H = \hbar \sum_n \Omega_n |n\rangle\langle n| + \sum_{n \neq m} J(n-m)|n\rangle\langle m| \tag{6.125}$$

where we have written the interaction operator J_{nm} as depending upon the distance between sites n and m. In this representation, the density matrix is given by $\rho_{nm} = \langle n|\rho|m\rangle$. At this point we change variables to relative and center of mass variables by writing

$$n = r - s/2$$

and

$$m = r + s/2$$

with $\rho(r, s) = \langle r - s/2|\rho|r + s/2\rangle$. $\rho(r, 0$ is a diagonal element of the density matrix and gives the probability for the exciton's being located at lattice position r at a given time. Consequently, the $\rho(r, s \neq 0)$ carries the phase coherence information between two sites separated by distance s on the lattice. In this representation the time evolution of the density matrix is given by

$$i\dot\rho(r, s) = \sum_a J(a)(\rho(r + a/2, s - a) - \rho(r + a/2, s + a))$$
$$+ (\Omega_{r-s/2} - \Omega_{r+s/2})\rho(r, s) \tag{6.126}$$

where the a summation index runs over all displacements in the lattice. The second term will vanish if all the sites have the same energy. We can now apply the Wigner transformation to both sides of the equation.

$$\rho(r, s) = \frac{1}{\sqrt{N}} \sum_p W(r, p)e^{ips} \tag{6.127}$$

As above, the resulting Liouville equation has both kinetic and potential energy contributions

$$\dot{W} = \mathscr{T}W + \mathscr{V}W \tag{6.128}$$

For the potential term, we derive this by first taking the discrete sine transform of the energy differences

$$V(r, p) = \frac{2}{\sqrt{N}} \sum_s \sin(ps)(E_{r-s/2} - E_{r+s/2}) \tag{6.129}$$

and then taking the convolution with the Wigner transformed density matrix,

$$\mathscr{V}W = \frac{1}{\hbar} \sum_{p'} V(r, p - p')W(r, p') \tag{6.130}$$

Taking the continuum limit, we can write this in the Groenewold form as[59]

$$\frac{1}{\hbar} \sum_{p'} V(r, p - p')W(r, p') = \frac{i}{\hbar} \left(E\left(r + \frac{i\hbar}{2}\frac{\partial}{\partial p}\right) - E\left(r - \frac{i\hbar}{2}\frac{\partial}{\partial p}\right) \right) W(r, p)$$
$$\tag{6.131}$$

Here, again, we see that dynamics can be interpreted as that of a particle that scatters onto some site where it receives a random momentum kick according to the momentum distribution at that site. Quantum effects (that is, constructive and destructive interference) occur when we sum over all scattering events.

The kinetic term arises from the hopping terms in our original Hamiltonian and can be directly evaluated by inserting Equation 6.127 into Equation 6.126

$$\mathscr{T}W = 2 \sum_a J(a)\sin(pa)W(r + a/2, p) \tag{6.132}$$

Pulling everything together yields

$$\dot{W}(r, p) = 2 \sum_a J(a)\sin(pa)W(r + a/2, p)$$
$$+ \frac{1}{\hbar} \sum_{p'} V(r, p - p')W(r, p') \tag{6.133}$$

Taking the continuum limit for the first term requires us to write $J(a)$ as a symmetric function $J(a) = J(-a)$. If it is sufficiently short ranged, then the first term becomes simply

$$2 \sum_a J(a) \sin(pa) W(r + a/2, p) = \frac{p}{m^*} \frac{\partial}{\partial r} W(r, p)$$

where m^* is the effective mass of the exciton given by $m^* = \hbar/(2Jl^2)$ where l is the lattice spacing.[61]

6.7.2 Enforcing Fermi–Dirac Statistics

So far we have dealt with single-particle systems. However, we can use the Wigner representation to develop many-body theories. As an example, let us consider a lattice system with more than one exciton present but restrict the system in such a way that any given site can have at most a single exciton present on it at a given time. This is essentially the Pauli exclusion rule that must be applied to systems with half-integer spin (such as electrons or protons). However, since excitons are generally prepared optically, they are typically singlet or triplet in their spin multiplicity and hence would be more aptly described as bosons. The justification for enforcing the exclusion rule in the case of excitons is that typically the energy required to doubly excite a given site is generally more than the energy carried by two excitons on separate sites. Rather than putting in the double exciton states by hand and carrying the extra baggage, we simply put a constraint on the system excluding two excitons from being on the same site.

Fermion systems obey the antisymmetrization rule, and as a result, Fermion operators obey an anticommutation rule $\{B_m^\dagger, B_n\} = \delta_{nm}$ rather than the usual $[B_m^\dagger, B_n = \delta_{nm}]$, which is fine for Bose particles. We can write the commutation relation in a general form using

$$\left[B_m^\dagger, B_n \right] = \delta_{nm}(1 - \zeta W_n)$$

where $W_n = 2B_n^\dagger B_n$ and $\zeta = 1$ for fermions and $\zeta = 0$ for bosons. This, along with $B_n^2 = (B_n^\dagger)^2 = 0$, yields the following Heisenberg equations of motion for the exciton operators:[60,62]

$$i \dot{B}_n = \Omega_n B_n + \sum_{m \neq n} J_{mn}(1 - W_n) B_m$$

$$i \dot{B}_n^\dagger = -\Omega_n B_n^\dagger - \sum_{m \neq n} J_{mn}(1 - W_n) B_m^\dagger \qquad (6.134)$$

Furthermore, the Heisenberg equations for the binary product $B_m^\dagger B_n$ are given by

$$i \frac{d}{dt} B_n^\dagger B_m = (\Omega_n - \Omega_m) B_n^\dagger B_m + \sum_k J_{mk} B_n^\dagger B_k - J_{nk} B_k^\dagger B_m + \Gamma_{nm}^{collision} \qquad (6.135)$$

where the collision term is

$$\Gamma_{nm}^{collision} = \sum_{k \neq m, n} J_{mk} W_m B_n^\dagger B_k - J_{nk} W_n B_k^\dagger B_m \qquad (6.136)$$

This term represents all possible bimolecular collisions between excitons. Ignoring this term, which is allowable in the limit of few excitations in the system, brings us back to the expressions above.*

Notice that this expression involves the product of B_m and $B_m^\dagger$ with the W_n number operator. Consequently (and alas), we are again faced with a hierarchy of equations since we need to deduce the equations of motion for operator products and their expectation values. Various approximations are possible, the simplest being to factor all many-operator terms into single-operator terms, viz,

$$\langle W_k B_n^\dagger B_m \rangle = \langle B_k^\dagger \rangle \langle B_k \rangle \langle B_n^\dagger \rangle \langle B_m \rangle$$

This is essentially a local-field approximation whereby an exciton at site n moves in the mean-field of excitons at all other sites. We can, however, also approximate the $\langle W_k B_n^\dagger B_m \rangle$ product with the factorized product $\langle W_k \rangle \langle B_n^\dagger B_m \rangle$ where $\langle W_k \rangle$ is the expected occupation number at site k.

The connection to the analysis above is that the density matrix elements are given by $\rho_{nm} = \langle B_n^\dagger B_m \rangle$. Consequently, we can make the same substitutions as previously and use the convolution theorem to write the collisional term as

$$\Gamma_{collision}(r, p) = \frac{1}{N} \sum_{r', p'} J(r, p') \langle W(r) \rangle W(r', p - p')$$
$$- J(r', p') \langle W(r') \rangle W(r, p - p') \qquad (6.137)$$

where the first term represents scattering to site r from all other sites and the second represents scattering from site r to any other site each with a momentum change of $p - p'$.

We can also make a couple of simplifications to the collisional term. First, we can invoke Boltzmann's *Stosszahl Ansatz* and assume that prior to collision at time t, the two excitons were uncorrelated. If we take the hopping term in as a momentum transfer, then the collision operator can be written as

$$\Gamma_{boltz}(r, p) = \iint J(p - p', q)(W(r, p + q)W(r, p' - q)$$
$$- W(r, p)W(r, p'))dp'dq \qquad (6.138)$$

Alternatively, one can use the Bhatnagar–Gross–Krook (BGK) approximation[63] used in the lattice Boltzmann method for simulating fluid flow and plasmas

$$\Gamma_{BGK}(r, p) = -\gamma(W(r, p) - W_{eq}(r, p)) \qquad (6.139)$$

where $W_{eq}(r, p)$ is the equilibrium (stationary) distribution. Here, γ is the collision frequency and it is assumed that the resulting momentum distribution after each

---

* It is also possible to introduce the Pauli exclusion principle as a dynamical constraint on the system using the Dirac bracket method discussed earlier. Here, one appends to the Hamiltonian series of constraints on the canonical variables such that the dynamics occurs on a surface defined by $\phi_n = B_n^\dagger B_n + B_n B_n^\dagger - 1 = 0$. Thus, the constraint generates the collision term in Eq. (6.136).

collision is equal to the equilibrium momentum distribution. If we assume that all momenta are equally likely and the site energies are all the same, then $\langle W_{eq}(r, p)\rangle = 1/\sqrt{N}$.

As we have seen, the primary source of quantum mechanical effects in the Wigner representation is any term in the potential listed as x^3 or greater in the displacement coordinate. For the homogeneous lattice or even the harmonic oscillator, the dynamics in the Wigner representation is identical to that of a classical ensemble of particles. This has a number of advantages because it allows us to treat excitons in a large system as a classical gas with essentially hard-sphere interactions that limit the number of excitons on a given site to just one per site.

Just as Boltzmann's collision-number approximation is valid when the collision frequency is long compared with the thermal relaxation time, the approximate forms for the exciton/exciton interaction given here are only valid if we have few excitons in the system. We have also ignored the possibility that exciton-exciton collisions may not necessarily conserve total exciton numbers. In fact, if we use a more generalized Frenkel exciton model that includes $B_n^\dagger B_m^\dagger$ and $B_n B_m$, then we open the possibility for exciton-exciton annihilation or production of additional excitons on the lattice. Because the dynamics is still described by a Hermitian Hamiltonian, total energy will be conserved even if the total particle number is no longer a constant of the motion. Exciton-exciton annihilation is an especially important process for triplet excitons and can lead to a loss of quantum efficiency in photovoltaic devices.

6.7.3   The k, Δ Representation

Although the Wigner distribution has proven to be quite useful in developing semiclassical representations of quantum mechanics, other representations are possible. One such is the so-called k, Δ representation given by

$$\rho(k, \Delta) = Tr[\rho e^{i(k\hat{x}+\Delta\hat{p})}] = Tr[D\rho] = \langle D\rho \rangle \tag{6.140}$$

where we introduce the last term to simplify our notation and to distinguish $\rho(k, \Delta)$ from $\rho(x, x')$. Since the density matrix is Hermitian, $\rho(k, \Delta)$ has the following symmetry: $\rho^*(k, \Delta) = \rho(-k, -\Delta)$.

The transformation is similar to the Wigner transform in that it involves the Fourier transform of the density matrix

$$\rho(k, \Delta) = \int dx e^{ikx} \rho(x + \Delta/2, x - \Delta/2) \tag{6.141}$$

and is related to the Wigner function via

$$W(x, p) = \left(\frac{1}{2\pi}\right)^{1/2} \int dk \int d\Delta e^{-i(kx+\Delta p)} \rho(k, \Delta) \tag{6.142}$$

or taking the inverse

$$\rho(k, \Delta) = \int \int e^{i(kx+p\Delta)} W(x, p) dx dp = \langle e^{i(kx+p\Delta)} \rangle_W \tag{6.143}$$

A characteristic function is the expected value of $e^{it \cdot x}$ for a given distribution, assuming that t is real. As such, $\rho(k, \Delta)$ is the *characteristic function* for the Wigner distribution. Characteristic functions are useful in deriving the moments of a given distribution. For example, the characteristic function for the normal (or Gaussian) distribution is

$$c(\mu, \sigma, t) = e^{it\mu} e^{-t^2 \sigma^2 / 2} \tag{6.144}$$

Taking the derivative of c with respect to t, then setting $t = 0$, generates the moments of the distribution:

$$(-i)^n \frac{d^n c}{dt^n} \bigg|_{t=0} = \langle x^n \rangle. \tag{6.145}$$

For example, the first two moments of the normal distribution read

$$m = \{\mu, \mu^2 + \sigma^2\} \tag{6.146}$$

For a Gaussian distribution, all subsequent moments can be related to these first two moments.

Likewise, for the Wigner function we can obtain

$$\langle x^n p^m \rangle = \lim_{k, \Delta \to 0} (-i)^{n+m} \frac{\partial^{n+m}}{\partial k^n \partial \Delta^m} \rho(k, \Delta) \tag{6.147}$$

which are the expectation values for the operator $x^n p^m$. Moreover, if we know the time derivative of $\rho(k, \Delta)$, we can derive the Heisenberg equations of motion for operators composed of x and p:

$$\frac{\partial \langle x^n p^m \rangle}{\partial t} = \lim_{k, \Delta \to 0} (-i)^{n+m} \frac{\partial^{n+m}}{\partial k^n \partial \Delta^m} \frac{\partial \rho(k, \Delta)}{\partial t} \tag{6.148}$$

We can also use the characteristic functions to derive the cumulants of a distribution as well. These are related to the moments but only include the "connected" parts. In other words, they cannot be reduced into a sum of other moments. For example, the cumulants of the Gaussian distribution are simply the μ and σ specifying the center and width of the Gaussian function. Cumulants are given by the log-derivative of the characteristic function:

$$c_n = (-i)^n \frac{1}{G(t)} \frac{\partial G}{\partial t} \bigg|_{t=0} \tag{6.149}$$

Thus, the cumulants of the Wigner function are the log-derivatives of the $\langle D\rho \rangle$ characteristic function

$$c_{n,m} = (-i)^{n+m} \frac{1}{\langle D\rho \rangle} \frac{\partial^{n+m} \langle D\rho \rangle}{\partial k^m \partial \Delta^n} \bigg|_{k, \Delta=0} \tag{6.150}$$

The advantage of the k, Δ representation is that one can make considerable use of commutation relations to simplify the equations of motion. For example, the D operator can be written as

$$D = e^{i(kx + \Delta p)} = e^{ik\Delta/2} e^{ikx} e^{ip\Delta} \tag{6.151}$$

Taking derivatives of D with respect to the k and Δ variables,

$$\frac{\partial D}{\partial k} = i\left(\frac{\Delta}{2} + \hat{x}\right)D \tag{6.152}$$

and

$$\frac{\partial D}{\partial \Delta} = i\left(\frac{k}{2} + \hat{p}\right)D \tag{6.153}$$

which can be rearranged to

$$\hat{x}D = -\left(\frac{D}{2} + i\frac{\partial}{\partial k}\right)D \tag{6.154}$$

$$\hat{p}D = -\left(\frac{k}{2} + i\frac{\partial}{\partial \Delta}\right)D \tag{6.155}$$

Finally, we have the commutation relations

$$[\hat{p}, D] = kD \tag{6.156}$$

and

$$[\hat{x}, D] = -\Delta D \tag{6.157}$$

We can work out the time-evolution equation for ρ in the k, Δ representation by taking the time derivative of $\rho(k, \Delta) = tr[D\rho]$ and applying Equations 6.154 to 6.157. For a particle with mass m in a harmonic well with frequency ω the time evolution for the density matrix is

$$i\dot{\rho} = \frac{1}{2m}[p^2, \rho] + \frac{m\omega^2}{2}[x^2, \rho] \tag{6.158}$$

Multiplying through by D and taking the trace, we arrive at

$$\frac{d}{dt}\langle D\rho\rangle = \left(\frac{k}{m}\frac{\partial}{\partial \Delta} - m\omega^2\Delta\frac{\partial}{\partial k}\right)\langle D\rho\rangle. \tag{6.159}$$

It is interesting to point out that we have replaced a complex second-order elliptical partial differential equation in x, x' with a real first-order partial differential equation in k, Δ. This presents us with an equation that (in principle at least) should be easier to solve than the original Liouville–von Neumann equation. The physics carried by $\rho(k, \Delta)$ has not changed; we have simply reduced the effort required to derive (or solve for) its time evolution.

Finally, consider the equations of motion for the moments $\langle x\rangle$ and $\langle p\rangle$ corresponding to the expected values of the position and momentum. Using the characteristic equation and the equation of motion for $\langle D\rho\rangle$ above, we find

$$\frac{\partial\langle x\rangle}{\partial t} = \frac{1}{m}\frac{\partial}{\partial \Delta}\langle D\rho\rangle\bigg|_{k,\Delta=0} = \frac{\langle p\rangle}{m} \tag{6.160}$$

$$\frac{\partial\langle p\rangle}{\partial t} = -m\omega^2\frac{\partial}{\partial k}\langle D\rho\rangle\bigg|_{k,\Delta=0} = -m\omega^2\langle x\rangle \tag{6.161}$$

which are what we expect to find for the Ehrenfest equations of motion for a harmonic oscillator. In Problem 6.3 we derive more general equations of motion for a particle in a potential.

6.8 PROBLEMS AND EXERCISES

Problem 6.1 Starting from the equations of motion for the density matrix in Equation 6.54, verify, using Laplace transform techniques, that Equation 6.69 is correct.

Problem 6.2 Show that the Wigner distribution for the harmonic oscillator is given by

$$W(x, p) = \frac{-1^n}{\pi} e^{-2H(p,x)/\omega} L_n(4H/\omega)$$

where L_n is a Laguerre polynomial and H is the Hamiltonian for a classical oscillator.

Problem 6.3 Show that the relations given in Equations 6.154 to 6.157 are correct. Using these, derive the equations of motion for a particle with mass m in a general polynomial potential

$$V = \sum_{n=0}^{\infty} \frac{x^n}{n!} \left(\frac{d^n V}{dx^n} \right)_{x=0} \tag{6.162}$$

Hint: Evaluating the commutators is straightforward; however, you need to be careful that the p and x in the exponent are still operators and do not commute:

$$[\hat{p}, D] = e^{ik\Delta}[p, e^{ikx} e^{ip\Delta}] \tag{6.163}$$

$$= e^{ik\Delta}[p, e^{ikx}] e^{ip\Delta} \tag{6.164}$$

$$[p, e^{ikx}]f = i \left[\frac{d}{dx}, e^{ikx} \right] f = i(f' e^{ikx} - \frac{d}{dx} e^{ikx} f) \tag{6.165}$$

$$= -i(ik) e^{ikx} f \tag{6.166}$$

thus, $[\hat{p}, D] = kD$.

The second is a bit trickier since it involves putting p in the exponent when evaluating $[x, e^{ip\Delta}]$. For this use, expand the exponent

$$[\hat{x}, e^{ip\Delta}] = \sum_{n=0}^{\infty} \frac{(i\Delta)^n}{n!} [\hat{x}, \hat{p}^n] \tag{6.167}$$

then use the relation $[A, B^n] = nB^{n-1}[A, B]$,

$$[\hat{x}, e^{ip\Delta}] = \sum_{n=1}^{\infty} \frac{(i\Delta)^n}{n!} np^{n-1} [\hat{x}, \hat{p}] \tag{6.168}$$

and $[x, p] = i$ (keeping $\hbar = 1$, pulling out an $i\Delta$ and changing range of summation index since the $n = 0$ case vanishes),

$$[\hat{x}, e^{ip\Delta}] = -\Delta \sum_{n=1}^{\infty} \frac{(i\Delta)^{n-1}}{(n-1)!} p^{n-1} \tag{6.169}$$

again, changing summation index and doing the sum

$$[\hat{x}, e^{ip\Delta}] = -\Delta e^{ip\Delta} \tag{6.170}$$

Thus, we arrive at the equations above.

Problem 6.4   When a dissipative bath is included under a certain set of assumptions, the equations of motion for the density matrix can be written as

$$i\dot{\rho} = \frac{1}{2m}[p^2, \rho] + \frac{m\omega^2}{2}[x^2, \rho] - \Lambda[x, [x, \rho]] + \gamma[x, \{p, \rho\}] \tag{6.171}$$

where Λ and γ are constants and $\{A, B\}$ denotes the anticommutation relation: $\{A, B\} = AB + BA$.

1. Derive the equivalent equation of motion for $\rho(k, \Delta)$.
2. Taking a Gaussian form for $\rho(k, \Delta)$

$$\rho(k, \Delta) = \exp(-(c_1 k^2 + c_2 \Delta^2 + c_3 k\Delta + ic_4 k + ic_5 \Delta + c_6)) \tag{6.172}$$

where $c_1 \cdots c_6$ are time-dependent coefficients, derive the appropriate equations of motion for $\dot{c}_1 \cdots \dot{c}_6$ for a particle in a dissipative environment.

3. Using *Mathematica* or other means, find numerical solutions for the coefficients given some appropriate initial conditions, and plot these vs. time. Comment upon how the system relaxes as you vary γ and Λ.
4. Using the numerical solutions, make contour plots of $\rho(k, \Delta)$ at various times as it relaxes.
5. Using the Gaussian form

$$\rho(k, \Delta) = \exp(-(c_1 k^2 + c_2 \Delta^2 + c_3 k\Delta + ic_4 k + ic_5 \Delta + c_6)) \tag{6.173}$$

derive the corresponding Wigner function and make either contour or three-dimensional plots of $W(x, p)$ at the same time steps as in your plot of $\rho(k, \Delta)$. How does this compare with how you would expect a classical system to behave under dissipative conditions?

Partial Solutions:

2. You should arrive at the following equations of motion:

$$\dot{c}_1 = c_2/m \tag{6.174}$$

$$\dot{c}_2 = 2c_3/m - 2m\omega^2 c_1 - \gamma c_2 \tag{6.175}$$

$$\dot{c}_3 = \Lambda - m\omega^2 c_2 - 2\gamma c_3 \tag{6.176}$$

$$\dot{c}_4 = c_5/m \tag{6.177}$$

$$\dot{c}_5 = -m\omega^2 c_4 - \gamma c_5 \tag{6.178}$$

$$\dot{c}_6 = 0 \tag{6.179}$$

5. The Wigner function corresponding to the Gaussian form of $\rho(k, \Delta)$ is

$$W(x, p) = \frac{1}{\sqrt{2\pi}} \sqrt{\frac{1}{4c_1 c_3 - c_2^2}} \, e^{\frac{c_3(c_4 - x)^2 + (c_5 - p)(c_1(c_5 - p) + c_2(x - c_4))}{c_2^2 - 4c_1 c_3}} \tag{6.180}$$

Representative plots are given in the *Mathematica* notebooks.

Problem 6.5 Consider the time evolution of the Wigner function for a free particle,

$$\frac{\partial W}{\partial t} + \frac{p}{m} \nabla_r W = 0$$

Using the k, Δ representation, derive and solve the equations of motion for $\rho(k, \Delta)$ assuming that at time $t = 0$ the initial Wigner function is given by $W(x, p; 0)$. Using $\rho(k, \Delta; t)$, derive expressions for $\langle(x - \langle x\rangle)^2\rangle(t)$ and $\langle(p - \langle p\rangle)^2\rangle(t)$. Are these the same as you would expect for the time evolution of a free particle using the Schrödinger equation?

Problem 6.6 Let ρ be the density operator for an arbitrary system where $|\chi_l\rangle$ and π_l are its eigenvectors and eigenvalues. Write ρ and ρ^2 in terms of the $|\chi_l\rangle$ and π_l. What do the matrices representing these operators look like in the $\{|\chi_l\rangle\}$ basis—first in the case where ρ describes a pure state and second where ρ describes a mixed state. Begin by showing that in the pure case, ρ has a single nonzero diagonal element equal to 1, while for a statistical mixture, it has a several diagonal elements between 0 and 1. Show that ρ corresponds to a pure case if and only if $tr[\rho^2] = 1$.

Problem 6.7 Consider a system with density matrix ρ evolving under Hamiltonian $H(t)$. Show that $tr[\rho^2(t)]$ does not change in time. Can the system evolve to be successively a pure state and a statistical mixture of states?

Problem 6.8 Let (1) and (2) be a global ensemble consisting of two subspaces (1) and (2). A and B denote operators acting in the state space $\mathscr{E}(1) \otimes \mathscr{E}(2)$. Show that the partial traces $tr_1(AB)$ and $tr_1(BA)$ are equal only if A or B acts only in space $\mathscr{E}(1)$. That is, A or B can be written as $A = A(1) \otimes I(2)$ or $B = B(1) \otimes I(2)$.

Note: $tr_1[]$ means that you take the trace ONLY over space (2). For example, take the case where we have states $|a, i\rangle$ spanning $\mathscr{E}(1) \otimes \mathscr{E}(2)$. Then

$$tr_1[A] = \sum_i A_{ai,a'i}$$

where $A_{ai,a'i} = \langle ai|A|a'i\rangle$ are the matrix elements of operator A.

7 Excitation Energy Transfer

A photoexcited molecule is rarely a stable species. The fact that we have just pumped in excess of 1 to 3 eV of energy into a small molecule through the interaction with a visible or UV photon means that this energy is likely to be rapidly dissipated to other degrees of freedom, to phonons in the form of heat, to other electronic states of the system via intersystem crossing or nonradiative decay, or through emission of photons. Typically we think of photoemission as leading to some observable spectroscopic signal. However, if in fact there are neighboring molecules that can absorb the emitted photon, the excitation that started off localized on one molecule may be transferred reversibly or irreversibly to the next. Typically, this is an irreversible process since the time scale for emission is roughly a thousandfold slower than the time scale for intramolecular vibrational relaxation and reorganization of the surrounding media. Thus, at each energy transfer event, some energy is lost to heat.

Figure 7.1 shows the various energy transfer and relaxation events that can occur following photo-excitation.

In this chapter, we explore the basis for excitation energy transfer between molecules. We begin with a discussion of irreversibility in a quantum mechanical system. We shall leave the molecular-level details of what causes this irreversibility for later, focusing our attention upon a phenomenological treatment in which we introduce in a rather ad hoc way the requisite decay times. Following this, we will consider how to compute the exciton coupling matrix element between molecular species using modern quantum chemical approaches.

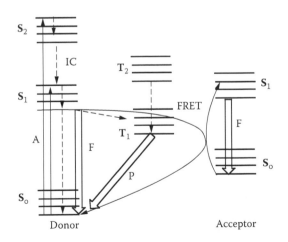

FIGURE 7.1 Possible photochemical pathways following excitation. The dashed lines indicate non-radiative processes whereas the solid lines indicate radiative. A = excitation of the donor molecule from its singlet ground state to one of its singlet excited states: S_1 and S_2 (IC = internal conversion, F = fluorescence, P = phosphorescence). The FRET process corresponds to the transfer of the excitation from one molecule to the next.

7.1   DIPOLE–DIPOLE INTERACTIONS

Molecules interact with each other at a distance via Coulomb forces determined by the shape and polarizability of the electronic density surrounding each of them. In general, we work in the limit that a given pair of molecules are far enough apart that electron exchange and correlation contributions can be safely ignored. Thus, the interaction can be written as

$$V_{ab} = \frac{1}{2} \int d^3 r_a \int d^3 r_b \frac{e^2 \rho_a(r_a) \rho_b(r_b)}{|r_a - r_b|} \tag{7.1}$$

where ρ_a and ρ_b are the transition densities of molecules A and B, respectively, between the initial and final electronic states. In loose terms, the transition density can be thought of as the induced charge oscillations in the ground-state electronic density in response to a linear oscillating driving force (that is, the electromagnetic field) at the transition frequency. If the distance, R, between A and B is large compared to the size of either molecule, a, we can safely expand the integrand in terms of its multipole moments and write the interaction in terms of the transition dipole moments of each molecule

$$M = \frac{1}{R^3} \left(\vec{p}_A \cdot \vec{p}_B - \frac{3}{R^2} (\vec{p}_A \cdot \vec{R})(\vec{p}_B \cdot \vec{R}) \right) \tag{7.2}$$

where $\vec{R}$ is a vector extending from the charge center of A to the charge center of B. Setting this to be the z axis, we can write M as a function of the angles (see Figure 7.2)

$$\chi(\theta_a, \theta_b, \phi) = \sin \theta_a \sin \theta_b \cos \phi - 2 \cos \theta_a \cos \theta_b \tag{7.3}$$

If all angles are statistically possible, we obtain the mean value

$$\overline{\chi}^2 = 2/3 \tag{7.4}$$

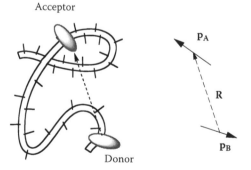

FIGURE 7.2 Schematic of dipole–dipole interaction between a donor-acceptor pair. Here p_A and p_B indicate the relative orientation of the transition dipoles associated with the donor or acceptor species. The molecular framework represents either a solvent matrix or the actual backbone of a polymer or biomolecule in which the chromophore pairs are intercalated.

Before moving forward, let us make a back-of-the-envelope estimate of the transfer rate between two molecules treated as a two-level system. Taking the molecules to be randomly oriented, the matrix element squared that would go into a golden rule expression becomes

$$M_{ab}^2 = \frac{2}{3} \frac{|p_A|^2 |p_B|^2}{R^6} \tag{7.5}$$

The transition dipoles can be replaced by their oscillator strengths, viz,

$$p_A^2 = \frac{\hbar e^2}{2m\omega} f_A \tag{7.6}$$

Assuming only radiative transitions are allowed, the lifetime of A of B is given by

$$\frac{1}{\tau_A} = \frac{f_A}{\tau_{cl}} \tag{7.7}$$

where, as we discussed previously, τ_{cl} is the decay time for a classical electronic oscillator $\tau_{cl} = 3(mc^3/e^2\omega^2)/2$.

Now, consider the ratio of the transfer rate W to the radiative rate

$$W\tau_A = \frac{2}{\hbar^2} |M_{ab}|^2 T_2 \tau_A \tag{7.8}$$

Inserting our expression for M_{ab} from above,

$$W\tau_A = \frac{3}{8\pi} f_B \left(\frac{\lambda}{R}\right)^6 \frac{T_2}{\tau_{cl}} \tag{7.9}$$

It is an important practice to learn to insert numbers into equations such as this in order to determine their range of validity for molecular-scale systems. For example, if the radiative linewidth is on the order of 100 cm^{-1}, then $T_2 \approx 50$ fs. Taking $\tau_{cl} \approx 10$ ns, then for $f_b = 1$ the characteristic distance for which $W\tau_A = 1$ is $R_o \approx 0.02\lambda$. For typical molecules with electronic transitions in the UV/visible region, $\lambda \approx 300$ nm, so $R_o \approx 6$ nm or 60 Å. This is consistent with experimental values of around 50 Å for most molecular systems. For small aromatic rings, $a \approx 10$ Å, so about $W\tau_A = 1$ for molecules separated by about 10 molecular radii. Notice that this estimate is independent of the oscillator strength of A.

At what distance can the interaction be considered strong? Consider the distance for which

$$\frac{2}{\hbar^2} |V_{ab}|^2 T_2 \tau_b \approx 1 \tag{7.10}$$

From what we have just seen above, it is the distance for which

$$R \approx R_o (\tau_B/\tau_A)^{1/6} \tag{7.11}$$

If the two molecules are identical or even similar, $\tau_B \approx \tau_A$, and only when $R \approx R_o$, does the interaction become sufficiently strong.

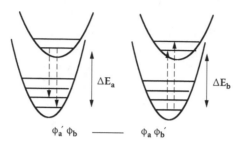

FIGURE 7.3 Resonant energy transfer between donor (a) and acceptor (b) energy levels.

From the discussion above, it should be clear that if we can measure the radiitive transfer rate between two distinct species, we have the means of measuring the instantaneous distance of separation between the two, provided the transfer rate is fast compared to the time scale for the relative motion between A and B. Because of this and the advent of single-molecule spectroscopic techniques that can selectively excite and collect photons from what amounts to single chromophores, we can effectively monitor experimentally the dynamics of complex molecular reactions through a technique termed "Förster resonant energy transfer" or FRET. A schematic of this process is shown in Figure 7.3

7.2 FÖRSTER'S THEORY

The model presented above is highly instructive and enables us to develop a theoretical "feel" for energy transfer between donor and acceptor chromophores. It is far from complete but can be easily extended to include more complex models, for example, which take into account the intramolecular vibrational motions.

An alternative approach, first developed by German physicist Theodor Förster [64–67] in 1948, proves to be highly useful in the case where the emission and absorption spectra of the donor and acceptor species are broad and diffuse. In this limit, we cannot make the assumption that the transverse relaxation time is short or that we are dealing with a two-level system.

Förster suggested that one should treat the spectra as being continous and applied the golden rule/first-order perturbative result for the case of a continuos spectra

$$dW = \frac{2\pi}{\hbar}|V_{ab}|^2 \delta(\Delta E_a - \Delta E_b) \tag{7.12}$$

where $\Delta E_a - \Delta E_b = (E'_a - E_a) - (E'_b - E_b)$ is the difference in energys as shown in Figure 7.3. This ensures that energy is conserved upon performing the final integration over energies.

Following Förster's approach, we write the initial and final wave functions as

$$\Psi_1 = \phi'_a \phi_b \, \Phi'_a(E'_a) \Phi_b(E_b) \tag{7.13}$$

$$\Psi_2 = \phi'_b \phi_a \, \Phi_a(E_a) \Phi'_b(E'_b) \tag{7.14}$$

where the ϕ's are the ground and excited electronic states of the system and the Φ's are vibrational wave functions. We assume here that the Born–Oppenheime approximation is valid so that the Φ's represent vibrational motion on a given potential energy surface associated with either the ground or excited state of either molecule. We denote the energy origin of each surface with E_a, E'_a, and so forth. Taking these as our states, we can write the coupling matrix element as

$$V_{12} = \langle \Psi_1 | V | \Psi_2 \rangle \tag{7.15}$$

$$= V_{ab} \langle \Phi'_a(E'_a) | \Phi_a(E_a) \rangle \langle \Phi_b(E_b) | \Phi'_b(E'_b) \rangle \tag{7.16}$$

$$= V_{ab} S_a(E'_a, E_a) S_b(E_b, E'_b) \tag{7.17}$$

where S_a and S_b are the matrices of overlap integrals between vibrational wave functions and

$$V_{ab} = \chi |p_a| |p_b| / R^3 \tag{7.18}$$

is the dipole–dipole coupling matrix element.

Notice that in writing this expression we have made the assumption that the electronic transition is effectively decoupled from the vibrational dynamics and takes place in a fixed frame of the nuclei. This is often termed the "Condon approximation" after Eugene Condon. Its justification is similar to that of the Born–Oppenheimer approximation.

Now, let $g'(E'_a)$ be the energy distribution for molecule A when it is in its excited electronic state and $g(E_b)$ be the energy distribution for B in its ground electronic state. If we assume that intramolecular vibrational relaxation is fast compared with the transfer rate, then the mechanism suggested by Figure 7.3 where, upon excitation, the donor (A) relaxes in its excited state to some lowest vibrational state and then transfers its remaining electronic energy to B. Assuming thermalization is fast compared with the transfer rate, then both of these represent thermal populations.

Pulling all this together yields

$$W = \frac{2\pi}{\hbar} \int V^2_{ab} \int g'(E'_a) S^2_a(E'_a, E'_a - E) dE'_a$$

$$\times \int g(E_b) S^2_b(E_b, E_b + E) dE_b dE \tag{7.19}$$

$$= \frac{2\pi \chi^2}{R^6} \int \left[p^2_a \int g'(E'_a) S_a(E'_a, E'_a - E) dE'_a \right]$$

$$\times \left[p^2_b \int g(E_b) S_b(E_b, E_b - E) dE_b \right] d\omega \tag{7.20}$$

The terms in square brackets are directly related to experimentally observable quantities, namely, the first is the normalized emission spectra of the donor (A) and the second is the normalized absorption spectra of the acceptor (B):

$$\left[p^2_a \int g'(E'_a) S_a(E'_a, E'_a - E) dE'_a \right] = \frac{1}{\hbar} p^2_a G_a(\omega) \tag{7.21}$$

and

$$\left[p_b^2 \int g(E_b) S_b(E_b, E_b - E) dE_b \right] = \frac{1}{\hbar} p_b^2 G_b(\omega) \tag{7.22}$$

where $\int G_{a,b}(\omega) d\omega = 1$.

There is a well-known relation between the lifetime τ, the absorption index $\mu(\omega)$, and the transition dipole moment p_a:

$$F_a(\omega) = \frac{4\omega^3}{3\hbar c^3} p_a^2 \tau_a G_a(\omega) \tag{7.23}$$

$$\mu_b(\omega) = \frac{4\pi^2 \omega}{3\hbar c} N_b p_b^2 G_b(\omega) \tag{7.24}$$

where $F_a(\omega)$ is the normalized radiation spectrum of A given as the number of quanta per unit frequency range. The second, $\mu_b(\omega)$, is the absorption coefficient as per the Beer–Lambert relation

$$I(z) = I_o e^{-\mu(\omega)z} \tag{7.25}$$

where N_b is the number of acceptor molecules per cm^3 and z is the thickness of the sample. Finally, we arrive at a well-known result that

$$W = \frac{9\chi^2 c^4}{8\pi N_b \tau_a R^6} \int F_a(\omega)\mu_b(\omega)\omega^{-4} d\omega \tag{7.26}$$

whereby the rate is obtained by taking the overlap integral between the emission spectrum of A and the absorption spectrum of B, multiplied by the appropriate scaling factors. The advantage of this formula is that both spectra can be determined independently by simple spectroscopic techniques.

The first general requirement for efficient energy transfer is a good degree of spectral overlap between the emission spectrum of the donor species and the absorption spectrum of the acceptor species. This is determined by the integral in Equation 7.26, which is often written as J:

$$J = \int F_a(\omega)\mu_b(\omega)\omega^{-4} d\omega$$

Herein, though, lies one of the experimental paradoxes of FRET. The spectral profiles of the FRET pair cannot be so separated that they have poor overlap, yet we want to avoid "cross-talk" between the two imaging channels—that is, ideally the donor emission filter set must collect only the light from the donor and none from the acceptor, and vice versa for the acceptor. In practice, this can be somewhat realized by employing short bandpass filters that collect light from only the shorter-wavelength side of the donor emission and the longer-wavelength side of the acceptor emission. This can limit somewhat the photon flux from both donor and acceptor during a typical exposure, especially when we bear in mind that these measurements are best performed under conditions of reduced excitation power, such that we do not accelerate the rates of bleaching.

Secondly, the rate scales as $1/R^6$ due to the dipole–dipole nature of the coupling matrix element. Consequently, we can define a distance in which the transfer rate W is equal to the radiative rate τ_a by

$$R_o^6 = \frac{9\chi^2 c^4}{8\pi N_b R^6} J \tag{7.27}$$

Thus,

$$W = \frac{1}{\tau_a}\left(\frac{R_o}{R}\right)^6$$

where R_o is the "Förster radius." At this distance, energy transfer is 50% efficient.

Often the FRET technique is combined with imaging microscopy techniques to monitor the proximity of two fluorophores. Since fluorophores can be employed to specifically label biomolecules and the distance condition for FRET is of the order of the diameter of most biomolecules, FRET is often used to determine when and where two or more biomolecules, often proteins, interact within their physiological surroundings. Since energy transfer occurs over distances of 1–10 nm, a FRET signal corresponding to a particular location within a microscope image provides an additional distance accuracy surpassing the optical resolution (≈ 0.25 mm) of the light microscope.

Furthermore, the transfer rate depends critically upon the relative orientation of the two transition dipoles. Above, we have expressed χ in terms of the relative dihedral angles between the two dipoles. Furthermore, assuming the donor and acceptor species are randomly oriented, $\chi^2 = 2/3$. However, if the two molecules are tethered to a common backbone, the instantaneous orientation factor χ^2 will reflect the instantaneous relative orientation of the two dipoles and as such may provide a sensitive probe of the dynamics of the backbone provided the time scale of the motion is long compared with the experimental time scale.

Finally, we can express all these factors in a general equation in spectroscopic units (per mol):

$$W = \frac{\chi^2 J}{n^4 \tau_o}\frac{1}{R^6}8.785 \times 10^{-23}$$

where χ^2 is the orientation factor, n the refractive index of the medium, τ_o the radiative lifetime of the donor, R the distance (in cm) between the donor and acceptor, and J the spectral overlap (in coherent units $cm^6 mol^{-1}$) between the donor fluorescence spectrum and acceptor absorbance spectrum. We can also write the Förster radius (in cm) as

$$R_o^6 = \frac{\chi^2 \Phi_D J}{n^4}8.785 \times 10^{-5}$$

where Φ_D is the quantum efficiency of the donor. The efficiency of the transfer may be evaluated by comparing the fluorescence lifetime of the donor in the presence of τ_a and in the absence of the acceptor τ_a^o or by the quantum yield in the presence Φ_D and in the absence Φ_D^o of the acceptor:

$$E = 1 - \frac{\tau_a}{\tau_a^o} = 1 - \frac{\Phi_D}{\Phi_D^o}$$

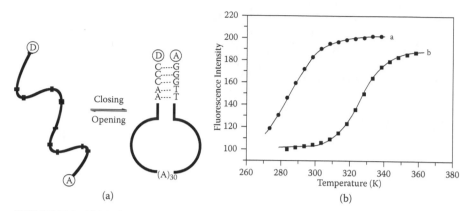

FIGURE 7.4 (a) Model open/closed loop structures for DNA hairpin. (b) Equilibrium thermal melting curves for the DNA hairpin loop. The closed-to-open transition was monitored by the ratio of fluorescence intensity of TMR in a double-labeled sample to that of a TMR-only-labeled sample. (a) 10 mM Tris/1 mM EDTA (pH 7.5) and 100 mM NaCl; (b) 10 mM Tris and 20 mM MgCl$_2$. Solid lines are fits to a two-state model (from Ref. 69).

Because the energy transfer rate is highly sensitive to the distance of separation between the donor and acceptor pairs, we can use resonant energy transfer to accurately measure inter- and intramolecular distances. Over 30 years ago, Hass et al. showed that FRET techniques could be used to monitor the end-to-end chain diffusion of a tagged biopolymer over a range of 2–10 nm.[68] In Figure 7.4 we show an example of FRET measurements that can be used to monitor the melting of a DNA hairpin loop where fluorescence from the donor is quenched by the proximity of a tagged acceptor. Here, in the work by Wallace et al.,[69] FRET techniques were used to determine the thermodynamic parameters of the closed-to-open transition of a model DNA oligomer in which the ends were terminated by the dye molecules carboxytetramethylrhodamine (TMR), which is a fluorescence donor, and indodicarbocyanine (Cy5), here the fluorescence acceptor. DNA hairpin-loop structures fluctuate between different conformations and are involved in various biological functions including gene expression and regulation.[70,71] Loops have also been used in biotechnology as biosensors and molecular beacons.[72]

In Figure 7.4b are the equilibrium melting curves for a model DNA hairpin as determined by comparing the FRET intensities for the donor-acceptor labeled sample to that of the donor-only labeled sample. The high-temperature/high-fluorescence intensity limit corresponds to the case where the two ends are farthest apart. Consequently, the fluorescence from the TMR is not quenched or transferred to the Cy5 as efficiently as in the low-temperature case.

7.3 BEYOND FÖRSTER

Förster's approach is only valid for distances that are large on the length scale of the individual chromophores involved. However, in many systems, particularly those involving π-conjugated molecules, the distance of separation between chromophore units is oftentimes comparable to the actual size of the molecule. In such cases, the Förster approach is incapable of providing an accurate estimate of the energy transfer rate. The problem stems from the fact that at sufficiently short ranges, the donor molecular "feels" segments of the acceptor species more strongly than others. Hence, one needs to account for the inhomogeneities in the transition densities about the donor and acceptor.

One improvement on the Förster scheme was proposed by Beenken and Pullerits[73] and given much more rigorous justification by Barford[74] whereby the total transition dipole moment for a polymer chain is projected onto individual monomeric units and the total interaction is summed as a line-dipole:

$$M = \sum_{ij} \frac{1}{R_{ij}^3} \left(\vec{p}_{Ai} \cdot \vec{p}_{Bj} - \frac{3}{R_{ij}^2} (\vec{p}_{Ai} \cdot \vec{R}_{ij})(\vec{p}_{Bj} \cdot \vec{R}) \right) \qquad (7.28)$$

where the $\vec{p}_{Ai}$ and $\vec{p}_{Bj}$ are fractional transition dipoles that obey the sum rule

$$\vec{p}_A = \sum_i \vec{p}_{Ai}$$

As seen in Figure 7.5, the line-dipole approach does a far better job of approaching the Coulomb coupling limit than the point-dipole approximation for linear polymers. Only when the distance of separation between the charge centers of the two chains is slightly larger than the actual chain lengths (in this case of 64 Å) do the three approaches agree. Notice, also, that the point-dipole approach consistently overestimates the coupling. For typical packing distances of $R \approx 4$–10 Å, the point-dipole approach can be as much as 2 to 4 orders of magnitude too large. For parallel polyene chains, it can be shown analytically that the Coulomb coupling integral between donor and acceptor species V_{DA} scales as the chain length L when L is smaller than the separation distance and $V_{DA} \propto 1/L$ when L is larger than their separation within a plane-wave approximation of the excitonic wave functions:[74]

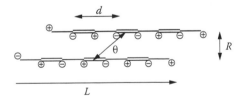

The scaling of $V_{DA} \propto L$ for short chain lengths is a reflection of the fact that at these length scales the point-dipole approximation may be applied to the entire chain, implying that $V_{DA} \propto L$. Similarly, the scaling of $V_{DA} \propto L^{-1}$ for collinear chains in the plane-wave approximation is easy to understand for chain lengths that are large compared to their separation. In this case, the exciton dipoles are uniformly distributed along both chains of length L. As a result, the double line integral of r^{-3} yields the L^{-1} scaling.[74]

The scaling of V_{DA} with L for long parallel chains is somewhat less intuitive since it implies that the probability for exciton transfer between neighboring chains is

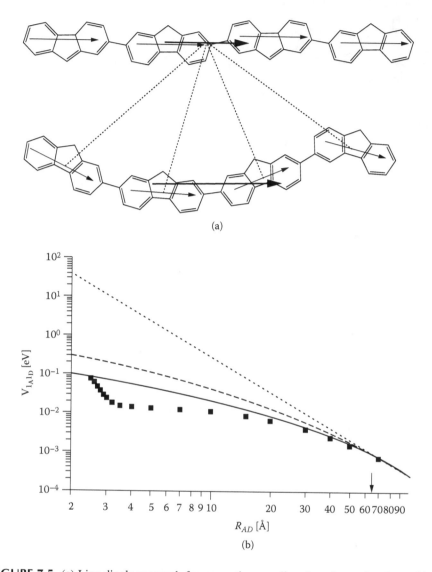

(a)

(b)

FIGURE 7.5 (a) Line-dipole approach for computing couplings based upon local transition dipole approximation between two polyfluorene oligomers with slightly different comformations. Bold arrows are the total transition dipole moment for each polymer chain while the small arrows indicate the projection of the total moment onto the individual repeating unit. (b). Excitonic coupling V_{DA} for the lowest singlet excited states between two transoid sedecithophene oligomers in a skew-line arrangement comparing the point-dipole (dotted), the line-dipole (dashed), and Coulomb intergral (solid). The squares indicate the half-splitting energies from ZINDO calculations of the dimer. The arrow at 64 Å indicates the length of a single 16-ring sedecithophene chain (figure from Ref. 73).

a decreasing function of the chain length. The scaling can be understood in one of two ways. First, if the distance of separation between the two polyenes, R, is on the order of a monomer length, as in the case of π stacked polyenes, V_{DA} becomes a periodic function of the relative alignment (or shift) of the two chains. For polyene-polyene dimers, one has "in-phase" or "out of phase" stacking configurations depending upon whether or not the C=C bonds on one chain are aligned with the C=C bonds on the other polyacetylene chain. This results in a modulation of V_{DA} as the two chains are displaced relative to each other. As the chains become farther and farther apart, this periodic variation vanishes due to interference effects from increasingly longer-ranged local transition densities. Ultimately, the coupling integral will vanish in the asymptotic limit even for chain separations greater than a few monomers. Alternatively, one can imagine that as two finite-length dipoles side longitudinally relative to each other, the sign of the dipole–dipole coupling changes once $\cos\theta = 1/\sqrt{3}$. Consequently, any small variation in the alignment results in a vanishing of V_{DA} as L becomes large.

7.4 TRANSITION DENSITY CUBE APPROACH

Thus far we have assumed that the interactions can be computed with the dipole–dipole approximation that assumes that the donor and acceptor molecules are far enough apart so that the distance of separation, R, is much larger than their size. However, as experimental techniques advanced, especially in the 1990s, it became apparent that in photosynthetic and other light-harvesting systems, the rapid time scales involved in the energy transfer between pigment molecules was much larger than what one would expect based upon a dipole–dipole coupling. This is especially evident for systems with extended electronic states, such as conjugated polymers (as in the case of optical electronic devices) and caratinoids in LH1 (for a review, see, for instance, V. Sundström, et al., *J. Phys. Chem. B*, 1999, 103, 2327–2346).

As noted earlier in this chapter, in the standard approach developed by Förster, one uses the lowest-order term in the multipole expansion of the Coulomb matrix between $|D^*A\rangle$ and $DA^*\rangle$ where D and D^* denote ground- and excited-state wave functions for the donor (D) or acceptor (A) molecules

$$V_{DA} = \kappa \frac{|\mu_D||\mu_A|}{R_{DA}^3}$$

where κ is an orientation factor

$$\kappa = \vec{\mu}_D \cdot \vec{\mu}_A - 3(\vec{\mu}_D \cdot \vec{n})(\vec{\mu}_A \cdot \vec{n})$$

where $\vec{\mu}_{D,A}$ is the unit vector giving the direction of the transition dipole moment for the donor or acceptor species and $\vec{n}$ is a unit vector pointing from the charge center of the donor to the charge center of the acceptor.

As noted, this approximation is no longer valid when $R \approx a$, that is, when the donor and acceptors are within a few molecular radii. At short range, higher-order multipoles must be taken into account to properly describe the charge density associated with the transition. Secondly, we have ignored the direct overlap between the wave functions on each molecule. Consequently, the exchange interaction must also be included once the two molecules become very close. In fact, if the two species are too close, the assumption that the two molecules are "independent" is simply too severe and one should really use a full quantum chemical treatment for the whole system.

The most robust approach aside from a full quantum chemical treatment is to compute the Coulomb matrix element directly from the donor and acceptor wave functions:[75]

$$V_{DA} \approx \left\langle D^*A \left| \frac{e^2}{|r_a - r_d|} \right| DA^* \right\rangle \tag{7.29}$$

where r_a denotes the coordinates of electrons associated with the acceptor molecule and r_d the coordinates for the electrons associated with the donor molecule. Under this assumption, the above integral can be recast as an integral over two densities,

$$M_D(r) = \langle r|D \rangle \langle D^*|r \rangle \tag{7.30}$$

and

$$M_A(r) = \langle r|A \rangle \langle A^*|r \rangle \tag{7.31}$$

where $|D\rangle\langle D^*|$ is the excitation operator (or projection operator) constructed by taking the outer product between the ground- and excited-state wave functions of the donor molecule (integrating over the spin coordinate) and likewise for the acceptor molecule. Both of these quantities can be computed using *separate* excited-state quantum chemical calculations involving the donor and acceptor species. The advantage, then, is that the accuracy of the exciton-exciton coupling is determined entirely by the accuracy of the quantum chemical approach used in determining the excited states of the donor and acceptor molecules.

Numerically, this is implemented by approximating the transition densities as

$$M_D(ijk) = \delta x \delta y \delta z \int ds \int_{x_i}^{x_i+\delta x} \int_{y_j}^{y_j+\delta y} \int_{z_k}^{z_k+\delta z} \langle r_{ijk}|D \rangle \langle D^*|r_{ijk} \rangle \, dx \, dy \, dz \tag{7.32}$$

where the integration is only over a small "voxel" or volume cell of dimension $\{\delta x, \delta y, \delta z\}$. Such volume renderings are very useful for analysis by external programs since they are essentially independent of the choice of basis functions used in quantum chemical routines used to generate the data. The final integral is constructed by taking

$$V_{DA} = \int dr_D \int dr_A \frac{M_D(r_D)M_A(r_A)}{|r_A - r_D|} \tag{7.33}$$

where the integrals are over the full three-dimensional volume. The de facto data format for the transition densities is the format used by the Gaussian quantum chemical code.[76] This format is also used by the Orca code[77] and Qchem.

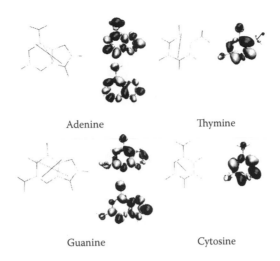

Adenine   Thymine

Guanine   Cytosine

FIGURE 7.6 Transition densities for $S_o \to S_1$ excited states of the four DNA bases (from Ref. 78).

As an example of the techniques, consider the electronic couplings between DNA bases. The transition densities for the four DNA bases are shown in Figure 7.6 for the $S_o \to S_1$ transitions for the pyrimidines (thymine and cytosine) and the $S_o \to S_1$ and $S_o \to S_2$ transitions for the purines (adenine and guanine).[78] To calculate these, the geometries of the DNA bases—adenine, guanine, cytosine, and thymine—in their most common tautomeric forms were optimized at the MP2/TZVP level of theory in chloroform using the GAUSSIAN03 suite of programs.[76] The optimized geometries were subsequently used to calculate the singlet excitation energies in gas phase at the TD-DFT level using PBE0 functional and TZVP basis set augmented with the diffusion functions on all atoms as implemented in Orca.[77]

In order to compare the transition density cube method to the simple point-dipole approximation, we show here the values of the Coulombic couplings between the lowest energy $\pi\pi^*$ transitions of the adenine and thymine and two π-stacked thymines as a function of distance between the bases (Figure 7.7). The comparison of the coupling elements obtained with the two methods, point-dipole approximation and transition density cube (Figure 7.7), shows a good agreement at a separation between the bases larger than 5 and 6 Å for the AT pair and two stacked thymines, respectively. At a shorter separations, in the range of 3–4 Å, which is typical for DNA structures, the agreement between point-dipole approximation and transition density cube is very poor with the differences between calculated couplings larger than 100% in case of AT pair. The aforementioned good agreement between point-dipole approximation and transition density cube at larger and poor agreement at shorter separations between nucleobases indicates that the shape and spatial extent of transition density (Figure 7.7) become important and cannot be neglected at distances between the bases typical for double helices DNA. The agreement between the two methods becomes very good in the limit of very large separation (> 8 Å).

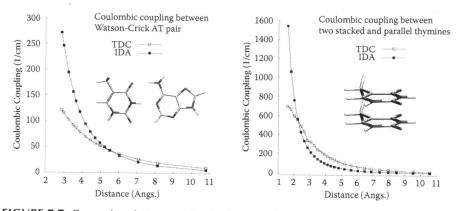

FIGURE 7.7 Comparison between point-dipole approximation (left) and exact (numerical) evaluation (right) of coupling between two DNA bases (from Ref. 78).

For the stacking and pairing distances corresponding to the idealized B-DNA geometry, the coupling elements calculated with the point-dipole approximation result with several-fold larger absolute values compared with the corresponding values calculated using the transition density cube method. The largest differences between the two methods are obtained for the couplings between the π-stacked adenines. For the idealized B-DNA geometry, the coupling between two adenines located on the same strand calculated using point-dipole approximation, 872 cm^{-1}, is more than five fold larger compared with the value obtained using transition density cube, 161 cm^{-1}. The differences in the calculated couplings using the same two methods for two stacked thymines are much smaller. For this base pair, the Coulombic coupling calculated using point-dipole approximation is equal to approximately 230 cm^{-1}, more than twice the value of 101 cm^{-1} obtained with transition density cube. Bouvier et al.[79] reported the magnitudes of Coulombic coupling calculated using atomic transition charges model.[80] The corresponding values for the intrastrand nearest neighbors in a standard B-DNA geometry are 170 and 217 cm^{-1} for the lowest energy $\pi\pi^*$ transitions of adenine and thymine, respectively. The absolute values of the coupling elements between the second-nearest neighbors located on the same strand are much smaller. At the point-dipole level of approximation, the coupling between the two adenines is only 57 cm^{-1} compared with 9 cm^{-1} calculated for the same base pairs using transition density cubes. The coupling between the two thymine bases on the same strand is even smaller— approximately 3 and 1 cm^{-1} for point-dipole approximation and transition density cube methods, respectively. We conclude then that while the point-dipole approximation provides a simple and robust way of estimating the electronic coupling between chromophores that are well separated, the simple approximation decidedly breaks down once the donor and acceptor speeds are brought to within close proximity.

SUGGESTED READING

1. *Quantum Mechanics in Chemistry*, G. C. Schatz and M. A. Ratner (Mineola, NY: Dover Books, 2002).
2. *Principles of Nonlinear Optics and Spectroscopy*, S. Mukamel (Oxford: Oxford University Press, 1995).
3. *Optical Resonance and Two-level Atoms*, L. Allen and J. H. Eberly (Mineola, NY: Dover Books, 1974).
4. *Principles of Nuclear Magnetism*, A Abragam (Oxford: Oxford University Press, 1961).
5. *The Quantum Theory of Light*, R. Loudon (Oxford: Oxford University Press, 1973).
6. *Laser Theory*, H. Haken, *Handbuch der Physik*, Vol. XXV/2 (Springer-Verlag, Berlin, 1970).
7. *Fundamentals of Quantum Electronics*, R. H. Pantell and H. E. Puthoff (New York: Wiley, 1969).

8 Electronic Structure of Conjugated Systems

The underlying physical laws necessary for the mathematical theory of a large part of physics and the whole of chemistry are thus completely known, and the difficulty is only that the exact application of these laws leads to equations much too complicated to be solvable

Paul Dirac

. . . that is solvable by a person armed only with pencil and paper.

8.1 π CONJUGATION IN ORGANIC SYSTEMS

In organic chemistry, a conjugated system is one in which there is a chain of unsaturated alternating single and double bonds as in $CH_2{=}CH{-}CH{=}CH{-}CH{=}CH_2$. In such polyene chains, the valence orbitals about the carbon atoms are in the sp^2 hybrid electronic configuration that forms the σ-bonding frame for the chain. The remaining $2p_z$ orbitals are aligned perpendicular to the σ bonding plane, and electrons within these orbitals are more or less delocalized over the entire molecule. In general, this delocalization decreases the overall kinetic energy of the electrons and lowers the total electronic energy of the system. Conjugated systems absorb readily in the ultraviolet and visible regions of the spectrum due to excitations of π bonding to π^* antibonding orbitals. Generally, chains with fewer than eight conjugated bonds absorb in the UV region while increasingly longer chains absorb at longer wavelengths.

A complete description of the electronic properties of conjugated molecules requires that we consider all the electrons in the system, and certainly modern quantum chemical codes and computational hardware are up to the challenge for even moderately large systems. However, with a few judicious approximations we can arrive at fairly robust models that can oftentimes outperform more exact treatments. Moreover, having a simple theory that works well allows us to develop a deeper understanding of the underlying physics of these systems that lead to many of their spectroscopic and optical-electronic properties.

The crucial approximation we make is that the total electronic wave function can be factored into two parts:

$$\Psi = \Psi_\sigma \Psi_\pi \Psi_{core} \tag{8.1}$$

where Ψ_σ describes the σ orbitals, Ψ_π describes the π orbitals, and Ψ_{core} are core orbitals that do not participate in chemical bonding. The justification for this is that, by and large, there is very little overlap between the σ orbitals that are generally localized along lines directly connecting the nuclear centers and the π orbitals that are generally delocalized and lie above and below the σ-bonding plane. Certainly for many systems there is some degree of mixing between the σ and π manifolds. However, for the

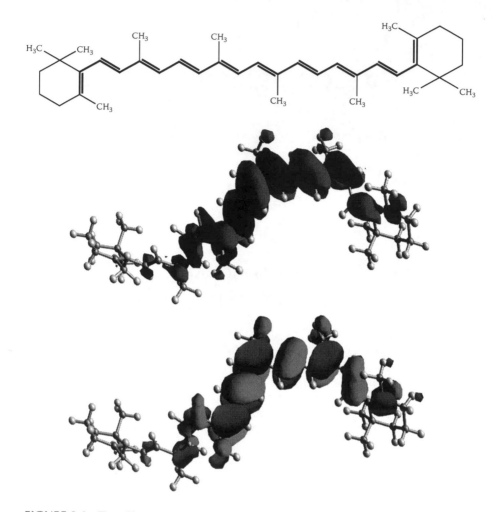

FIGURE 8.1 (Top) Chemical structure of betacarotene. The conjugated domain is indicated in bold. LUMO (middle) and HOMO (bottom) orbitals of betacarotene superimposed on its 3D structure.

orbitals energetically nearest the highest occupied and lowest unoccupied molecular orbitals (HOMO and LUMO) as well as the first few excited electronic states, this is fairly good approximation.

In this chapter we will develop a description of the π electronic structure of conjugated organic systems. We will start with a simple free-electron model and finish with a brief description of modern quantum chemical techniques. A simple model for understanding this trend is the "free-electron model" where we assume that the electrons within the π bonding network are more or less free particles and we ignore any electron/electron interaction. If the average C–C bond length is a from carbon center to carbon center then an electron within the π orbital in the $C_N H_{N+2}$ polyene is confined to a "box" of length L. From elementary quantum mechanics,

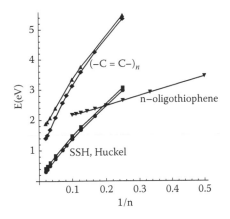

FIGURE 8.2 Variation of optical energy gap with number of repeating units for polythiophene and polyacetylene oligomers as computed various the semiempirical models.

such a system has energy levels

$$E_n = \frac{\hbar^2}{2m_e} \frac{\pi^2 n^2}{L^2} \tag{8.2}$$

since each C atom contributes one electron to the π system. So for a system with k C=C double bonds, we have a total of $2k$ electrons with the $n = k$ level being the highest occupied and the $k = n + 1$ being the lowest unoccupied levels. Assuming the optical transitions are between these two levels,

$$\Delta E = \frac{\hbar^2 \pi^2}{2m_e L^2}(2k + 1) \tag{8.3}$$

The length of the "box" is actually somewhat arbitrary since the π orbital extends a bit beyond the terminal C atoms, say, 1/2 a C–C bond length ($a = 1.4$ Å), in which case we get a length of $L = (2k + 1)a$. Thus,

$$\Delta E = \frac{\hbar^2 \pi^2}{2m_e} \frac{1}{2k + 1} = \frac{1.54 \times 10^5 \text{cm}^{-1}}{2k + 1} \tag{8.4}$$

after inserting values for the other physical constants. A comparison of the free electron prediction and the observed UV/VIS absorption maxima for polyenes of various lengths is given in Table 8.1. The agreement is not so good; however, this simple model does give the correct variation of energy gap with increasing chain length, $\Delta E \propto 1/L$.

The free electron model also predicts that for infinitely long chains, $\Delta E \to 0$ as $L \to \infty$, implying that a linear fit of the data to $a + b/(2k + 1)$ should give $a = 0$ as a y intercept. However, such a fit gives $a = 16\,775.9$ cm^{-1} and $b = 1.396 \times 10^5$ cm^{-1}. There are a number of likely reasons for the deviation. The most obvious is that we have ignored the interactions between electrons. However, this actually turns out not to be the primary reason for the deviation. The problem is that the C atoms in

TABLE 8.1
Electronic Spectra of Linear Polyenes versus
Number of C=C Double Bonds

k	ν_{free}	ν_{huck}	ν_{obs}
1	51333.3 cm^{-1}	63009.7 cm^{-1}	62000 cm^{-1}
2	30800.0	45119.1	46000
3	22000.0	37016.5	38000
4	17111.1	32438.3	33000
5	14000.0	29503.1	30000
6	11846.2	27463.0	27500
7	10266.7	25963.4	26000
8	9058.82	24814.9	24000
10	7333.33	23172.0	22000

polyacetylene are not evenly spaced and in fact alternate their bond lengths between double (C=C) and single (C–C) bonds. As we shall show later in this chapter, this gives rise to a nonzero energy gap for the infinitely long chain. Finally, we have neglected the fact that realistic polyene molecules are not straight chains. They can have twists and kinks and other contortions that can limit the extent of π conjugation. However, the fact that the energy gap does follow the predicted $\Delta E \propto 1/L$ behavior indicates that the electrons in the π orbitals are moving ballistically as free particles more or less unaware of the geometry of the molecule. Consequently, even the π electronic structure for a molecule such as betacarotene shown in Figure 8.1 can be well understood within a free-electron model.

8.2 HÜCKEL MODEL

In the free electron model, we assumed that as far as the electron was concerned there was no energetic cost in moving from one C atom to the other. However, a more systematic approach can be developed if we assume that there is an energy α associated with placing an electron in a C $2p_z$ orbital and an energy β associated with transferring that electron from one C $2p_z$ to another neighboring C $2p_z$ orbital. In other words, in a basis of C $2p_z$ orbitals localized about the C atoms in the chain,

$$\langle\phi_i|H|\phi_j\rangle = \beta \quad \text{and} \quad \langle\phi_i|H|\phi_i\rangle = \alpha \tag{8.5}$$

We shall also assume that the overlap integral between neighboring C $2p_z$ orbitals is exactly zero, $\langle\phi_i|\phi_j\rangle = \delta_{ij}$, and provide a sufficient basis to expand the electronic wave functions:

$$|\Psi\rangle = \sum_{j=1}^{N} c_j|\phi_j\rangle \tag{8.6}$$

Finally, we also will neglect the Coulombic interaction between the electrons. Our task, then, is to determine the coefficients and the energy eigenvalues. Since we have ignored interactions between the electrons, we need to solve the one-electron Schrödinger equation:

$$H|\Psi\rangle = E|\Psi\rangle \tag{8.7}$$

In general, H will have nondiagonal elements whenever there is a π bond linking adjacent C atoms. For a linear chain, only nearest neighbors are linked, H becomes a tridiagonal matrix, the Schrödinger equation in matrix form reads:

$$
\begin{bmatrix}
\alpha & \beta & 0 & 0 & \cdots & 0 \\
\beta & \alpha & \beta & 0 & \cdots & 0 \\
0 & \beta & \alpha & \beta & \cdots & 0 \\
\vdots & \vdots & \vdots & \vdots & \ddots & \vdots \\
0 & \cdots & 0 & \beta & \alpha & \beta \\
0 & \cdots & 0 & 0 & \beta & \alpha
\end{bmatrix}
\begin{bmatrix}
c_1 \\ c_2 \\ c_3 \\ \vdots \\ c_{N-1} \\ c_N
\end{bmatrix}
= E
\begin{bmatrix}
c_1 \\ c_2 \\ c_3 \\ \vdots \\ c_{N-1} \\ c_N
\end{bmatrix}
\tag{8.8}
$$

Introducing a dimensionless energy $E' = (E - \alpha)/\beta$, the equations for $j \neq 1$ and $j \neq N$ become

$$c_{j-1} + c_{j+1} = E'c_j \tag{8.9}$$

If we write

$$c_j = Ae^{ikj} + Be^{-ikj} \tag{8.10}$$

then

$$E' = e^{ik} + e^{-ik} = 2\cos(k) \tag{8.11}$$

where A and B are constants and k is a parameter. Imposing the boundary condition

$$c_2 = E'c_i \quad \text{and} \quad c_{N-1} = E'c_N \tag{8.12}$$

we arrive at two simple equations:

$$A + B = 0 \tag{8.13}$$

$$Ae^{ik(N+1)} + Be^{-ik(N+1)} = 0 \tag{8.14}$$

which allow us to deduce the allowed values of k:

$$e^{2ik(N+1)} = 1 \tag{8.15}$$

which leads to

$$k = \frac{\pi n}{N+1}, \quad n = 1, 2, \ldots, N \tag{8.16}$$

We must disallow the case for $n = 0$ since it leads to the trivial solution of $c_j = 0$ for all values of j. The constant A can be determined by the normalization condition

$$\sum_j c_j^2 = 1 \tag{8.17}$$

Thus, the allowed energy levels for the discretized polymer lattice are given by

$$E_n = \alpha + 2\beta \cos\left(\frac{\pi n}{N+1}\right) \tag{8.18}$$

As N increases, the width of the energy spectrum tends to $4|\beta|$. Moreover, in the infinite limit, the energy spacing between each successive energy level shrinks to zero and k becomes a continuous variable. The coefficients for the eigenstates are given by (including normalization)

$$c_{nj} = \sqrt{\frac{2}{N+1}} \sin(nj\pi/(N+1)) \tag{8.19}$$

8.2.1 JUSTIFICATION FOR THE HÜCKEL MODEL

The Hückel model rests upon the following assumptions. First, as noted above, we discount any electron-electron interaction within the π system itself. Thus, the matrix elements of our π electron Hamiltonian reduce to

$$h_{aa} = \int \phi_a \hat{h} \phi_a dr \tag{8.20}$$

where ϕ_a is a C $2p_z$ orbitals located on the carbon atom a. The operator $\hat{h}$ is really an effective operator since we are assuming that it contains all core-level interactions. The diagonal elements of h_{aa} include (tacitly) an integration over all core electrons not included in the π system. Since we are talking about C atoms participating in σ bonding, these matrix elements should be roughly the same for all similar C atoms. Hence we set $h_{aa} = \alpha$ as a constant. For heteroatoms, one needs to use other values of α.

Integrals of the form

$$h_{ab} = \int \phi_a \hat{h} \phi_b dr \tag{8.21}$$

with $a \neq b$ are termed *resonance* integrals. If we allow each $2p_z$ to be given by a Slater-type orbital (STO) with a radial function

$$R_n(r) = Nr^{n-1}e^{-\zeta r} \tag{8.22}$$

where n is the principle quantum number with values $n = 1, 2, \ldots$; N is the normalization; and ζ is related to the effective charge of the nucleus. The normalization is given by

$$\int_0^\infty x^n e^{-\alpha x} dx = \frac{n!}{\alpha^{n+1}} \tag{8.23}$$

which gives

$$N = (2\zeta)^n \sqrt{\frac{2\zeta}{(2n)!}} \tag{8.24}$$

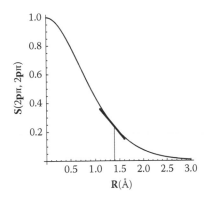

FIGURE 8.3 Overlap integral between two adjacent C $2p\pi$ orbitals.

The angular terms are given by the real form of the spherical harmonics. Clearly, given the fact that R_n decays exponentially with radial distance r, the resonance integrals will be nonvanishing only between C atoms that are close to each other. Hence, for adjacent C atoms we set $h_{ab} = \beta$. Since orbital overlap varies with bond length, one expects some systematic variation in β with bond length. Mullikin suggested that $\beta(r)$ should vary as the overlap integral between two $2p$ STOs,

$$S(2p\pi, 2p\pi) = e^{-p}(1 + p + (2/5)p^2 + p^3/15) \tag{8.25}$$

where $p = 1.625R/a_o$ for C atoms separated by distance R (in Bohr radii). [?] This function is plotted in Figure 8.3 where we note that for $R > 3$ Å, the overlap integral is nearly vanishing. Taking the typical C–C bond length to be 1.39 Å and expanding about this point gives

$$\beta(r) = \beta_o(1 - 1.72(R - 1.39 \text{ Å})) \tag{8.26}$$

as an approximate variation of β with C–C bond distance where β_o is the resonance integral at 1.39 Å. We shall later use this approximately linear variation in the resonance integral as a means for including electron–phonon coupling in these systems.

Finally, the remaining assumption that the orbitals localized on one C atom are orthogonal to orbitals localized on different C atoms requires us to write

$$S_{ab} = \int \phi_a \phi_b \, dr = 0 \tag{8.27}$$

unless $a = b$, in which case $S_{aa} = 1$. In the simplest case, this is a rather extreme approximation as we can see in Figure 8.3 where at the typical C–C bond length $S \approx 0.25$. We can improve upon the simple Hückel model by solving the generalized eigenvalue equation

$$(h - \varepsilon S)\psi = 0 \tag{8.28}$$

where S is the overlap matrix. For example, Roald Hoffman's extended Hückel approach includes the overlap integral

$$S_{ij} = \langle i | j \rangle \tag{8.29}$$

between basis functions on different atomic centers as a way to account for bond bends and torsions using STOs on each atom. The remaining parameters are given by the ionization potentials, electron affinities, and core charges of the atomic sites.

8.2.2 EXAMPLE: 1,3 BUTADIENE

We begin by drawing a sketch of the molecule:

where each C atom is labeled and the adjacency is indicated by a solid line. We then write the Hamiltonian for the π electrons as

$$
H = \begin{bmatrix} \alpha & \beta & 0 & 0 \\ \beta & \alpha & \beta & 0 \\ 0 & \beta & \alpha & \beta \\ 0 & 0 & \beta & \alpha \end{bmatrix}
\tag{8.30}
$$

The eigenvalues and eigenvectors can be readily determined either numerically or algebraically by solving the secular determinant equation

$$
\begin{vmatrix} x & 1 & 0 & 0 \\ 1 & x & 1 & 0 \\ 0 & 1 & x & 1 \\ 0 & 0 & 1 & x \end{vmatrix} = 0
\tag{8.31}
$$

where $x = (\alpha - \varepsilon)/\beta$. Expanding the determinant gives

$$
x^4 - 3x^2 + 1 = 0
\tag{8.32}
$$

which has four roots corresponding to the orbital energies as shown in Figure 8.4.

The total energy is then

$$
E_\pi = \sum_j n_j E_j
\tag{8.33}
$$

where n_j is the occupancy of the jth energy level. For the case of 1-3-butadiene, each C atom contributes one electron to the π system. Taking the Pauli principle into account, only the lowest two levels are fully occupied and the total π energy is

$$
E_\pi = 4\alpha + 4.472\beta
\tag{8.34}
$$

If the π electrons in butadiene were not delocalized but rather formed two isolated double bonds, the total energy would be simply $4\alpha + 4\beta$, that is, twice the energy of ethylene. By delocalizing the electrons, the total energy is lowered by 0.472β.*

---

* Note that the resonance integral β is a negative value since delocalization should lower the energy of the system.

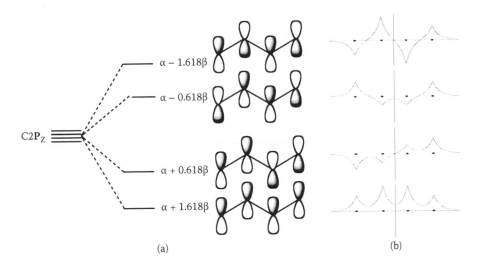

(a)   (b)

FIGURE 8.4 Energy levels and orbitals for 1, 3-butadiene from the Hückel model.

An alternative way to derive the eigenvalues for the linear chain is to consider the roots of the determinant equation

$$D_n(x) = \begin{Vmatrix} x & 1 & 0 & 0 & \cdots & 0 \\ 1 & x & 1 & 0 & \cdots & 0 \\ 0 & 1 & x & 1 & \cdots & 0 \\ \vdots & & & \ddots & \ddots & \ddots \end{Vmatrix} \qquad (8.35)$$

where $D_n(x)$ is an nth-order polynomial and $x = (\alpha - \lambda)/\beta$. Expanding the determinant in terms of its cofactors, we find a recursion relation

$$D_n(x) = x D_{n-1}(x) - D_{n-2}(x) \qquad (8.36)$$

This is the recursion relation for the Tchebychev polynomials with $D_0(x) = 1$, $D_1(x) = 1$, or in general

$$D_n(x) = 2^{1-n} \cos(n \cos^{-1}(x)) \qquad (8.37)$$

Setting $x = 2 \cos \theta$,

$$D_n(x) = \sin((n + 1)\theta)/ \sin \theta = 0 \qquad (8.38)$$

Consequently, the roots occur whenever $\theta = i\pi/(n + 1)$ for $i = 1, 2, \ldots, n$ or $x = 2 \cos(i\pi/(n + 1))$. Thus

$$\lambda_i = \alpha + 2\beta \cos(i\pi/(n + 1)) = \alpha + 2\beta T_i(\cos(\pi/(n + 1))) \qquad (8.39)$$

are the roots and $T_i(x)$ are the Tchebychev polynomials.

The eigenfunctions are obtained as linear combinations of the C $2p_z$ orbitals

$$\psi_n = \sum_i c_{ni} \phi_i \qquad (8.40)$$

In general it is a tedious procedure to determine the expansion coefficients by hand. For anything with more than four–six atoms, this is best done numerically using various eigenvector routines. In fact, the problem we have just solved is equivalent to that of the Schrödinger equation for a square well potential on a mesh of $N + 2$ points with $\psi(N = 0) = \psi(N + 1) = 0$ as boundary conditions and $t = \hbar^2/(2ma^2)$. The eigenvalue coefficients for a linear Hückel chain are the amplitudes for standing waves on a grid of points located at $x_j = ja$ where a is the interatomic spacing. Thus, the normalized eigenstate corresponding to the nth energy level for a chain of N identical atoms is given by

$$\psi_n = \sqrt{\frac{2}{N+1}} \sum_{i=1}^{N} \sin(ni\pi/(N+1))\phi_i \qquad (8.41)$$

Furthermore, using the x_j points to sample a function for integration is equivalent to performing the Gauss-Tchebychev quadrature with weightings given by $2/(N + 1)$. Such grids are the basis for the Tchebychev version of the discrete variable representation (DVR) used extensively in grid-based quantum dynamical calculations.[81,82]

The Hückel model can also be used to predict and interpret electronic $\pi - \pi^*$ transitions. Assuming that the primary optical transition is between the HOMO and LUMO orbitals, the energy gap for linear polyenes with k C=C bonds can be written as

$$\Delta E = 4\beta \sin(\pi/(4k + 2)) \qquad (8.42)$$

In Table 8.1 we compare the experimental UV/VIS absorption maximum for polyenes of various lengths to an empirical fit to the Hückel model

$$\nu_{huck} = a + 4b \sin(\pi/(4k + 2)) \qquad (8.43)$$

with $a = 16\,171.6$ cm^{-1} and $b = 2.34 \times 10^5$ cm^{-1}. The fact that the y intercept does not vanish implies that no single value of β can give a good fit for all the data. However, the fact that the energy gap varies as $4\sin(\pi/(4k + 2))$ implies that the first excited states of similar molecules are correlated in about the same way as their respective ground states.

While this energy gap law generally holds for polymethines (odd-numbered C chains), it does not hold for polyenes (even-numbered C chains), which have a finite gap as n grows to be large. The reason for this is that the transfer integral β is not the same between every pair of neighboring C atoms due to bond-length alternation. Longer bonds have slightly smaller β values while short bonds will have slightly larger β values (in magnitude).

8.2.3 Cyclic Systems

For cyclic systems, the Hamiltonian is essentially the same as above except that atoms 1 and N are now linked and $H_{1N} = H_{N1} = \beta$. Because of the periodicity, $\Psi(x + N) = \Psi(x)$ so that

$$e^{iNk} = 1 \rightarrow k = \frac{2\pi n}{N}, \quad n = 0, 1, 2, \ldots, N - 1 \qquad (8.44)$$

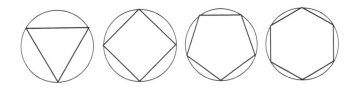

FIGURE 8.5 Hückel levels for cyclic molecules.

The eigenvalues are then

$$E = \alpha + 2\beta \cos(2n\pi/N) \tag{8.45}$$

where $n = 0, 1, 2, 3, \ldots, N-1$. In the limit of an infinite ring, k becomes a continuous variable between 0 and 2π. Since the cosine function is a periodic function, we can redefine the range of k to be between $-\pi$ and $+\pi$ to highlight the degeneracy between $+k$ and $-k$. Thus, the spectrum of an infinite system with open boundaries is equivalent to that of a system with periodic boundary conditions.

Secondly, we note a useful way to remember the level structure of a cyclic system with N sites. Notice that the argument of the cosine in the energy expression can be written as

$$E = \alpha + 2\beta \cos(n\theta) \tag{8.46}$$

where $\theta = 2\pi/N$ is one angular step going about a circle. These define the vertices of a regular polygon circumscribed by a circle of radius 2β centered about α with the $\theta = 0$ vertex pointing straight down. The polygons for the C_nH_n for $n = 3$ to 6 are shown in Figure 8.5. From this, one can more or less immediately read off the Hückel energy levels without having to rederive the energy equation. For the case of cyclobutadiene, the total π energy is $E_\pi = 4\alpha + 4\beta$. Recall that E_π for two ethylene molecules was also $4\alpha + 4\beta$. Hence, according to the Hückel model at least, there is no additional energy stabilization gained by delocalizing the π electrons in cyclobutadiene.

While cyclobutadiene is not stabilized by delocalization, other cyclic rings are stabilized. Table 8.2 compares the experimental resonance energies of a number of common cyclic aromatic systems to their Hückel delocalization energies along with an estimate for the resonance integral. The fact that β is fairly constant over a number of

TABLE 8.2
Experimental Resonance Energies and Hückel Delocalization Energies

Molecule	Experimental Resonance Energy (kcal/mol)	Hückel Delocalization Energy	Apparent β (kcal/mol)
Benzene	36	2β	18
Napthalene	75	3.7β	20
Anthracene	105	5.3β	20
Phenanthrene	111	5.45β	20
Biphenyl	65	4.4β	15

chemical compounds lends credence to its use as an empirical parameter for predicting the stability and other electronic properties for related compounds.

8.2.4 SUMMARY OF RESULTS FROM THE HÜCKEL MODEL

The Hückel model is both extremely simple and extremely powerful. Its success relies on the fact that one of the most important deciding factors in the energy and shape of a molecular orbital is its nodal structure. This is determined by the connectivity of the π electronic system and hence reflects the topology of the molecule. The Hückel matrix can written as

$$H = \alpha I + \beta M \tag{8.47}$$

where I is the $N \times N$ identity matrix and M is the topology matrix with elements $M_{ij} = 1$ if atoms i and j are neighbors and $M_{ij} = 0$ otherwise. Clearly, since H and M commute, they share the same set of eigenvectors, c_k. Likewise, their eigenvalues are related

$$\varepsilon_k = \alpha + \beta \lambda_k \tag{8.48}$$

Consequently, everything about the electronic structure can be deduced by knowing λ_k and c_k without any regard or information about the site energy α or transfer integral β.

As we have seen above, there are a number of cases for which we can obtain closed-form solutions for the energies and eigenvectors. We summarize four important cases here.

1. Polyenes and polymethines $C_n H_{n+2}$.[83,84]

$$\lambda_k = 2\cos(k\pi/(n+1)), \quad k = 1, 2, \ldots n \tag{8.49}$$

2. Cyclic polyenes (annulenes) $C_n H_n$ (example: benzene)

$$\lambda_k = 2\cos(2k\pi/n), \quad k = 1, 2, \cdots n \text{ and } \lambda_k = \lambda_{n-k} \tag{8.50}$$

3. Polyacenes, including benzene ($N = 1$), napthalene ($N = 2$), and anthracene ($N = 3$).[85]

$$\lambda_k^{(s)} = 1 \tag{8.51}$$

$$\lambda_{k\pm}^{(s)} = +\frac{1}{2}\left[1 \pm \sqrt{9 + 8\cos(k\pi/(N+1))}\right] \tag{8.52}$$

$$\lambda_k^{(a)} = -1 \tag{8.53}$$

$$\lambda_{k\pm}^{(s)} = -\frac{1}{2}\left[1 \pm \sqrt{9 + 8\cos(k\pi/(N+1))}\right] \tag{8.54}$$

where the (a) and (s) superscripts denote symmetric and antisymmetric states and $k = 1, \ldots N$.

4. Radialenes $(CCH_2)_n$ (n membered rings with exocyclic $-CH_2$ groups[86]

$$\lambda_{k\pm} = 2\cos(2k\pi/n) \pm \sqrt{\cos^2(2k\pi/n) + 1}, \quad k = 1, 2, \ldots, n \tag{8.55}$$

There are also a number of close expressions for the delocalization (or resonance) energy. This is the difference between the total π electron energy and the energy corresponding to a set of localized bonds.

1. Polyenes (even number of C atoms)

$$E_{res}/\beta = 2\csc(\pi/(2(n+1))) - n - 2 \tag{8.56}$$

2. Polymethines (n is odd)

$$E_{res}/\beta = 2\cot(\pi/(2(n+1))) - n - 1 \tag{8.57}$$

3. Hückel annulenes ($n = 4N + 2$)

$$E_{res}/\beta = 4\csc(\pi/n) - n \tag{8.58}$$

4. Anti-Hückel annulenes ($n = 4N$)

$$E_{res}/\beta = 4\cot(\pi/n) - n \tag{8.59}$$

In all cases, as $n \to \infty$,

$$\lim_{n\to\infty} \frac{E_{res}}{\beta} = \frac{4}{\pi} - 1 = 0.2723 \tag{8.60}$$

The corresponding limit for polyacenes[86] is given by

$$\lim_{n\to\infty} \frac{E_{res}}{n\beta} = \frac{1}{2\pi} \int_0^\pi \sqrt{1 + 16\cos^2\theta}\, d\theta - 1 = 0.40284 \tag{8.61}$$

This can be compared with the values obtained for benzene, 0.33333, 0.36832 for napthalene, and 0.37954 for anthracene. Graphene has the highest resonance energy of 0.575.[87]

The bond order between neighboring C atoms in Hückel-type annulene is

$$p_{r,r+1} = \frac{2}{n}\csc(\pi/n) \tag{8.62}$$

which in the limit of an infinite system becomes

$$\lim_{n\to\infty} p_{r,r+1} = \frac{2}{\pi} = 0.6366\ldots \tag{8.63}$$

8.2.5 ALTERNANT SYSTEMS

The model works well for systems that can be classified as *alternant*. An alternant system is such that one can divide the carbon atoms in the molecule into two sets such that members of the first set are only bonded to members of the second set. All linear chains and branched linear chains are alternant as are rings with an even number of carbon atoms as shown in Figure 8.6. In the case of rings with at least one odd-membered ring, the system is nonalternant.

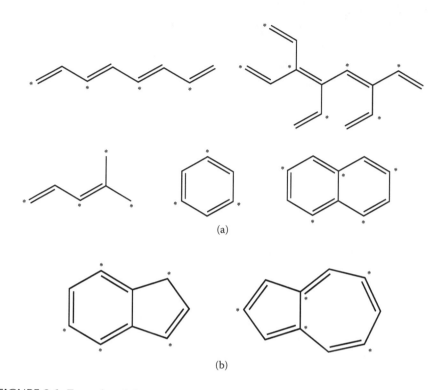

FIGURE 8.6 Examples of alternant (a) and nonalternant (b) hydrocarbon systems. The asterisks labeling different carbons indicate C atoms belonging to one of the alternant sets.

The alternant property imparts a number of important topological implications on both the electronic energy levels and the shape of the wave functions. The first is the *Coulson–Rushbrooke pairing theorem*, which states

THEOREM 8.1

For every Hückel molecular orbital energy $\alpha + \beta x$ in an alternant hydrocarbon system, there exists another orbital with energy $\alpha - \beta x$. In other words, the roots of $||H_\pi - \lambda|| = 0$ appear in pairs. In addition, for linear or branched-linear chains with an odd number of C atoms, there will be one root with $x = 0$.

The proof follows from the parity properties of the Tchebychev polynomials as shown in Figure 8.5. Under the parity operation $x \mapsto -x$, we see that $T_n(x) = (-1)^n T(-x)$. Thus, systems with an even number of sites will have roots corresponding to the even polynomials while systems with an odd number of sites will have odd polynomial roots.

Before moving on, we introduce the bond-charge density matrix ρ

$$\rho = \sum_{i=1}^{N} n_i |\phi_i\rangle\langle\phi_i| \tag{8.64}$$

where $\{\phi_i\}$ are molecular orbitals and n_i is the occupancy. In the basis of local C $2p_z$ orbitals, the diagonal elements $q_m = \rho_{mm}$ represent the number of π electrons localized about the mth carbon atom while the off-diagonal terms, $p_{mn} = \rho_{mn}$, indicate the number of π electrons shared between the mth and nth atom, that is, the bond order. In terms of the orbital coefficients, we can define

$$q_r = \rho_{rr} = \sum_i n_i |c_{ri}|^2 \tag{8.65}$$

$$p_{rs} = \rho_{rs} = \sum_i n_i c_{ri} c_{si}^* \tag{8.66}$$

The total π energy is, thus,

$$E_\pi = Tr[\rho H_\pi] \tag{8.67}$$

For systems with $2n$ carbons and (hence) $2n$ electrons in the π system one finds

$$E_\pi = 2 \sum_i^n \varepsilon_i = 2 \sum_{rs} \rho_{rs} (H_\pi)_{rs} \tag{8.68}$$

$$= \alpha \sum_r q_r + 2\beta \sum_{r<s} p_{rs} \tag{8.69}$$

Returning to the example of 1,3-butadiene above, the molecular orbitals are given by

$$\psi_1 = 0.3718(\phi_1 + \phi_4) + 0.6015(\phi_2 + \phi_3)$$
$$\psi_2 = 0.6015(\phi_1 - \phi_4) + 0.3718(\phi_2 - \phi_3)$$
$$\psi_3 = 0.6015(\phi_1 + \phi_4) - 0.3718(\phi_2 + \phi_3)$$
$$\psi_4 = 0.3718(\phi_1 - \phi_4) - 0.6015(\phi_2 - \phi_3)$$

with $\varepsilon_{1,4} = \alpha \pm 1.618\beta$ and $\varepsilon_{2,3} = \alpha \pm 0.618\beta$. Experimentally, the **trans** diastereomer of 1,3 butadiene is more stable than the **cis** diastereomer. Furthermore, it belongs to the C_{2h} point-group and as such its orbitals can be classified according to the irreducible representations of this group. One finds that the four π orbitals span the reducible representation of $2A_u \oplus 2B_g$ with ϕ_1 and ϕ_3 being of A_u symmetry and the ϕ_2 and ϕ_4 being of B_g symmetry. It is interesting to note that the energy levels linked by the Coulson-Rushbrooke pairing theorem span irreducible representations with opposite symmetry character with respect to inversion and rotation about the C_2 axis.

Next we note that the charge densities on each atom is identical since by Eq. (8.65)

$$q_1 = 2(0.3718)^2 + 2(0.6015)^2 = q_4 = 1 \tag{8.70}$$
$$q_2 = 2(0.6015)^2 + 2(0.3718)^2 = q_3 = 1 \tag{8.71}$$

This brings us to two other important theorems.

THEOREM 8.2
Within the Hückel molecular orbital approximation for alternant systems, the total electron density on each site for an N-electron N-site system is 1.

The proof of the first part of the theorem can be seen by examining Equation 8.65. For an N-electron/N-site system, the sum in Equation 8.65 is identical to taking

$$q_r = 2 \sum_{i=1}^{N/2} \langle r|\psi_i\rangle\langle\psi_i|r\rangle \qquad (8.72)$$

Here the $|\psi_i\rangle\langle\psi_i|$ acts like a projection operator and we are summing over exactly 1/2 the total states in the system. Hence,

$$q_r = 2\langle r|r\rangle \times \frac{1}{2} = 1 \qquad (8.73)$$

In other words, for a neutral alternant system, no charge transfer or charge localization occurs within the molecule.

THEOREM 8.3
The orbital coefficients of paired molecular orbitals are the same for "starred" atoms and have opposite sign for "unstarred" atoms.

This we can see by inspection of the orbital coefficients themselves.

An important special case of the pairing theorem occurs when there is an odd number of sites in the system, as in the case of the allyl radical. Since the energy eigenvalues are determined by the roots of an Nth-order polynomial, any system with an odd number of sites will have a characteristic polynomial that must pass through $x = 0$. The corresponding orbital will be a nonbonding molecular orbital with nodes passing through unstarred alternant sites as shown in Figure 8.7.

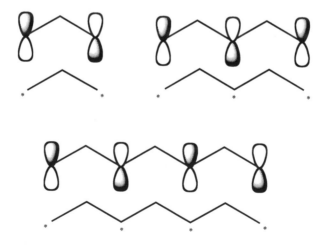

FIGURE 8.7 Nonbonding Hückel molecular orbitals with energy $E = \alpha$ for chains of length $N = 3, 5,$ and 7.

8.2.6 WHY BOTHER WITH THE HÜCKEL THEORY?

The Hückel model represents historically one of the first steps towards a quantum mechanical description of chemical bonding and electronic structure. Even by the 1950s it was certainly being supplanted by more sophisticated models that took electron-electron interactions into account. Today, given the incredible advances and proliferation of quantum chemical methods, we wonder why we should bother with Hückel theory at all beyond its role as a pedagogic tool.

First, because the model depends only upon the topology and hence symmetry of the molecule, it provides an important way to interpret the rigorous and more detailed results we obtain from ab initio calculations. Certain effects and trends can be accounted for within the context of the model. Certainly, for π electronic systems, the insight gained from a Hückel model is very valuable. In fact, this is what separates a "theory" or model from an exact result. A good theory should be able to reproduce a physical result with a modest number of parameters. These parameters should be valid within some range such that the parameters for one physical system should be transferable to another related system. Finally, a good theory should be simple enough in its form to be readily understandable and based upon a logical set of arguments. The Hückel model has all these characteristics and this I believe is why, even after 70 or so years, the model remains central to our discussion of electronic structure.

8.3 ELECTRONIC STRUCTURE MODELS

The goal of electronic structure theory is to solve the Schrödinger equation for a set of interacting electrons and nuclei. Typically, this is done within the framework of a fixed geometry of the nuclei as justified by the Born–Oppenheimer approximation. Essentially, we assume that since the mass of an electron is on the order of 10^{-4}, that of a typical nucleus, we can separate the nuclear motions described by Q from the electronic motions q and write

$$H_{ele}(q; Q)\Psi(q; Q) = E_{ele}(Q)\Psi(q; Q) \qquad (8.74)$$

where Q in this equation reminds us that the electronic energy levels and associated wave functions depend parametrically upon the nuclear positions. Within a fixed frame, we can equivalently write this as

$$H_{ele} = \sum_{\sigma}\sum_{lm} h_{lm}a_{l\sigma}^{\dagger}a_{m\sigma} + \frac{1}{2}\sum_{klmn}\sum_{\sigma\rho}\langle kl|v|mn\rangle a_{k\sigma}^{\dagger}a_{l\rho}^{\dagger}a_{m\rho}a_{n\sigma} \qquad (8.75)$$

where $\{a_{i\sigma}^{\dagger}, a_{j\rho}\} = \delta_{ij}\delta_{\sigma\rho}$ are fermion operators that add or remove electrons from single-electron basis functions.

$$a_{k\sigma}^{\dagger}|0\rangle = |k_{\sigma}\rangle \quad \text{and} \quad a_{k\sigma}||k_{\sigma}\rangle = |0\rangle \qquad (8.76)$$

Also, since we can put only one electron with a given spin into a given basis function,

$$a_{k\sigma}^{\dagger}a_{k\sigma}^{\dagger}|0\rangle = 0 \qquad (8.77)$$

The first term in H_{ele} represents the single-electron terms and includes the kinetic energy and the electron-nuclear interactions. These are independent of spin,

$$h_{lm} = \langle l | \left(-\frac{1}{2}\nabla^2 - e^2 \sum_{n=1}^{N_a} \frac{Z_n}{|q - Q_n|} \right) | m \rangle \qquad (8.78)$$

The electron-electron repulsion is introduced by the second term. Here the $\langle ij|v|kl \rangle = \langle ij|kl \rangle$ bracket denotes the Coulomb integral

$$\langle ij|kl \rangle = \int dr_1^3 \int dr_2^3 \phi_i^*(1)\phi_j^*(2)\frac{e^2}{r_{12}}\phi_k(1)\phi_l(2) \qquad (8.79)$$

Be careful! A number of authors use different conventions for this. Here, adopt the notation more prevalent in the many-body physics literature that is consistent with the creation/annihilation operator formalism we are following. In the quantum chemical literature, the bracket $\langle ij|kl \rangle$ is taken to mean

$$\langle ij|kl \rangle = \int dr_1^3 \int dr_2^3 \phi_i(1)\phi_j(1)\frac{e^2}{r_{12}}\phi_k(2)\phi_l(2) \qquad (8.80)$$

and is sometimes denoted with a square bracket $[ij|kl]$ when working with spin orbitals. The connection is that

$$\langle ij|kl \rangle \equiv [ik|jl] \qquad (8.81)$$

for real orbitals. The reason the two notations have evolved is historical, with one camp adopting one notation and another camp adopting the other notation. The appendix explains the difference (but not the cause).

We can interpret each term in the electronic Hamiltonian as follows:

- $h_{22}a_{2\mu}^\dagger a_{2\mu}$ = energy to place an electron with spin μ into the ϕ_2 basis function
- $h_{23}a_{2\mu}^\dagger a_{3\mu}$ = energy to remove an electron from ϕ_3 and place it into ϕ_2
- $\langle 22|22 \rangle a_{2\alpha}^\dagger a_{2\beta}^\dagger a_{2\beta} a_{2\alpha}$ = Coulomb repulsion between a spin-up α electron and a spin-down β electron when both are in the ϕ_2 basis function

Notice that, in general, if we have N basis functions, we would need to perform N^4 six-dimensional integrals to completely account for all the electron-electron interactions.

8.3.1 HARTREE–FOCK APPROXIMATION

Writing the Hamiltonian and solving the requisite integrals is the first step. We are still a long way from a solution. By and large, most quantum chemical treatments use the Hartree–Fock (HF) approximation as solved by the self-consistent field (SCF) approach. The basic SCF procedure was first developed (independently) by Clemens Roothaan[88–90] and G. G. Hall[91] in the early 1950s. There are a number of derivations for these equations; one particularly elegant derivation involves the use of

Heisenberg equations of motion[92] where we write the time derivative of each electron operator as

$$i\hbar \frac{d}{dt} a_{s\mu} = [a_{s\mu}, H] \qquad (8.82)$$

This can be approximated by

$$i\hbar \frac{d}{dt} a_{s\mu} \approx \sum_u f_{su}^{\mu} a_{u\mu} \qquad (8.83)$$

If we allow the equality, then we can write

$$[a_{s\mu}, H] = \sum_u f_{su}^{\mu} a_{u\mu} \qquad (8.84)$$

Now, multiplying on left and on the right by $a_{t\mu}^{\dagger}$ and adding the two together,

$$\{a_{t\mu}^{\dagger}, [a_{s\mu}, H]\} = \sum_u f_{su}^{\mu} \{a_{t\mu}^{\dagger}, a_{u\mu}\} \qquad (8.85)$$

where we have assumed we are working in a orthonormal basis. Taking the anticommutator on the right-hand side and averaging over the electronic ground state of the system,

$$\langle \{a_{t\mu}^{\dagger}, [a_{s\mu}, H]\} \rangle = f_{st}^{\mu} \qquad (8.86)$$

f_{st}^{μ} is called the Fock operator. It is Hermitian and its eigenvalues and eigenvectors correspond to the energies and single-particle orbitals,

$$
\begin{aligned}
f_{st}^{\mu} &= \langle \{a_{t\mu}^{\dagger}, [a_{s\mu}, H]\} \rangle \\
&= -\langle \{a_{t\mu}^{\dagger}, [H, a_{s\mu}]\} \rangle \\
&= -\left\langle \left\{ a_{t\mu}^{\dagger}, \left[\sum_{\sigma} \sum_{lm} h_{lm} a_{l\sigma}^{\dagger} a_{m\sigma}, a_{s\mu} \right] \right\} \right\rangle \\
&\quad - \frac{1}{2} \left\langle \left\{ a_{t\mu}^{\dagger}, \left[\sum_{klmn} \sum_{\sigma\rho} \langle kl|nm \rangle a_{k\sigma}^{\dagger} a_{l\rho}^{\dagger} a_{m\rho} a_{n\sigma}, a_{s\mu} \right] \right\} \right\rangle \qquad (8.87)
\end{aligned}
$$

Working through the operator algebra yields the Fock operator:

$$f_{st}^{\mu} = h_{st} + \sum_{lm} \left\{ \sum_{\sigma} \langle ls|mt \rangle \langle a_{l\sigma}^{\dagger} a_{m\sigma} \rangle - \langle ls|tm \rangle \langle a_{l\mu}^{\dagger} a_{m\mu} \rangle \right\} \qquad (8.88)$$

The density matrix or "bond-charge" matrix is

$$\gamma_{ml}^{\mu} = \langle a_{l\mu}^{\dagger} a_{m\mu} \rangle \qquad (8.89)$$

The diagonal elements are electrons in a given basis function of a given spin, and the off-diagonal elements are how those electrons are shared between the different basis functions.

For a closed shell system in which all electrons are paired, we can equate the spin-up (α) densities with the spin-down (β) densities

$$\langle a_{l\alpha}^{\dagger} a_{m\alpha} \rangle = \langle a_{l\beta}^{\dagger} a_{m\beta} \rangle \tag{8.90}$$

Using this and the symmetry relations of the two-body matrix elements, we arrive at

$$f_{st}^{\mu} = h_{st} + \sum_{lm} (2\langle sl|tm\rangle - \langle sl|mt\rangle)\langle a_{l\mu}^{\dagger} a_{m\mu} \rangle \tag{8.91}$$

Notice that the Fock operator depends upon its own eigenstates since we can expand the density matrix in terms of the eigenvectors of the Fock operator

$$\psi_k^{\mu} = \sum_m c_{km}^{\mu} \phi_m^{\mu} \tag{8.92}$$

as

$$\gamma_{ml}^{\mu} = \sum_k \left(c_{kl}^{\mu}\right)^* c_{km}^{\mu} n_k^{\mu} \tag{8.93}$$

where the coefficients satisfy

$$\sum_m \left(f_{sm}^{\mu} - E_k S_{sm}\right) c_{km}^{\mu} = 0 \tag{8.94}$$

where S is the overlap matrix between different basis functions and n_k^{μ} is the occupancy of the kth eigenstate.

The total ground-state energy is not the sum over the energies of the occupied HF energy levels. In calculating the energy for orbital 1, E_1, we include the interaction between electrons 1 and 2, 1 and 3, and so on. Likewise for electron 2, we average over the interactions between 2 and 1, 2 and 3, 2 and 4, and so on. In other words, by summing

$$E = \sum_k E_k n_k \tag{8.95}$$

where $n_k = 0$ or 1 is the occupation number of state k, we have included the Coulombic interaction $\langle ij|ij\rangle$ term twice, so we need to subtract this out in writing the total ground-state energy,

$$E_{gs}^{HF} = \sum_k n_k (E_k - \sum_{j>k} \langle kj|kj\rangle) \tag{8.96}$$

Operationally, we "guess" at γ_{ml}^{μ} either by doing a low-level calculation or, simply by using Hückel theory, construct the Fock operator as a matrix in the basis of choice, and diagonalize this to obtain an improved set of orbital energies and single-particle orbitals. We then reconstruct the bond-charge matrix using these new orbitals and repeat this process until neither the energies, orbitals, nor bond-charge matrix changes to within a suitable tolerance. In performing either ab initio or semi empirical calculations, this part of the calculation usually takes the most time and one has no real guarantee on how many iterations are required to converge the Hartree–Fock equations. Depending upon the size of the system, the complexity of the basis set, and the speed of our computer, this can take anywhere from a few seconds to months or years. A good strategy is to start off with a low-level basis set, get a good guess at the bond-charge matrix, then repeat the procedure with increasingly more accurate basis sets. Most modern quantum chemical codes allow us to import the results (checkpoint file) from a previous calculation as a starting point.

8.3.2 VARIATIONAL DERIVATION OF THE HARTREE–FOCK APPROACH

We shall now rederive the Hartree–Fock equation using a more standard approach. The assumption made is that the ground-state wave function consists of a single Slater determinant, $|\phi\rangle$, and that the ground-state energy is obtained by minimizing $\langle\phi|H|\phi\rangle$ with respect to variations in the density matrix (that is, bond-charge matrix). Our generic electronic structure Hamiltonian has the form

$$H = \sum_{ij} h_{ij} a_i^\dagger a_j + \frac{1}{4} \sum_{ijkl} \langle ij|v|kl\rangle a_i^\dagger a_j^\dagger a_l a_k \tag{8.97}$$

where i, j, k, l label single-particle basis functions that we shall take to be orthogonal. If we assume the ground state is composed of a single-determinant $|\phi\rangle$, the energy expectation value is

$$E[\rho] = \langle\phi|H|\phi\rangle \tag{8.98}$$

$$= \sum_{ij} \langle i|h|j\rangle \rho_{ij} + \frac{1}{2} \sum_{ijkl} \langle ij||kl\rangle \rho_{ki} \rho_{lj} \tag{8.99}$$

where ρ is the single-particle density matrix with elements $\rho_{ij} = \langle\phi|a_i^\dagger a_j|\phi\rangle$. The density matrix satisfies the conditions $\rho^2 = \rho$ and $Tr[\rho] = N$. With this in mind, we can minimize $E[\rho]$ with respect to the density under the constraint that $\rho^2 = \rho$ remain satisfied,

$$\delta(E[\rho] - Tr\,\lambda(\rho^2 - \rho)) = 0$$

where Λ is a matrix of Lagrange multipliers. Expanding the variation

$$\left(\frac{\delta E[\rho]}{\delta\rho} - \rho\Lambda - \Lambda\rho + \Lambda \right) \delta\rho = 0$$

This must hold for all $\delta\rho$, so we send

$$F - \rho\Lambda - \Lambda\rho + \Lambda = 0 \tag{8.100}$$

and define Fock matrix F with elements:

$$F_{ij} = \frac{\delta E[\rho]}{\delta\rho}$$

The Λ's can be eliminated by multiplying Eq. (8.100) on the right and left by ρ and subtracting the two (taking $\rho^2 = \rho$ into account). One thus obtains

$$[F, \rho] = 0$$

In other words, the density matrix that minimizes the energy commutes with the Hartree–Fock Hamiltonian F.

We can generalize this by asking what happens to ρ if $[F, \rho] \neq 0$ or if H is a function of time. Let us consider the case where the system at all times is described by a single Slater determinant $|\psi(t)\rangle$ composed of orthonormal orbitals $\{|\phi_i\rangle\}$ such that the time-dependent density matrix is given by

$$\rho(t) = \sum_{i=1}^{N} |\phi_i(t)\rangle\langle\phi_i(t)| \tag{8.101}$$

We write the action in terms of the orbitals as

$$S = \int_0^t ds \left[i\hbar \sum_{i=1}^{N} \langle\phi_i|\dot{\phi}_i\rangle - E[\rho] - \sum_{ij} \lambda_{ij}\langle\phi_i|\phi_j\rangle \right] \tag{8.102}$$

Here, the dot denotes the time derivative, $E[\rho] = \langle\psi|H(t)|\psi\rangle$ is the Hartree–Fock energy functional, and the λ_{ij}'s are Lagrange multipliers introduced to ensure orthogonality of the orbitals.

We now minimize S with respect to the variations in the orbitals

$$\frac{\delta S}{\delta|\phi_i\rangle} = -i\hbar\langle\dot{\phi}_i| - \langle\phi_i|F - \sum_j \lambda_{ij}\langle\phi_j| = 0 \tag{8.103}$$

$$\frac{\delta S}{\delta\langle\phi_i|} = i\hbar|\dot{\phi}_i\rangle - F|\phi_i\rangle - \sum_j \lambda_{ij}|\phi_j\rangle = 0 \tag{8.104}$$

where $F = \delta H/\delta\rho$ is the Hartree–Fock Hamiltonian operator. Again, we can eliminate the λ's and find the equations of motion for the density matrix

$$i\hbar\frac{\partial\rho}{\partial t} = [F, \rho] \tag{8.105}$$

This is the *time-dependent Hartree–Fock equation*. In fact, the stationary Hartree–Fock equation can be viewed as a limiting case where $\dot{\rho} = 0$, describing a system at equilibrium. The time-dependent Hartree–Fock approach is useful for studying systems displaced from equilibrium or the response of a system to a time-dependent driving field and obtain transition amplitudes.

8.4   NEGLECT OF DIFFERENTIAL OVERLAP

The electron-electron interactions can be approximated in a number of ways as well. First, we note that the two-electron integral depends upon the overlap between two electron densities:

$$\langle ik|jl\rangle = \int dr_1^3 \int dr_2^3 [u_i(1)u_j(1)]\frac{e^2}{r_{12}}[u_l(2)u_k(2)] \qquad (8.106)$$

The integral will more or less vanish unless u_i and u_j are the same basis function. Thus, we can write:

$$[u_i(1)u_j(1)] = \delta_{ij}u_i^2(1) \qquad (8.107)$$

This approximation, termed CNDO for "complete neglect of differential overlap," reduces the total number of such two-electron integrals to $N(N+1)/2 \propto N^2$ from $(N(N+1)/2)(N(N+1)/2+1)/2 \propto N^4)$ that are included in ab initio calculations. Note that in making this approximation, we now have only electron–electron repulsions between an electron in basis u_i and an electron in basis u_j. Also, there are no spin-dependent interactions. Methods, such as the Pariser–Parr–Pople method (PPP) and CNDO/2 use the zero-differential overlap approximation. One can relax the zero-differential overlap approximation to some extent by instead writing

$$[u_i(1)u_j(1)] = \delta_{ci}\delta_{cj}u_i(1)u_j(1) \qquad (8.108)$$

where $\delta_{ci}\delta_{cj} = 0$ unless the two orbitals are on the same atomic center. Such intermediate neglect is used in the MNDO, PM3, and AM1 methods while methods such as the INDO, MINDO, ZINDO, and SINDO approaches do not apply the rule when all the orbitals in the integral are on the same atomic site.

The notion of neglect of differential overlap also plays a key role in the development of tight-binding treatments used in solid-state physics. In many ways, we can use the terms *tight-binding* and *semiempirical* interchangeably (although certain purists will argue otherwise). For solid-state systems we can often take advantage of the periodic nature of the lattice to develop recursion relations or treat the problem in reciprocal space to compute the band structure of a given system. For more complete treatments, any number of excellent texts on solid-state physics may be consulted.

If we further assume that

$$u_i(1)u_i(1)u_j(2)u_j(2) = \delta_{ij}(u_i(1)u_j(2))^2 \qquad (8.109)$$

and assume that $h_{ij} = 0$ unless $i = j$ or i is on atoms adjacent to j, we have the Pariser–Parr–Pople Hamiltonian:

$$H_{PPP} = \sum_\sigma \sum_i h_{ii}a_{i\sigma}^\dagger a_{i\sigma} + \sum_\sigma \sum_{ij}' t_{ij}a_{i\sigma}^\dagger a_{j\sigma} + \sum_i' U_i a_{i\uparrow}^\dagger a_{i\uparrow} a_{j\downarrow}^\dagger a_{j\downarrow}$$

$$= \sum_\sigma \sum_i h_{ii}n_{i\sigma} + \sum_\sigma \sum_{ij}' t_{ij}a_{i\sigma}^\dagger a_{j\sigma} + \sum_i' U_i \left(n_{i\uparrow} - \frac{1}{2}\right)\left(n_{i\downarrow} - \frac{1}{2}\right)$$

$$(8.110)$$

where $U_i = \langle ii|ii\rangle$, the prime on the second summation reminds us that we sum only atoms participating in the π-electron network, and $n_{i\uparrow}$ or $n_{i\downarrow}$ indicate the number of spin-up or spin-down electrons in a given basis orbital. The original purpose of the approach was to predict the electronic properties of organic dye molecules. In fact, when combined with a Hartree–Fock treatment of the electronic ground state, the PPP model does a remarkably good job of predicting the position and oscillator strengths of the lowest singlet transitions in many π-conjugated systems.

Unlike most semiempirical treatment, π-electron theories have a rigorous ab initio underpinning. H_{PPP} is in fact an approximate effective operator acting on the π-electronic subspace. Likewise, its parameters include effective electron correlation effects between the π system and the core. The connection between the PPP model and more rigorous approaches was explored by Freed and coworkers using diagrammatic techniques for solving multireference perturbation theory.[93-95]

Now, if all the sites are equivalent, we can write $t_{ij} = t$, $h_{ii} = \varepsilon$, and $U_i = U$ and arrive at the Hubbard Hamiltonian:[96]

$$H = \varepsilon \sum_{\sigma} \sum_{i} a_{i\sigma}^{\dagger} a_{i\sigma} + t \sum_{\sigma} \sum_{ij}' a_{i\sigma}^{\dagger} a_{j\sigma} + U \sum_{i}' a_{i\uparrow}^{\dagger} a_{i\uparrow} a_{j\downarrow}^{\dagger} a_{j\downarrow} \quad (8.111)$$

This can be further reduced by noticing that the first term is simply the sum over the number operators and as such is equal to εN. This can be removed by choosing our energy origin so that $\varepsilon = 0$. Next, the interaction term can be also written in terms of the electron number operators and we arrive at

$$H = t \sum_{\sigma} \sum_{ij}' a_{i\sigma}^{\dagger} a_{j\sigma} + U \sum_{i}' \left(n_{i\uparrow} - \frac{1}{2} \right) \left(n_{j\downarrow} - \frac{1}{2} \right) \quad (8.112)$$

This particular model is seldom used in the chemical field but widely used in the solid-state physics community for describing electrons in narrow band-gap materials and has been applied to problems such as high-T_c superconductivity, band magnetism, and the metal-insulator transition. For small systems, one can perform numerically exact calculations, and the code for doing so is included on the disk accompanying this text with some representative results shown in Figure 8.8 for the case of a spin-paired four-electron/four-site model of 1,3-butadiene.

The Hubbard model can be solved exactly in one dimension as shown by Lieb and Wu[97] in 1968 using the Bethe Ansatz technique. This solution was later shown to be the complete solution by Essler et al. in 1991.[98] One of the most significant predictions of the model is that there is an absence of a "Mott transition" between conducting and insulating states as the strength of the interaction U is increased.

8.5   AN EXACT SOLUTION: INDO TREATMENT OF ETHYLENE

There are very few systems that can be treated exactly—most of which are poor representations of the actual physical systems; however, we can use models to test the limits and range validity of our theoretical approaches. Oftentimes, one can learn more about a broad class of physical systems by using models and then performing more detailed calculations to get a better match to the experimental data.

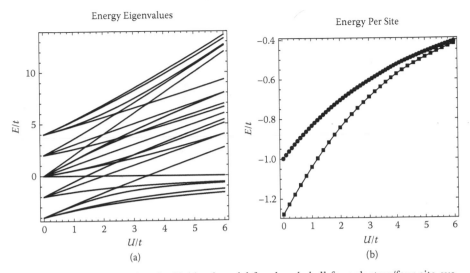

FIGURE 8.8 Exact solution for Hubbard model for closed-shell four-electron/four-site system. (a) Energy eigenvalues for the 36×36 configuration system. (b) Total ground-state energy per site.

Let us consider a solvable two-electron problem to illustrate what we have developed thus far. Consider a two-atom/two-electron system described by the Hubbard Hamiltonian as would be the case for the *pi* system for ethene. The orbital basis ϕ_1 and ϕ_2 we take as the C $2p_z$ orbitals centered about each carbon atom and assume that we can work within the NDO approximation and treat the basis functions as orthogonal.

We have a total of six possible two-electron configurations:

$$\phi_1 = a_1^\dagger \overline{a_2}^\dagger |0\rangle$$
$$\phi_2 = a_2^\dagger \overline{a_1}^\dagger |0\rangle$$
$$\phi_3 = a_1^\dagger \overline{a_1}^\dagger |0\rangle$$
$$\phi_4 = a_2^\dagger \overline{a_2}^\dagger |0\rangle$$
$$\phi_5 = a_1^\dagger a_2^\dagger |0\rangle$$
$$\phi_6 = \overline{a_1}^\dagger \overline{a_2}^\dagger |0\rangle \qquad (8.113)$$

where ϕ_1 is the case where we have a spin-up electron on atom 1 and a spin-down electron on atom 2 (as denoted by the overbar), ϕ_2 is the reverse where the spin-down electron is on atom 1 and the spin-up electron is on atom 2. ϕ_3 and ϕ_4 are the cases where we have both spin-paired electrons on either atom 1 or atom 2. Lastly, ϕ_5 and ϕ_6 are the cases where the two spins are parallel but on different atoms as indicated in the diagram below.

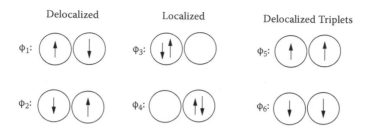

Using this as a basis, we can construct the following Hamiltonian matrix:

$$
H = \begin{bmatrix}
2\varepsilon & 0 & t & t & 0 & 0 \\
0 & 2\varepsilon & t & t & 0 & 0 \\
t & t & 2\varepsilon + U & 0 & 0 & 0 \\
t & t & 0 & 2\varepsilon + U & 0 & 0 \\
0 & 0 & 0 & 0 & 2\varepsilon & 0 \\
0 & 0 & 0 & 0 & 0 & 2\varepsilon
\end{bmatrix}
\tag{8.114}
$$

where we have included ε as the local site energy. Since we are working with fermion operators already, antisymmetry is already enforced within our basis states so we do not need to construct separate exchange and Coulomb contributions to the total energy.

Notice that only in the case where there is spin pairing does the electron/electron Coulomb coupling term enter into H. Furthermore, notice that the two parallel spin configurations are completely decoupled from the other antiparallel configurations and H can be block-diagonalized along these lines with ϕ_5 and ϕ_6 being eigenstates of H each with energy 2ε. The remaining eigenstates are symmetric and antisymmetric linear combinations of the remaining four basis states. Again, we notice that ϕ_1 and ϕ_2 are decoupled, as are ϕ_3 and ϕ_4. Thus, we can form a suitable basis by taking even and odd linear combinations of the two types of states,

$$
\phi_{ag} = (\phi_1 + \phi_2)/\sqrt{2}
\tag{8.115}
$$

$$
\phi_{bg} = (\phi_3 + \phi_4)/\sqrt{2}
\tag{8.116}
$$

$$
\phi_{au} = (\phi_1 - \phi_2)/\sqrt{2}
\tag{8.117}
$$

$$
\phi_{bu} = (\phi_3 - \phi_4)/\sqrt{2}
\tag{8.118}
$$

where u and g indicated odd (ungerade) and even (gerade) linear combinations. Within the gerade basis, H is again block-diagonal with eigenvalues

$$
E_{\pm}^{s} = 2\varepsilon + U/2 \pm \sqrt{U^2 + 16t^2}/2
\tag{8.119}
$$

The ungerade eigenstates of H with energies 2ε and $2\varepsilon + U$, respectively. These two states are clearly antibonding and we notice that ϕ_{a-} is of the triplet states.

We are typically interested in the ground electronic state of a given system. For a closed shell system such as this, the ground electronic state is the totally symmetric gerade state

$$\Psi_{gs} = (\cos\theta\phi_{ag} + \sin\theta\phi_{bg}) \tag{8.120}$$

where θ is the mixing angle that mixes delocalized (ϕ_{ag}) and localized (ϕ_{bg}) electronic configurations. The ground-state energy is then

$$E_{gs} = 2\varepsilon + U/2 - \sqrt{(U/2)^2 + 4t^2} \tag{8.121}$$

Recalling Chapter 2, we can write the mixing angle as

$$\tan 2\theta = \frac{4t}{U} \tag{8.122}$$

For the case of small U, the Coulomb repulsion between electrons is very small relative to the hopping, and the delocalized configuration ϕ_{a+} dominates. In the limit of no interaction, $E_{gs} = 2\varepsilon - 2|t|$, which is what we expect from a Hückel treatment. As U becomes large compared to t, hopping is insignificant and the ground state is dominated by the localized configurations ϕ_{b+} with an asymptotic energy of $2\varepsilon + U$, which is the energy necessary to have two electrons on the same atom with Coulomb interaction U. In the exact case, we have perfect correlation between the two asymptotic limits. If you imagine that t is a function of the bond distance, then we have a smooth potential energy surface connecting the bound molecular system to the dissociated atom limit. This is not always the case when we make approximations to the system.

8.5.1 HF Treatment of Ethylene

Within the Hückel model, the single-electron term is given by

$$h = \begin{bmatrix} h_{11} & h_{12} \\ h_{21} & h_{22} \end{bmatrix} \tag{8.123}$$

with $h_{11} = h_{22} = \alpha$ and $h_{12} = h_{21}^* = t$ as per the Hückel notation. In the localized basis, matrix elements of the Fock operator are

$$f_{11}^\mu = h_{11} + \sum_{lm} \langle a_{l\mu}^\dagger a_{m\mu}\rangle(2\langle l1|m1\rangle - \langle l1|1m\rangle \tag{8.124}$$

$$f_{22}^\mu = h_{22} + \sum_{lm} \langle a_{l\mu}^\dagger a_{m\mu}\rangle(2\langle l2|m2\rangle - \langle l2|2m\rangle \tag{8.125}$$

$$f_{12}^\mu = h_{12} + \sum_{lm} \langle a_{l\mu}^\dagger a_{m\mu}\rangle(2\langle l1|m2\rangle - \langle l1|2m\rangle \tag{8.126}$$

$$f_{21}^\mu = f_{12}^\mu \tag{8.127}$$

While we can work in the localized basis, it is more convenient to take the Hückel ground-state molecular orbital as a trial wave function for performing the Hartree–Fock calculation. This we can find by diagonalizing h and forming the molecular orbitals

$$\phi_\pm = \frac{1}{\sqrt{2}}(\phi_1 \pm \phi_2) \tag{8.128}$$

with energies

$$\varepsilon_\pm = \alpha \mp |t| \tag{8.129}$$

Thus, we form the trial solution to the Hartree–Fock equations by writing the ground-state wave function as

$$\psi_{gs} = a_{+\beta}^\dagger a_{+\alpha}^\dagger |0\rangle \tag{8.130}$$

Expanding this in the local orbital basis,

$$\psi_{gs} = \frac{1}{2}\left[\left(a_{1\beta}^\dagger a_{1\alpha}^\dagger + a_{2\beta}^\dagger a_{2\alpha}^\dagger\right) + \left(a_{1\beta}^\dagger a_{2\alpha}^\dagger + a_{2\beta}^\dagger a_{1\alpha}^\dagger\right)\right]|0\rangle \tag{8.131}$$

where the first term represents a symmetric linear combination of the two possible ionic configurations and the second term is a symmetric linear combination of the two covalent configurations. Writing the Fock matrix in the molecular orbital basis results in

$$F = \begin{bmatrix} \varepsilon_+ + \langle ++ | ++ \rangle & 0 \\ 0 & \varepsilon_- + 2\langle -+ | -+ \rangle - \langle -+ | +- \rangle \end{bmatrix} \tag{8.132}$$

No further refinement is needed, and we find the single-electron (Hartree–Fock) orbital energies to be

$$\varepsilon_1^{hf} = \varepsilon_+ + \langle ++ | ++ \rangle \tag{8.133}$$

$$\varepsilon_2^{hf} = \varepsilon_- + 2\langle -+ | -+ \rangle - \langle -+ | +- \rangle \tag{8.134}$$

Finally, we write the total ground-state energy as

$$E_{gs} = 2h_{++} + \langle ++ | ++ \rangle = 2\varepsilon - 2|t| + \langle ++ | ++ \rangle \tag{8.135}$$

Evaluating the electron-electron (e-e) interaction, we find that the only terms that survive are those involving the Coulomb interactions in the ionic configurations:

$$\langle ++ | ++ \rangle = \frac{1}{4}(\langle 11|11 \rangle + \langle 22|22 \rangle) = \frac{1}{2}U \tag{8.136}$$

which we take to be the same for both atoms:

$$U = \int\int dr_1 dr_2 |\phi_1(r_1)|^2 |\phi_1(r_2)|^2 \frac{e^2}{r_{12}} \tag{8.137}$$

Hartree–Fock limit. We previously derived this result:

$$E_{hf} = 2\varepsilon + U/2 - 2|t| \tag{8.138}$$

Here we see a linear variation in the ground-state energy with increasing e-e interaction. When U/t is small (large t or small U) the Hartree–Fock energy approaches the exact energy from above. While the HF energy does asymptotically approach the

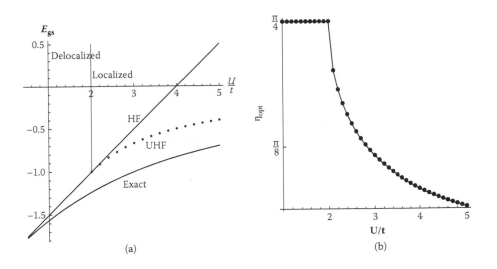

FIGURE 8.9 (a) Comparison between Hartree–Fock, unrestricted Hartree–Fock, and the exact ground-state energy for a diatomic model using the Hubbard model. For $U/t < 2$, the UHF and HF results are identical. (b) UHF mixing angle η vs. coupling. At $U/t = 2$ the derivative of the UHF energy with respect to the coupling is discontinuous, indicating a sudden transition from localized to delocalized states.

exact value in the limit of $U \to 0$, it fails to reproduce the dissociated atom limit as plotted in Figure 8.9.

We can extend this a bit further by adopting an *unrestricted Hartree–Fock* approach by constructing a variational wave function mixed spin configuration:

$$\phi_\alpha = \sin\eta|1\uparrow\rangle + \cos\eta|2\uparrow\rangle \tag{8.139}$$

$$\phi_\beta = \cos\eta|1\downarrow\rangle + \sin\eta|2\downarrow\rangle \tag{8.140}$$

The coefficients are such that the segregation of the electrons can be controlled by varying η. Using these one-electron states as a trial wave function, the UHF energy is

$$E_{uhf}(\eta) = 2\varepsilon + (2\varepsilon + U)\cos^2\eta\sin^2\eta + t\sin(2\eta) \tag{8.141}$$

which reduces to the HF value for $\eta = \pi/4$. The UHF energy is the case where $E_{uhf}(\eta)$ has a minimum value. Solving $dE_{uhf}(\eta)/d\eta = 0$ for various values of U yields the variation of the UHF energy with U. The resulting E_{uhf} energy is plotted in Figure 8.9 along with a plot of the optimal mixing angle. Here we notice that for $U/t \leq 2$, the HF and UHF energies are identical with $\eta_{opt} = \pi/4$, indicating that only delocalized configurations contribute to the ground state. At $U/t = 2$ the first derivative of E_{uhf} is discontinuous with respect to the coupling. This is indicative of a sudden and discontinuous change in the ground state to include localized configurations. Asymptotically, as $\eta \to 0$, the UHF converges to the exact ground-state energy. While the UHF does incorporate the physically correct behavior of charge segregation, it does it in a discontinuous way.

8.6   AB INITIO TREATMENTS

The goal of ab initio quantum chemistry is to evaluate the three- and six-dimensional integrals properly by assuming some finite set of spatial basis functions, $\{u_j\}$. Typically the basis functions are chosen to be a set of atomic-centered Slater or Gaussian-type orbitals. Slater-type orbitals were among the first basis sets to be introduced in the early development of quantum chemistry and are based upon the hydogenic atomic orbitals[99] centered about the atomic sites. They have the radial form

$$R_n(r) = Nr^{n-1}e^{-\zeta r} \tag{8.142}$$

as given above and are generally considered to be a more accurate basis for a given number of basis functions. However, because the radial parts are based upon exponential forms, it becomes compuationally costly to compute the various multicentered integrals required to evaluate H_{ele}.

In the early 1950s Boys[100] introduced the use of Gaussian-type orbitals—although there is some evidence that Roy McWeeny was using them as early as 1946. Gaussian-type orbitals have a radial part

$$R_n(r) = Nr^{n-1}e^{-\zeta r^2} \tag{8.143}$$

Although the Gaussian-type orbitals (GTO) are not as closely matched to the actual atomic orbitals as the STOs and generally more GTOs are needed to achieve comparable accuracy, the advantage gained is a tremendous speedup in the evaluation of multicentered integrals. This is gained by the use of the "Gaussian product theorem" whereby the integral over two Gaussian orbitals centered on different atoms is a finite sum of Gaussians centered on a point along the axis connecting them. Hence, any four-centered integral reduces to a sum of two-centered integrals and any two-centered integrals become a finite number of single-centered integrals. This facilitates a speedup on the order of 4–5 orders of magnitude.

Currently there are literally hundreds of basis sets composed of Gaussian-type orbitals. A minimum basis set is one in which, on each atom in the molecule, a single basis function is used for each orbital in a Hartree–Fock calculation on the free atom. However, often additional basis functions are added. For example for the Li atom, which would require the 1s and 2s orbitals, one adds a complete set of 2p orbitals. Thus, for the first row on the periodic table, one has a basis of 2s functions and 3p functions for a total of 5 basis functions per atom. The most common minimal bases is the STO-nG basis where the n integer represents the number of Gaussian primitive functions comprising a single basis function. It is usually a good idea to start off using a minimal basis and then improve the calculation using a more accurate (hence larger) basis. Commonly used minimal basis sets in most ab initio codes are: STO-3G, STO-4G, STO-6G, and STO-3G*, the latter being a polarized version of the STO-3G basis.[101]

One of the problems with the minimal bases is that they are quite inflexible and are unable to adjust to the different electron densities encountered in forming chemical bonds. In chemical bonding, it is typically the outer or valence electrons that take part in forming bonds. Consequently, split-valence basis sets were developed to account for this fact by allowing each valence orbital to be composed of a fixed linear combination of primitive Gaussian functions. These are termed valence double, triple, or quadruple-zeta basis sets according to the number of Gaussians used. These are denoted using Pople's scheme via X-YZg where X is the number of primitive Gaussians composing each atomic core basis function and Y and Z denote how the valence orbitals are split. In this case, we would have a split-valence double-zeta basis. Split-valence triple- and split-valence quadruple-zeta basis sets are likewise termed X-YZWg and X-YZWVg, respectively.

Additionally, these basis sets may include polarization functions as denoted by an asterisk (*). Two asterisks (**) indicate that polarization functions are also added to the light atoms (H and He). In a minimal basis, only the 1s orbital would be used for these two atoms. In this case, adding polarization functions would involve including a p function about the light atom. The addition of polarization functions allows the electron density about these atoms to be more asymmetric about the atom center. Similarly d- and even f-type orbitals can be added. The more precise notation is to indicate exactly how many polarization functions were added such as (p,d).

One can also include diffuse functions to light atoms, as denoted by a + sign. Here one adds additional Gaussian functions that are very broad so as to more accurately represent the "tail" of the atomic orbits as one moves farther away from the atomic center. Such functions are necessary to include when performing calculations on anions or Rydberg states.

Finally, one can include correlation consistent basis functions. These were developed by Thom Dunning and co-workers and are designed to converge to the complete basis set limit using extrapolation techniques. These are the "cc-pVNZ" basis functions (for correlation consistent polarized) where $N = $ D,T, Q, 5, 6, ... for doubles, triples, and so on, and V denotes that they are valence-only basis sets. Such basis sets are currently considered state of the art and are widely used for post-Hartree–Fock calculations.

Needless to say, ab initio treatments are appealing since one strictly deals with the basic physical interactions between the electron and the nuclei. There is some art in designing basis functions, and certainly a lot of thoughtful computational design is required to both set up and solve the many-body problem. A number of standard implementations and codes are available, some at no cost and some at low cost for academic users. As better codes, better basis sets, and better theoretical techniques are developed, we also have nearly parallel progress in the amount of computational horsepower available. Consequently, armed with a modest modern workstation, we can perform quite accurate ab initio level calculations with rather large basis-set systems up to about 100 to 300 carbon atoms.

8.7 CREATION/ANNHILIATION OPERATOR FORMALISM FOR FERMION SYSTEMS

In this chapter, we have drawn heavily upon the use of fermion creation and annhiliation operators in describing the various models for electronic structure. The advantage of adopting such a formalism is that the Pauli exclusion principle and antisymmetrization rule are immediately enforced. Here we briefly summarize their properties.

An N-fermion wave function is a function of N coordinates describing the position of each fermion:

$$\langle r_1, \ldots, r_N | \psi \rangle = \psi(r_1, \ldots, r_N) \tag{8.144}$$

This is a normalizable function since

$$\int_{\infty}^{\infty} dr_1 \cdots dr_N |\psi(r_1, \ldots, r_N)|^2 < +\infty \tag{8.145}$$

and

$$P(r_1, \ldots, r_N) = dr_1 \cdots dr_N |\psi(r_1, \ldots, r_N)|^2 \tag{8.146}$$

gives the joint probability of finding electron 1 at r_1, electron 2 at r_2, and so on. Since the physics implied by this cannot depend upon how we arbitrarily assign the labels to the electrons (being identical particles), swapping labels should not affect the physics. Thus,

$$P(r_1, r_2, \ldots, r_N) = P(r_2, r_1, \ldots, r_N) \tag{8.147}$$

This introduces an ambiguity into the wave function since swapping particles can then only change the total phase of the wave function

$$P_{12}\psi(r_1, r_2, \ldots, r_N) = \psi(r_2, r_1, \ldots, r_N) = e^{i\delta}\psi(r_1, r_2, \ldots, r_N) \tag{8.148}$$

Repeated operations of the permutation operator will then introduce a phase change $n\delta$:

$$P_{ij}^n \psi(r_1, r_2, \ldots, r_N) = e^{in\delta}\psi(r_1, r_2, \ldots, r_N) \tag{8.149}$$

Consequently, one can readily conclude that δ can take one of two possible values: $\delta = \pi$ or $\delta = 0$. For the case of $\delta = \pi$, each binary permutation of indices changes the sign of the wave function while for $\delta = 0$, swapping particle indices has no effect on the sign of the wave function. From relativistic considerations, it can be shown that particles with half-integer spin (fermions) must have $\delta = \pi$ while particles with integer spin (bosons) must have $\delta = 0$. Thus, for electrons,

$$\psi(r_{p1}, r_{p2}, \ldots, r_{pN}) = (-1)^P \psi(r_1, r_2, \ldots, r_N) \tag{8.150}$$

where P is the binary transpositions required to return the permutation $\{p1, p2, p3, \ldots, pN\}$ to its original form. For a two-electron system,

$$\psi(1, 2) = (-1)\psi(2, 1) \tag{8.151}$$

and

$$\psi(1, 2, 3) = -\psi(2, 1, 3) = +\psi(2, 3, 1) \tag{8.152}$$

Since all of these wave functions are for all intents and purposes identical in terms of their physical content, we need to sum over all possible permutations in constructing the final fermionic state. Let us define $\hat{P}_F$ as the antisymmetrization operator that acts upon an N fermion state to produce an antisymmetric state

$$\hat{P}_F \psi(1, \ldots, N) = \frac{1}{N!} \sum_p (-1)^p \psi(p1, \ldots, pN) \tag{8.153}$$

This is a Hermitian operator since

$$P_F \psi(1, 2) = \frac{1}{2}(\psi(1, 2) - \psi(2, 1)) \tag{8.154}$$

It also acts as a projection operator since by operating twice it is equivalent to operating once:

$$P_F P_F \psi(1 \cdots N) = \frac{1}{N!} \frac{1}{N!} \sum_p \sum_{p'} (-1)^{p+p'} \psi(pp'1, pp'2, \ldots, pp'N) \tag{8.155}$$

$$= \frac{1}{N!} \sum_p \left(\frac{1}{N!} \sum_q (-1)^q \psi(q1, q2, \cdots qN) \right) \tag{8.156}$$

$$= \frac{1}{N!} \sum_p P_F \psi(1, \ldots, N) \tag{8.157}$$

$$= P_F \psi(1, \ldots, N) \tag{8.158}$$

This N-particle wave function is a vector in a N-dimensional Hilbert space $\mathscr{H}_N$, which is direct product space of N single-particle Hilbert spaces $\mathscr{H}$,

$$\mathscr{H}_N = \mathscr{H} \otimes \mathscr{H} \otimes \mathscr{H} \otimes \ldots \tag{8.159}$$

each spanned by an orthonormal basis of vectors $\{|\phi_i\rangle\}$. Thus, we can construct an appropriate basis for the N fermion state using

$$|\phi_1 \cdots \phi_N) = |\phi_1\rangle \otimes \cdots |\phi_N\rangle \tag{8.160}$$

For notation, we use the rounded ket,), to denote a nonsymmetrized direct product basis ket. Similarly, we use a curly bracket | } to denote a properly antisymmetrized and normalized ket

$$|\phi_1 \phi_2 \cdots \phi_N\} = \sqrt{N!} P_F |\phi_1 \phi_2 \cdots \phi_N) \tag{8.161}$$

$$= \frac{1}{\sqrt{N!}} \sum_p (-1)^p |\phi_{p1} \phi_{p2} \cdots \phi_{pN}) \tag{8.162}$$

For example, for a two-electron system, we write

$$|\phi_1, \phi_2\} = \frac{1}{\sqrt{N!}}(|\phi_1 \phi_2) - |\phi_2 \phi_1)) \tag{8.163}$$

This state obeys antisymmetry since applying the permutation operator yields

$$|\phi_1, \phi_2\} = -|\phi_2, \phi_1\} \tag{8.164}$$

Also, the Pauli principle is automatically enforced since

$$|\phi_1, \phi_1\} = \frac{1}{\sqrt{N!}}(|\phi_1\phi_1) - |\phi_1\phi_1)) = 0 \qquad (8.165)$$

In other words, putting two particles into the same orbital results in a vanishing wave function.

A second notational convention we shall adopt is that $|\phi_1\phi_2)$ reads "electron #1 is in single-particle orbital #1 and electron #2 is in single-particle orbital #2" while $|\phi_2\phi_1)$ reads "electron #1 is in single-particle orbital #2 and electron #2 is in single-particle orbital #1." That is,

$$(1, 2|\phi_1\phi_2) = \phi_1(1)\phi_2(2) \qquad (8.166)$$

while

$$(1, 2|\phi_2\phi_1) = \phi_2(1)\phi_1(2) \qquad (8.167)$$

Within the orbital basis we can write the antisymmetrtized state as a Slater determinant

$$|\phi_1 \cdots \phi_N\} = \frac{1}{\sqrt{N!}} \begin{vmatrix} \phi_1(1) & \phi_2(1) & \cdots & \phi_N(1) \\ \phi_1(2) & \phi_2(2) & \cdots & \phi_N(2) \\ \vdots & \vdots & \cdots & \vdots \\ \phi_1(N) & \phi_2(N) & \cdots & \phi_N(N) \end{vmatrix} \qquad (8.168)$$

Determinants automatically satisfy the antisymmetry requirement since swapping any two rows or any two columns results in a change of sign. Likewise, if any two (or more) columns or any two (or more) rows are identical to within a constant factor, the determinant vanishes.

Needless to say, the bookkeeping in this looks to be quite painful! However, we can simplify matters considerably by introducing operators that add or remove single electrons from the antisymmetrized state. Let

$$a_\lambda^\dagger|\lambda_1 \cdots \lambda_N\} = |\lambda\lambda_1 \cdots \lambda_N\} \qquad (8.169)$$

add an electron to basis orbital λ. Normalizing this (as per the harmonic oscillator operators)

$$a_\lambda^\dagger|\lambda_1 \cdots \lambda_N) = \sqrt{1 + n_\lambda}|\lambda\lambda_1 \cdots \lambda_N) \qquad (8.170)$$

where n_λ is the occupation number of orbital λ in $|\lambda_1 \cdots \lambda_N)$. Since the Pauli principle puts a limit that $n_\lambda = 0$ or 1, we have

$$a_\lambda^\dagger|\lambda_1 \cdots \lambda_N\} = \begin{cases} |\lambda\lambda_1 \cdots \lambda_N\} & \text{if } |\lambda) \text{ is not in } |\lambda_1 \cdots \lambda_N\} \\ 0 & \text{otherwise} \end{cases} \qquad (8.171)$$

Any basis vector $|\lambda_1 \cdots \lambda_N)$ or $|\lambda_1 \cdots \lambda_N\}$ can be constructed by repeated applications of the creation operator on the vacuum state $|0)$. Thus,

$$|\lambda_1 \cdots \lambda_N\} = a_{\lambda_1}^\dagger a_{\lambda_2}^\dagger \cdots a_{\lambda_N}^\dagger|0) \qquad (8.172)$$

and

$$|\lambda_1 \cdots \lambda_N\rangle = \frac{1}{\sqrt{n_1! \cdots n_N!}} a_{\lambda_1}^\dagger a_{\lambda_2}^\dagger \cdots a_{\lambda_N}^\dagger |0\rangle \qquad (8.173)$$

One must be careful in using these operators since

$$a_\lambda^\dagger a_\mu^\dagger |0\rangle = |\lambda\mu\} = -|\mu\lambda\} = -a_\mu^\dagger a_\lambda^\dagger |0\rangle \qquad (8.174)$$

Thus, we can conclude that

$$a_\mu^\dagger a_\lambda^\dagger + a_\lambda^\dagger a_\mu^\dagger = 1 \qquad (8.175)$$

or

$$\{a_\mu^\dagger, a_\lambda^\dagger\} = 1 \qquad (8.176)$$

In other words, the fermion creation operators anticommute.

The creation operators are also not self-adjoint, and we have the following relations:

$$a_\lambda |0\rangle = 0 \qquad (8.177)$$

$$\langle 0| a_\lambda^\dagger = 0 \qquad (8.178)$$

$$a_\lambda |\lambda\rangle = \sqrt{n_\lambda} |\lambda - 1\rangle \qquad (8.179)$$

In other words, the a_λ removes a fermion from orbital λ if this state is occupied to begin with. To see how this works on the antisymmetrized many-body state, consider

$$a_\lambda |\beta_1 \cdots \beta_N\} = \sum_P^\infty \frac{1}{P!} \sum_{\alpha_1 \cdots \alpha_P} \{\alpha_1 \cdots \alpha_P | a_\lambda |\beta_1 \cdots \beta_N\} |\alpha_1 \cdots \alpha_P\} \qquad (8.180)$$

$$= \sum_P^\infty \frac{1}{P!} \sum_{\alpha_1 \cdots \alpha_P} \{\lambda\alpha_1 \cdots \alpha_P | a_\lambda |\beta_1 \cdots \beta_N\} |\alpha_1 \cdots \alpha_P\} \qquad (8.181)$$

where P denotes the number of particles in a given state and the inner sums are over all P-particle basis states. Clearly, only terms with $P = N - 1$ and $(\lambda\alpha_1 \cdots \alpha_P)$ equal to some permutation of $(\beta_1 \cdots \beta_N)$ will contribute to the sum. Thus, we can write

$$a_\lambda |\beta_1 \cdots \beta_N\} = \sum_{i=1}^N (-1)^{i-1} \delta_{\lambda\beta_i} |\beta_1 \cdots \beta_{i-1}\beta_{i+1} \cdots \beta_N\} \qquad (8.182)$$

$$= \sum_{i=1}^N (-1)^{i-1} \delta_{\lambda\beta_i} |\beta_1 \cdots \hat{\beta}_i \cdots \beta_N\} \qquad (8.183)$$

where the hat denotes that an electron has been removed from orbital β_i. Thus, for fermions

$$a_\lambda |\beta_1 \cdots \beta_N\} = \begin{cases} (-1)^{i-1} |\beta_1 \cdots \hat{\beta}_\lambda \cdots \beta_N\} & \text{if } |\lambda\rangle \text{ is occupied} \\ 0 & \text{otherwise} \end{cases} \qquad (8.184)$$

Now to close the algebra. Consider the action of two operators on the vacuum state:

$$a_\lambda a_\mu^\dagger |0\rangle = \delta_{\lambda\mu} |0\rangle = a_\mu a_\lambda^\dagger |0\rangle \tag{8.185}$$

Thus, we can conclude

$$\{a_\lambda, a_\mu^\dagger\} = \delta_{\lambda\mu} \tag{8.186}$$

which gives the fermion anticommutation bracket. If we act upon an already occupied state, then

$$a_\mu^\dagger a_\lambda |\alpha_1 \cdots \alpha_N\rangle = \sum_{i=1}^{N} (-1)^{i-1} \delta_{\lambda\alpha_i} |\mu\alpha_1 \cdots \hat{\alpha}_i \cdots \alpha_N\} \tag{8.187}$$

Thus, we conclude

$$a_\lambda a_\mu^\dagger |\alpha_1 \cdots \alpha_N\} = \left(\delta_{\lambda\mu} - a_\mu^\dagger a_\lambda\right) |\alpha_1 \cdots \alpha_N\} \tag{8.188}$$

This last expression is extremely useful since typically we write operators in *normal ordered* form in which all creation operators are to the left of the annihilation operators. This then (typically) reduces the operator into terms involving occupation numbers $n_\lambda = a_\lambda^\dagger a_\lambda$ and integrals evaluated over the basis functions.

General rules: Let us summarize the general operational rules for the fermion operators:

- Many-particle state: $|ijkl\rangle$
- Antisymmetrized, many-particle state: $|ijkl\}$
- Normalized antisymmetrized, many-particle state: (that is, Slater determinant state) $|ijkl\rangle$
- Vacuum: $|0\rangle$ is defined as a state with no particles. It is normalized so that $\langle 0|0\rangle = 1$. Antisymmetrized states are generated by acting on the vacuum ket with the creation operator.

$$a_i^\dagger a_j^\dagger a_k^\dagger |0\rangle = |ijk\rangle \tag{8.189}$$

or by acting on the vacuum bra using the annihilation operator

$$\langle 0| a_i a_j a_k = \langle ijk| \tag{8.190}$$

The vacuum is recovered by the reverse operation

$$|0\rangle = a_k a_j a_i |ijk\rangle = a_k a_j a_i |A\rangle \tag{8.191}$$

and

$$\langle 0| = \langle A| (a_k a_j a_i)^\dagger = \langle A| a_i^\dagger a_j^\dagger a_k^\dagger \tag{8.192}$$

- When acting on an antisymmetrized state

$$a_i^\dagger |jkl \cdots\rangle = |ijkl \cdots\rangle \qquad (8.193)$$

and

$$a_i |ijkl \cdots\rangle = |\bar{i} jkl \cdots\rangle \qquad (8.194)$$

where the overline indicates an electron has been removed from orbital i.
If orbital i was occupied, the result is

$$a_i |\lambda jkl \cdots\rangle = |jkl \cdots\rangle \qquad (8.195)$$

otherwise

$$a_i |jkl \cdots\rangle = 0 \qquad (8.196)$$

- Order is important:

$$a_k^\dagger a_l^\dagger |0\rangle = |kl\rangle \qquad (8.197)$$

while

$$a_l^\dagger a_k^\dagger |0\rangle = |lk\rangle = -|kl\rangle \qquad (8.198)$$

- $a_i \times a_i = a_i^\dagger \times a_i^\dagger = 0$
- Anticommutation: $\{a_i, a_j^\dagger\} = \delta_{ij}$. This allows us to swap creation and annihilation operators for different orbitals $(i \neq j)$ provided we change the sign: $a_i^\dagger a_j = -a_j a_i^\dagger$. If we are dealing with the same orbital, then $a_i a_i^\dagger = 1 - a_i^\dagger a_i$.
- Overlap between two states $\langle A|B \rangle$ is evaluated by writing out creation/annihilation operators and rearranging to normal ordering. For example, if $|A\rangle = |ij\rangle$ and $|B\rangle = |kl\rangle$, $\langle A|B \rangle = \langle 0|(a_i^\dagger a_j^\dagger)^\dagger a_k^\dagger a_l^\dagger |0 >$. This we evaluate by using the symmetry and commutation rules as follows:

$$\langle A|B \rangle = \langle 0|a_j a_i a_k^\dagger a_l^\dagger |0\rangle \qquad (8.199)$$

$$= \langle 0|a_j(\delta_{ik} - a_k^\dagger a_i)a_l^\dagger |0\rangle$$

$$= \delta_{ik}\langle 0|a_j a_l^\dagger |0\rangle - \langle 0|a_j a_k^\dagger a_i a_l^\dagger |0\rangle \qquad (8.200)$$

$$= \delta_{ik}\langle 0|(\delta_{jl} - a_l^\dagger a_j)|0\rangle - \langle 0|(\delta_{jk} - a_k^\dagger a_j)(\delta_{il} - a_l^\dagger a_i)|0\rangle \qquad (8.201)$$

$$= \delta_{ik}\delta_{jl} - \delta_{jk}\delta_{il} \qquad (8.202)$$

- Normal order. Operators composed of fermion (or boson) operators are in "normal ordered form" if all creation operators are to the left of all annihilation operators. Operators in normal ordered form are denoted as

$$\mathcal{N}(O) =: O :$$

Single-particle operators: Single-particle operators—that is those involving either the coordinate or momentum of only a single electron—can always be written in the form

$$\hat{U} = \sum_{\lambda\mu} \langle \lambda|U|\mu\rangle a_\lambda^\dagger a_\mu \qquad (8.203)$$

where $\langle \lambda | U | \mu \rangle$ is the integral

$$\int dx \phi_\lambda^*(x) U(\hat{x}, \hat{p}) \phi_\mu(x) \tag{8.204}$$

Two-particle operators: Two-particle terms, in particular the electron-electron interaction, are such that

$$\hat{V} | \alpha \beta \rangle = V_{\alpha \beta} | \alpha \beta \rangle \tag{8.205}$$

For matrix elements between antisymmetrized states, we need to sum over all permutations

$$\{ \alpha_1 \cdots \alpha_N | \hat{V} | \alpha_1' \cdots \alpha_N' \} = \sum_P \zeta^P \frac{1}{2} \sum_{i \neq j} (\alpha_{Pi} \alpha_{Pj} | V | \alpha_1 \alpha_2) \tag{8.206}$$

$$= \left(\frac{1}{2} \sum_{i \neq j} V_{\alpha_i \alpha_j} \right) \{ \alpha_1' \cdots \alpha_N' | \alpha_1 \alpha_N \} \tag{8.207}$$

where the sum is over all distinct particle pairs in $| \alpha_1 \alpha_N \rangle$.

8.7.1   EVALUATING FERMION OPERATORS

The motivation in introducing an operator formalism is to simplify calculations involving operators composed of multiple fermion operators. More importantly, in developing electronic structure theories we need to evaluate commutators of operator products much like we did above in generating the Fock matrix or if we were interested in the time-dependent response of the system due to some external driving field. For this, we have two powerful allies given by antisymmetry. First,

$$\{ a_\lambda, a_\mu^\dagger \} = \delta_{\lambda \mu} \tag{8.208}$$

and second $a_\mu a_\lambda = -a_\lambda a_\mu$. For example, consider the commutator between a single fermion operator and a binary product (as would be encountered in the single-particle hopping terms):

$$\begin{aligned}
[a_k, a_l^\dagger a_m] &= a_k a_l^\dagger a_m - a_l^\dagger a_m a_k \\
&= \left(\delta_{kl} - a_l^\dagger a_k \right) a_m - a_l^\dagger a_m a_k \\
&= \left(\delta_{kl} - a_l^\dagger a_k \right) a_m - (-1) a_l^\dagger a_k a_m \\
&= \delta_{kl} a_m
\end{aligned} \tag{8.209}$$

Here we have shown the sequence of steps where we have used the two "tricks" in our arsenal. Obviously, more complicated examples can be given involving commutators of operators composed of two or more fermion operators each. A more complex example arises in the commutator between the density operator $a_k^\dagger a_l$ and the two-particle interaction operator $a_m^\dagger a_n^\dagger a_p a_q$,

$$\begin{aligned}
[a_k^\dagger a_l, a_m^\dagger a_n^\dagger a_p a_q] = {} &\delta_{lm} a_k^\dagger a_n^\dagger a_p a_q - \delta_{ln} a_k^\dagger a_m^\dagger a_p a_q \\
&+ \delta_{kp} a_m^\dagger a_n^\dagger a_q a_l + \delta_{kq} a_m^\dagger a_n^\dagger a_p a_l
\end{aligned} \tag{8.210}$$

A final example, this time involving the creation operator in the same sequence, results in

$$[a_k^\dagger, a_l^\dagger a_m^\dagger a_n a_m] = \delta_{kl} a_l^\dagger a_m^\dagger a_p - \delta_{kp} a_l^\dagger a_m^\dagger, a_n \qquad (8.211)$$

In all cases we manipulate the sequence of the fermion operators to bring it into the normal ordered form.

8.7.2 NOTATION FOR TWO-BODY INTEGRALS

One of the irritations of the quantum chemistry and the quantum many-body physics literature is that both branches of the same subject evolved over much the same period of time yet have developed slightly different notations for the same thing. Based upon our definitions above for many-body wave functions, we interpret the matrix element $\langle kn|v|lm \rangle$ to mean

$$\langle kn|v|lm \rangle = \int d1 \int d2 \phi_k^*(1)\phi_n^*(2)v(12)\phi_l(1)\phi_m(2) \qquad (8.212)$$

We also can abbreviate $\langle kn|v|lm \rangle \equiv \langle kn|lm \rangle$ and we have the following relations

$$\langle ij|kl \rangle = \langle ji|lk \rangle^* \qquad (8.213)$$

and

$$\langle ij|kl \rangle = \langle kl|ij \rangle^* \qquad (8.214)$$

This notation is mostly used in the many-body physics literature. We have adopted it here since it is easier to read the matrix element from the order of the creation/annihilation operators once they have been put into normal ordered form with the creation operators to the left of the annihilation operators. For example,

$$\langle kl|nm \rangle a_k^\dagger a_l^\dagger a_m a_n \qquad (8.215)$$

which appears in the electron-electron interaction term of the electronic Hamiltonian. *However*, it is an unfortunate fact of life that this same symbol can mean

$$\langle kn|lm \rangle = \int d1 \int d2 \phi_k(1)\phi_n(1)v(12)\phi_l(2)\phi_m(2) \qquad (8.216)$$

(assuming real orbitals) where the quantum numbers for particle #1 are in the left bracket and the quantum numbers for particle #2 are in the right. One also sees this written using the square brackets

$$[kn|lm] = \int dr_1 \int dr_2 \phi_k(r_1)\phi_n(r_1)v(12)\phi_l(r_2)\phi_m(r_2) \qquad (8.217)$$

where $\phi_n(r)$ is a spatial orbital (as opposed to spin orbital) and the integral is only over the spatial variables (as opposed to both spin and space). The rationale for this notation is convenient since for real orbitals

$$[ij|kl] = [ji, kl] = [ij, lk] = [ji|lk] \qquad (8.218)$$

For purposes of clarity, we shall denote integrals using the physics notation using round $(|)$ or angular $\langle|\rangle$ brackets and integrals using the quantum chemistry notation using the square $[|]$ brackets. The translation between physics and quantum chemistry is that

$$\langle ij|kl\rangle_{physics} = [ik|jl]_{q.chem} \tag{8.219}$$

Two important integrals are the Coulomb integral

$$J_{ij} = [ii|jj] = (ij|ij) \tag{8.220}$$

and the exchange integral

$$K_{ij} = [ij|ij] = (ij|ji) \tag{8.221}$$

When in doubt, double-check the notation the author is using. Also, it is a good idea to clearly specify which convention you are using.

8.8 PROBLEMS AND EXERCISES

Problem 8.1   Consider the two-atom/two-electron problem discussed in the chapter. Show that by using the mixed single-electron states

$$\phi_\alpha = \sin\eta|1\uparrow\rangle + \cos\eta|2\uparrow\rangle \phi_\beta = \cos\eta|1\downarrow\rangle + \sin\eta|2\downarrow\rangle \tag{8.222}$$

we arrive at the following expression for E_{uhf}

$$E_{uhf}(\eta) = 2\varepsilon + (2\varepsilon + U)\cos^2\eta\sin^2\eta + t\sin(2\eta) \tag{8.223}$$

Also show that in the limit of $U/t < 2$, E_{uhf} reduces to the Hartree–Fock limit.

Problem 8.2   Evaluate the following commutator identities:

$$\left[a_i, a_j^\dagger a_k^\dagger a_l a_m\right]$$

$$\left[a_i^\dagger a_j, a_k^\dagger a_l^\dagger a_m a_n\right]$$

Problem 8.3   If $|\psi\rangle$ is a single Slater determinant state, show that the following factorization holds:

$$\left\langle a_i^\dagger a_j^\dagger a_k a_l\right\rangle = \rho_{kj}\rho_{li} - \rho_{lj}\rho_{ki}$$

SUGGESTED READING

Below is a list of review articles and books concerning the electronic structure of π-conjugated systems.

1. "Light-emitting Diodes Based on Conjugated Polymers," J. H. Burroughes, D. D. C. Bradley, A. R. Brown, R. N. Marks, K. Mackay, R. H. Friend, P. L. Burns, and A. B. Holmes, *Nature, 347,* 539–541 (1990).

2. "An Organic Electronics Primer," G. Malliaras and R. H. Friend, *Physics Today, 58,* 53–58 (2005).

3. *Quantum Chemistry Aided Design of Organic Polymers: An Introduction to the Quantum Chemistry of Polymers and Its Applications,"* Jean-Marie Andre, Joseph Delhalle, and Jean-Luc Bredas (Singapore: World Scientific, 1991).

4. *Electronic Processes in Organic Crystals and Polymers,* 2nd ed, Martin Pope and Charles E. Swenberg (Oxford: Oxford University Press, 1999).

5. *Conjugated Polymers : The Novel Science and Technology of Highly Conducting and Nonlinear Optically Active Material,* J. L. Brédas and R. Silbey (Kluwer: Dordrect, 1991).

6. "An Overview of the First Half-Century of Molecular Electronics," Noel S. Hush, *Ann. N.Y. Acad. Sci., 1006,* 1 (2003).

7. "What I Like About the Hückel Model," W. Kutzelnigg, *J. Comp. Chem.* 28, 25–34 (2007). This is a recent essay by Prof. Werner Kutzelnigg based upon a lecture at the University of Marburg in 1996 celebrating the 100th birthday of Erich Hückel. It has a concise overview of Hückel's work and key references. Also, many of the results in Sec. 3.2 are presented in this paper.

The following are good texts concerning electronic structure theory of molecular systems:

1. *Quantum Chemistry,* 5th ed., I. Levine (Prentice Hall, 1999).

2. *Modern Quantum Chemistry: Introduction to Advanced Electronic Structure Theory,* A. Szabo and N. S. Ostlund (Dover, 1996).

3. *Quantum Mechanics in Chemistry,* George C. Schatz and Mark A. Ratner (Prentice Hall, 1993).

4. *Elementary Quantum Chemistry,* 2nd ed., Frank L. Pilar (McGraw-Hill, 1990).

9 Electron–Phonon Coupling in Conjugated Systems

9.1 SU–SCHRIEFFER–HEEGER MODEL FOR POLYACETYLENE

A significant distinguishing feature of π-electron systems is the relatively strong interaction between the electrons and phonons. For example, the carbon-carbon bond length in an olefinic C=C bond is generally taken to be 1.35 Å, whereas an olefinic C–C bond is somewhat larger at 1.45 Å. In aromatic systems, the C=C bond length is 1.40 Å. This is, of course, due to the modulation of bond order as one moves down the olefinic chain. For aromatic systems, we know that the resonance structures account for the homogenous delocalization of the electron density about the ring. Here we present the classic model developed by Su, Schrieffer, and Heeger (SSH) that accounts for semiconducting properties of olefinic chains.[102,103]

From the Hückel model, we found that if all the sites along the chain are equivalent, then the energy for the infinite chain is given by

$$E_k = \alpha - 2\beta \cos(ka) \tag{9.1}$$

where a is the bond distance between neighboring C atoms. For a π-electron system that is half-filled (that is, each C atom contributes one e- to the π system), then the band gap at the Fermi energy is exactly 0. This is the case if all the C–C bonds are the same length, as in aromatic rings.

However, as we just noted in olefinic chains, the C=C and C–C bonds alternate, so one expects the hopping term β to reflect this alternation. As the bond length increases, β should decrease in magnitude; while as the bond length is compressed, β should increase in magnitude. We can thus assign an order parameter, u, that reflects the expansion and compression of the bonds along the olefinic chain. Assuming all the sites are equivalent, we can write

$$H(u) = -\sum_{n\sigma} (\beta + (-1)^n 2\alpha u) * \left(a_{n\sigma}^\dagger a_{n+1\sigma} a_{n+1\sigma}^\dagger a_{n\sigma}\right) \tag{9.2}$$

For the sake of simplicity, take $\alpha = 0.05$ eV and $\beta = 1.40$ eV so as to correspond with the spectroscopic parameters of the PPP model. Thus, $u = 0$ corresponds to the uniform Hückel model and $u = \pm 1$ corresponds to the "dimerized" olefinic chain with alternating single and double bonds. (See Figure 9.1.)

For the undimerized $u = 0$ chain, we have two bands, the filled valence band

$$E_k^{0v} = -2\beta \cos(ka) = -\varepsilon_k \tag{9.3}$$

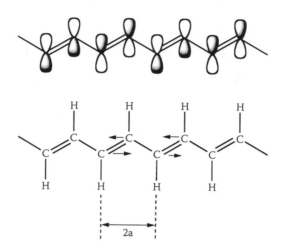

FIGURE 9.1 Polyacetylene chain showing (top) arrangement of C $2p_z$ orbitals forming a π system and (bottom) distortions of a C–C lattice from uniform lattice to dimerized lattice.

and the conduction band

$$E_k^{0c} = +2\beta \cos(ka) = +\varepsilon_k \tag{9.4}$$

with corresponding states

$$\chi_k^v = \frac{1}{\sqrt{N}} \sum_n e^{ikan} u_n \tag{9.5}$$

$$\chi_k^v = \frac{1}{\sqrt{N}} \sum_n e^{ikan} (-1)^n u_n \tag{9.6}$$

where the u_n are the localized basis functions. For a chain of length $L = Na$, we can define new operators for the valence and conduction bands as

$$c_{k\sigma}^v = \frac{1}{\sqrt{N}} \sum_n e^{ikan} a_{n\sigma} \tag{9.7}$$

$$c_{k\sigma}^c = \frac{1}{\sqrt{N}} \sum_n e^{ikan} (-1)^n a_{n\sigma} \tag{9.8}$$

Inverting these and reintroducing them back into the modulated Hückel Hamiltonian above, we have for the dimerized lattice

$$H(u) = \sum_{k\sigma} \varepsilon_k \left(c_{k\sigma}^{c\dagger} c_{k\sigma}^c - c_{k\sigma}^{v\dagger} c_{k\sigma}^v \right) + 4\alpha u \sin(ka) \left(c_{k\sigma}^{c\dagger} c_{k\sigma}^v + c_{k\sigma}^{v\dagger} c_{k\sigma}^c \right) \tag{9.9}$$

We now introduce the Bogolyubov transformation to bring $H(u)$ into diagonal form by mixing the valence and conduction bands,

$$\begin{pmatrix} a_{k\sigma}^v \\ a_{k\sigma}^c \end{pmatrix} = \begin{pmatrix} \alpha_k & \beta_k \\ \alpha_k^* & -\beta_k \end{pmatrix} \begin{pmatrix} c_{k\sigma}^v \\ c_{k\sigma}^c \end{pmatrix} \tag{9.10}$$

Since $|\alpha_k|^2 + |\beta_k|^2 = 1$, this is a unitary transformation. Now, inverting the transformation and requiring $H(u)$ to be diagonal in this new representation, we find

$$H(u) = \sum_{k\sigma} E_k \left(n_{k\sigma}^c - n_{k\sigma}^v \right) \tag{9.11}$$

where

$$E_k = \left(\varepsilon_k^2 + \Delta_k^2 \right)^{1/2} \tag{9.12}$$

and

$$\Delta_k = 4\alpha u \sin(ka) \tag{9.13}$$

The operators, $n_{k\sigma}^c$ and $n_{k\sigma}^v$ are the occupation numbers of the conduction and valence bands. If we restrict α_k to be real and positive, then

$$\alpha_k = \sqrt{\frac{1}{2}\left(1 + \frac{\varepsilon_k}{E_k}\right)} \tag{9.14}$$

$$\beta_k = \sqrt{\frac{1}{2}\left(1 - \frac{\varepsilon_k}{E_k}\right)} \tag{9.15}$$

which immediately gives us the conduction and valence band eigenstates. A plot of the valence and conduction bands for an olefinic chain is given in Figure 9.2a using the parameters suitable for polyacetylene. Notice at $ka = \pi/2$, the gap between the

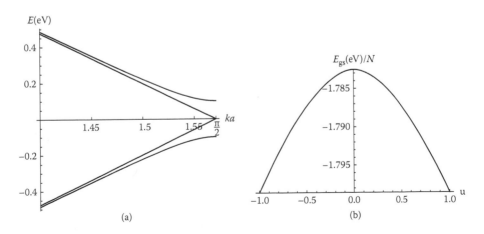

(a)   (b)

FIGURE 9.2 (a) Valence and conduction bands for dimerized ($u = 1$) and undimerized ($u = 0$) polyacetylene chains close to $ka = \pi/2$. (b) Variation of ground-state energy for chain of length Na versus the dimerization order parameter u.

valence bands ($E < 0$) and conduction bands ($E > 0$) opens as the order parameter varies from 0 to 1.

The ground-state energy (see Figure 9.2b) is given by summing over all occupied energy levels

$$E_{gs}(u) = \sum_k -2E_k \tag{9.16}$$

where the sum over k is over the entire Brillouin zone, $-\pi/2 \geq ka \leq \pi/2$. Taking the sum to an integral

$$E_{gs}(u) = -\frac{2L}{\pi} \int_0^{\pi/2a} [(2\beta \cos(ka))^2 + (4\alpha u \sin(ka))^2]^{1/2} dk \tag{9.17}$$

$$= -\frac{4N\beta}{\pi} \int_0^{\pi/2} (1 - (1 - z^2)\sin^2(ka))^{1/2} d(ka) \tag{9.18}$$

$$= -\frac{4N\beta}{\pi} E(1 - z^2) \tag{9.19}$$

The integral is the complete elliptic integral $E(1 - z^2)$ with argument $z = \beta_1/\beta = 2\alpha u/\beta$. For $z \in \Re$, the expansion of $E(1 - z^2)$ about small values of z produces

$$E(1 - z^2) \approx = 1 + \left(-\frac{\log(z)}{2} + \log(2) - \frac{1}{4}\right) z^2 + O(z^4) \tag{9.20}$$

from which we can conclude that the total ground-state energy E_{gs} is always lowered upon dimerization of the olefinic lattice and reaches a maximum for the perfectly uniform lattice. This analysis is, in fact, a demonstration of a very powerful theorem given by Peierls that states that one-dimensional metals cannot exist and so as such will always distort so as to lower their total energy and open their band gap.*

9.2 EXCITON SELF-TRAPPING

In the SSH model we included the effects of bond distortion to minimize the total energy of a linear chain by making the hopping integral a linear function of the distortion of atoms from uniform spacing. We can generalize this by writing the

---

* We can arrive at this conclusion by considering the fact that the period for the dimerized lattice has, in fact, doubled from a to $2a$ upon dimerization and as such the Brillouin zone now extends from $-\pi/(2a)$ to $\pi/(2a)$. If we now take the two atoms in the mth unit cell as having states u_{m1} and u_{m2}, then we can write the wave function for the lattice as

$$\psi_k = \sum_{m=-\infty}^{+\infty} e^{i2kma}(c_1(k)u_{m1} + c_2(k)u_{m2}) \tag{9.21}$$

Inserting this into the Schrödinger equation above we find

$$\begin{bmatrix} t_1 e^{ik2a} & t_2 \\ t_1 e^{-ik2a} & t_2 \end{bmatrix} \begin{bmatrix} c_1(k) \\ c_2(k) \end{bmatrix} = E_k \begin{bmatrix} c_2(k) \\ c_1(k) \end{bmatrix} \tag{9.22}$$

where $t_1 = \beta - \alpha u$ and $t_2 = \beta + \alpha u$. Solving for the energy produces the result given above.

Hamiltonian for a particle on a lattice as

$$H_{el}(x) = \sum_n \varepsilon_n a_n^\dagger a_n + \sum_n (\beta + \lambda(x_n - x_{n+1}))\left(a_n^\dagger a_{n+1} + a_{n+1}^\dagger a_n\right) \qquad (9.23)$$

where x_n denotes the displacements from the uniform lattice with a strain energy given by

$$E_{strain} = \sum_n \frac{\omega^2}{2}(x_n - x_{n+1})^2 \qquad (9.24)$$

Working within the adiabatic approximation, we can minimize the total energy of the system by requiring

$$-\langle \phi_o | \frac{\partial H_{el}}{\partial x_n} | \phi_o \rangle - \frac{\partial E_{strain}}{\partial x_n} = 0 \qquad (9.25)$$

where ϕ_o is the lowest energy eigenstate of H_{el} for a given lattice configuration. The first term is the Hellmann–Feynman force on the nth lattice site when the exciton is in the lowest energy eigenstate. The second is the strain force, which increases as the lattice is displaced from its uniform position. If β and λ are both negative, then increasing $|\phi_o|^2$ in a given region gives an attractive interaction between the lattice atoms. However, displacing the atoms from their uniform position increases the strain energy. The final equilibrium state is where the strain force and the Hellmann–Feynman forces are in balance.[104–109] In Figure 9.3 we show the effect of exciton self-trapping on a model 11-site lattice. The first (Figure 9.3a) is an order parameter showing the displacement of the lattice sites from their original position in the final relaxed state. The second (Figure 9.3b) shows how the lattice relaxes from its original uniform state (top) to the final relaxed or self-trapped state (bottom). In Figure 9.3c we show the exciton's probability density for the initial and final (relaxed) state.

We can extend the model for a continuum lattice by writing the Hamiltonian as

$$H = \sum_k \varepsilon(k) a_k^\dagger a_k + \sum_q \omega(q) B_q^\dagger B_q + \sum_{nq} g(n, q) a_n^\dagger a_n \left(B_q + B_q^\dagger\right) \qquad (9.26)$$

where $\{a_k, a_k^\dagger\}$ are exciton operators in k-space and $\{a_n, a_n^\dagger\}$ denote exciton operators in the lattice representation. The two are related by the Fourier relation. $\{B_q, B_q^\dagger\}$ are phonon operators with dispersion relation $\omega(q)$. The electron phonon coupling, $g(n, q)$, depends on the type of phonon in question. For longitudinal acoustic modes,

$$g(n, q) = \frac{1}{\sqrt{N}} e^{iqR_n} \frac{Cq}{2\rho\omega(q)a_o^3} \qquad (9.27)$$

where ρ is the density of the medium, a_o is the lattice constant, C is the deformation parameter, and $\omega(q) = vq$ with v as the velocity of sound. For optical (transverse) phonons,

$$g(n, q) = \frac{1}{\sqrt{N}} e^{iqR_n} \gamma_o \qquad (9.28)$$

where γ_o and ω are assumed to be constant.

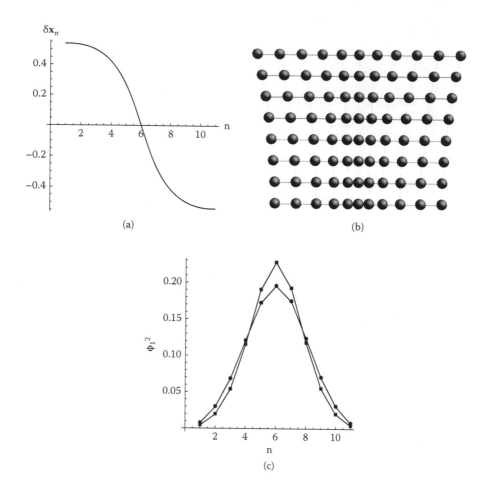

FIGURE 9.3 (a, b) Lattice distortion for an exciton self-trapping on a finite-sized lattice. (c) Comparing the unrelaxed to trapped exciton probability density for finite lattice.

By completing the square in H we arrive at

$$H = \sum_k \varepsilon(k)a_k^\dagger a_k + \sum_q \omega(q)\tilde{B}_q^\dagger \tilde{B}_q - \sum_{mnq} \frac{g(n, q)g(m, q)}{\omega(q)}\hat{n}_n\hat{n}_m \qquad (9.29)$$

where we have defined renormalized phonon operators as

$$\tilde{B}_q = B_q + \sum_n \hat{n}_n g(n, q)/\omega(q) \qquad (9.30)$$

and

$$\tilde{B}_q^\dagger = B_q^\dagger + \sum_n \hat{n}_n g(n, -q)/\omega(q) \qquad (9.31)$$

with $\hat{n}_n = a_n^\dagger a_n$ as the exciton number operator on site n. If we adiabatically minimize the total energy, the second term will always be a constant and can thus be ignored.

Thus, we write the adiabatic Hamiltonian as

$$H_{ad} = \sum_k \varepsilon(k) a_k^\dagger a_k - \sum_{mnq} \frac{g(n,q)g(m,q)}{\omega(q)} \hat{n}_n \hat{n}_m \qquad (9.32)$$

Irrespective of the phonon model, taking the phonons to be normal coordinates then

$$\sum_q \frac{g(n,q)g(m,q)}{\omega(q)} = \delta_{nm} C_o \qquad (9.33)$$

where

$$C_o = \begin{cases} \frac{C_o}{\rho} v^2 a_o^3 & \text{for acoustic modes} \\ \gamma_o^2/\omega & \text{for optical modes} \end{cases} \qquad (9.34)$$

Thus,

$$H_{ad} = \sum_k \varepsilon_k a_k^\dagger a_k - C_o \sum_n \hat{n}_n^2 \qquad (9.35)$$

So, as discussed above, increasing the exciton probability density at a given lattice site results in a net lowering of the exciton's energy.

We can push this analysis farther by assuming a Gaussian form for the exciton's wave function

$$a(r) = \sqrt{\frac{a_o}{\sigma}} \left(\frac{2}{\pi}\right)^{1/4} e^{-(r/\sigma)^2} \qquad (9.36)$$

Taking its Fourier transformation,

$$a(k) = \sqrt{\frac{\sigma}{a_o}} \left(\frac{2}{\pi}\right)^{1/4} e^{-k^2\sigma^2} \qquad (9.37)$$

These are both normalized so that

$$\sum_n a_n^2 = \frac{N}{L} \int dr a(r)^2 = 1 \qquad (9.38)$$

and

$$\sum_k a_k^2 = \frac{L}{N} \int dk a(k)^2 = 1 \qquad (9.39)$$

and $L/N = a_o$. Using these two relations,

$$\sum_n \hat{n}_n^2 = \frac{1}{\sqrt{\pi}} \frac{a_o}{\sigma} \qquad (9.40)$$

For a free particle on an infinite lattice, the energy dispersion is given by $\varepsilon(k) = 4\beta k^2 a_o^2$. Thus,

$$\sum_k \varepsilon(k) a_k^2 = \beta \left(\frac{\sigma}{a_o}\right) \qquad (9.41)$$

Putting this all together, one finds that the adiabatic energy is given by

$$E_{ad} = \beta \frac{a_o^2}{\sigma^2} - \frac{C_o}{\sqrt{\pi}} \frac{a_o}{\sigma} \tag{9.42}$$

Furthermore, extending this to the d-dimensional isotropic lattice, we find:

$$E_{ad} = \beta \frac{a_o^2}{\sigma^2} - \frac{C_o}{\pi^{d/2}} \left(\frac{a_o}{\sigma}\right)^d \tag{9.43}$$

Minimizing this with respect to the sole remaining parameter, σ, requires

$$\frac{aC_o d\pi^{-d/2} \left(\frac{a}{\sigma}\right)^{d-1}}{\sigma^2} - \frac{2a^2\beta}{\sigma^3} = 0 \tag{9.44}$$

Solving for σ, we find for the one-dimensional lattice

$$\sigma = \frac{2aJ\sqrt{\pi}}{C_o} \tag{9.45}$$

However, for the 2D isotropic lattice, we find that $dE_{ad}/d\sigma = 0$ occurs for the special case of $C_o = 2\beta/\pi$ for a continuum lattice. Sumi and Sumi reach a similar conclusion for finite-sized lattices[110] in which they conclude there is a phase boundary between the free and self-trapped (small-radius) exciton that depends upon the size of the system and the strength of the electron/phonon coupling.

More recent quantum/classical dynamical simulations by Kobrak et al.[107–109] and by Tretiak et al.[111–115] have examined the interplay between different types of vibrational motion in the trapping and relaxation of an exciton on a polymer chain. In particular for poly-phenylene vinylene (PPV), self-trapping of excitations on about six repeat units in the course of photoexcitation relaxation identifies specific slow (torsion) and fast (bond stretch) nuclear motions strongly coupled to the electronic degrees of freedom. Similar conclusions were drawn using semiempirical excited-state techniques by Karabunarliev and Bittner.[116,117]

9.3 DAVYDOV'S SOLITON

The exciton model is quite universal and can be invoked to study a wide variety of systems in which one has a highly quantized degree of freedom interacting with a harmonic reservoir. Moreover, the exciton need not be an electronic excitation. One particularly interesting application of the exciton model occurs in modeling energy transport along an alpha-helix chain in a protein. In a biological system, chemical energy is provided through the hydrolysis of adenosine triphosphate (ATP). An ATP molecule binds to the active site of a protein, reacts with water, and releases 0.49 eV of free energy. Roughly speaking, 1 eV corresponds to 12 000 K of thermal energy, so this 0.49 eV is about 20 times greater than the average energy available from the thermal background around 300 K. The crucial questions become "What happens to this energy?" and "How is it transported from the reaction site to where it is needed?" This is a highly nonequilibrium system and one would expect that the excess energy would be rapidly dissipated as heat to the surroundings. Classical molecular dynamics simulations confirm that this excess heat is rapidly distributed among the vast degrees of freedom of the protein and surrounding water molecules in a few picoseconds.

An alternative explanation is that the energy released in ATP hydrolysis is converted via resonant coupling to localized high-frequency modes of the protein, perhaps occurring through some intermediate vibrational coupling. One likely participant is the amide-I vibrational mode of a peptide group at 0.21 eV ($1660 \, \text{cm}^{-1}$). This is about half of the free energy released and almost resonant with the H-O-H bending mode at $1646 \, \text{cm}^{-1}$. The amide-I mode gives a strong peak in the IR and Raman spectra of proteins with little variation: $1665–1660 \, \text{cm}^{-1}$ for α-helices, $1660–1665 \, \text{cm}^{-1}$ for a β-sheet, and $1665–1680 \, \text{cm}^{-1}$ for a random coil. It is primarily composed of a C=O stretch weakly coupled to an N-H in-plane bend as shown in Figure 9.4a.

In an α helix, the peptide chain is wound into a right-handed helix whereby every third C=O is hydrogen-bonded to an amide hydrogen to form a quasi-linear chain, with each C=O separated by a distance $R = 4.5 \, \text{Å}$.

$$C=O \cdots H\text{-}N\text{-}C=O \cdots H\text{-}N\text{-}C=O$$

These chains form three quasi-linear spines along the α helix. These spines are not exactly linear and slowly wrap about the helical axis as shown in Figure 9.4b. Each C=O group carries a substantial permanent electric dipole moment μ directed from the $O^{\delta-}$ to the $C^{\delta+}$.* These dipoles run parallel to the spines. Thus, to a first approximation, neighboring C=O's along each spine are coupled to each other via dipole–dipole interactions

$$J_{i,i+1} = 2\frac{|\mu|^2}{R^3}$$

From this we can build a simple vibrational exciton model within a local basis where $|\phi_i\rangle$ represents a single stretching quantum on the ith C=O along a given spine. That is,

$$H_{ex} = \hbar\omega I + J \sum_i^N |\phi_i\rangle\langle\phi_{i+1}| \tag{9.46}$$

where $\varepsilon = \hbar\omega$ is the excitation energy for a C=O amide-I vibration and I is the $N \times N$ identity matrix. H_{ex} is the familiar tridiagonal Hamiltonian matrix we have seen previously. Thus, its eigenvalues and eigenvectors can be immediately deduced. For a sufficiently long helix, the eigenfunctions are plane waves with energy

$$\varepsilon(k) = \varepsilon + 2|J| \cos(kR)$$

Thus, for a perfect chain, a C=O vibrational exciton will be delocalized over the entire chain.

The α helix itself is free to undergo a variety of motions. A compression of the α-helix chain would change the local electrostatic environment about the C=O group.

---

* We use the "physics convention" where dipoles point from source $(-)$ to sink $(+)$ rather than the "chemistry convention" where dipoles point from $(+)$ to $(-)$.

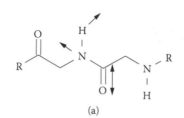

(a)

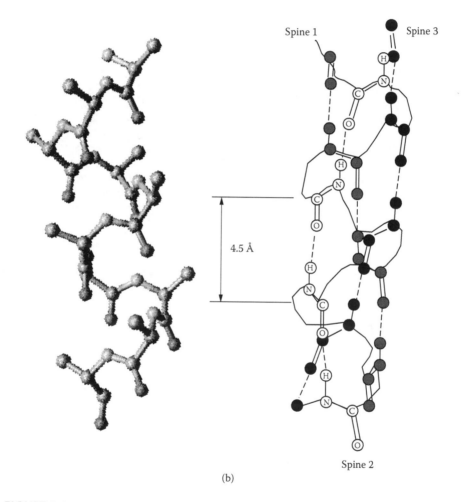

(b)

FIGURE 9.4 (a) Peptide group showing amide-I vibrational mode. The amide-I mode gives a strong peak in the IR and Raman spectra of proteins with little variation: 1665–1660 cm^{-1} for α helices, 1660–1665 cm^{-1} for a β sheet, and 1665–1680 cm^{-1} for a random coil. (b) Three-dimensional model of an alpha-helix coil with hydrogen bonds between linked C=O$\cdots$H-N groups.

Displacing a peptide changes the CO–CO distance R. Taking u_i to be the displacement of the ith peptide group, then the local site energy of the ith C=O becomes

$$\varepsilon_i = \varepsilon + \frac{\partial \varepsilon(\{u_i\})}{\partial R}(x_{i+1} - x_{i-1}) + \cdots$$

Secondly, a compression of the helix sets up a longitudinal sound wave along the chain. This wave travels at $v_s = R(\kappa/m)^{1/2}$ where κ is the bulk modulus or spring constant and m is the mass of an amide group. Assuming the chain to be harmonic,

$$H_{phonon} = \frac{1}{2m} \sum_i \hat{p}_i^2 + \frac{\kappa}{2} \sum_i (\hat{x}_i - \hat{x}_{i+1})^2$$

Combining these contributions, the total Hamiltonian takes the familiar form:

$$H = \sum_i \left(\varepsilon + \frac{\partial \varepsilon(\{x_i\})}{\partial R}(\hat{x}_{i+1} - \hat{x}_{i-1}) \right) |\phi_i\rangle\langle\phi_i|$$

$$+ J \sum_i^N |\phi_i\rangle\langle\phi_{i+1}|$$

$$+ \frac{1}{2m} \sum_i \hat{p}_i^2 + \frac{\kappa}{2} \sum_i (\hat{x}_i - \hat{x}_{i+1})^2 \tag{9.47}$$

where p_i is the momentum conjugate to x_i. Of course, the $\hat{x}_i$ and $\hat{p}_i$ are quantum mechanical operators and a rigorous solution demands that we take this into account. If we allow that the mass of an amide-I group is much larger than the effective mass of a C=O vibrational exciton $m^* = \hbar^2/2J$, then we may safely invoke a classical treatment for the u_i degrees of freedom. Davydov introduces this by making an ansatz for the state vector for the coupled exciton/lattice wave function:

$$|\psi(t)\rangle = \sum_i \phi_i(t)B_i^\dagger \exp\left[i\hbar \sum_j (u_j(t)\hat{p}_j - \pi_j(t)\hat{x}_j) \right] |0\rangle \tag{9.48}$$

where $|0\rangle$ is the ground-state vector and $B_i^\dagger = |\phi_i\rangle\langle 0|$ creates an excitation on site i. Davydov then makes what is essentially the Ehrenfest approximation by writing

$$u_i(t) = \langle\psi(t)|\hat{x}_i|\psi(t)\rangle \ \& \ \pi_i(t) = \langle\psi(t)|\hat{p}_i|\psi(t)\rangle \tag{9.49}$$

Thus, the fully quantum problem reduces to a mixed-quantum/classical problem with two coupled equations:

$$i\hbar\dot{\phi}_i = (\varepsilon + \chi(u_{i+1} - u_{i-1}))\phi_i - J(\phi_{i-1} + \phi_{i+1}) \tag{9.50}$$

and

$$m\ddot{u}_i - \kappa(u_{i+1} - 2u_i + u_{i-1}) = \chi(|\phi_{i-1}|^2 - |\phi_{i+1}|^2) \tag{9.51}$$

where ϕ_i is the quantum mechanical amplitude giving the probability $|\phi_i|^2$ for finding the ith C=O in its first vibrational excited state. Also, we have introduced χ as the linear coupling between adjacent C=O's. These last two equations are the main results of Davydov's original paper.[118-121] Let us next assume that the de Broglie wavelength of the exciton is large compared to R. Thus, we can rewrite these last two equations in continuum form as

$$i\hbar \frac{\partial \phi}{\partial t} = \left(\varepsilon_o + 2\chi \frac{\partial u}{\partial x} \right) \phi - J \frac{\partial^2 \phi}{\partial x^2} \tag{9.52}$$

and

$$\frac{\partial^2 u}{\partial t^2} - \frac{\kappa}{m} \frac{\partial^2 u}{\partial x^2} = \frac{2\chi}{m} \frac{\partial |\phi|^2}{\partial x} \tag{9.53}$$

where x now represents a location along the helical spine and $\varepsilon_o = \varepsilon - 2J$. The left-hand side of the second equation is that of the wave equation. The inhomogeneous term on the right-hand side is a source. What we see, then, is that the quantum mechanical motion of the C=O vibrational exciton acts as a source for the generation of longitudinal sound waves in the α helix.

With this in mind, let us seek traveling wave solutions to Equations 9.52 and 9.53 of the form

$$u(x,t) = u(x - vt)$$

where v is the velocity of propagation. Substituting this into Equation 9.53 yields

$$\frac{\partial u}{\partial x} = -\frac{2\chi}{\kappa(1 - s^2)} |\phi(x)|^2$$

where $s = v/v_s < 1$ is the ratio of the propagation velocity to the velocity of sound. Introducing this back into Equation 9.52 yields a nonlinear Schrödinger equation

$$i\hbar \frac{\partial \phi}{\partial t} = -J \frac{\partial^2 \phi}{\partial x^2} + \varepsilon_o \phi - \frac{4\chi^2}{\kappa(1 - s^2)} |\phi|^2 \phi$$
$$= H[\phi]\phi \tag{9.54}$$

This is the source of the nonlinear interactions in which the solution of the wave function depends upon the wave function itself! Similar nonlinear Schrödinger equations arise in various contexts, typically within the self-consistent field or Hartree approximation to the many-body problem. Here, however, the nonlinear interaction arises because we have a feedback mechanism between the vibrational motion of the helix and the quantum motion of the C=O exciton.

9.3.1 Approximate Gaussian Solution

As previously, let us make a simple ansatz to the form of the wave function solution to the stationary (time-independent) Schrödinger equation:

$$\phi(x) = \frac{1}{(a\pi)^{1/4}} e^{-x^2/(2a^2)}$$

and use the variational principle to minimize the total energy with respect to the width a

$$\frac{dE}{da} = \frac{d}{da} \langle \phi | H[\phi] | \phi \rangle = \frac{d}{da} \left(\varepsilon_o + -\frac{J}{2a} - \frac{1}{2} \frac{2\sqrt{2}\chi^2}{\sqrt{a\pi}\kappa(1-s^2)} \right) = 0 \qquad (9.55)$$

Note, the 1/2 appearing in front of the term arising from the exciton/lattice interaction is included to avoid the interaction between the exciton and itself. This produces the variational estimate of the width of the exciton wave function

$$a = \frac{J^2\kappa^2\pi(s^4 - 2s^2 + 1)}{8\chi^4}$$

Since $1 \geq 1 - 2s^2 + s^4 \geq 0$ for $s < 1$, $a > 0$ for all traveling wave solutions. Let us take the case where $s = 0$ to find the energy of a trapped exciton using

$$a = \frac{J^2\kappa^2\pi}{2\chi^4}$$

Introducing this into the variational estimate of the energy yields

$$E_{ste} = \varepsilon_o - \frac{3\chi^4}{\pi|J|\kappa^2}$$

9.3.2 Exact Solution

Zakharov and Shabat[122] have found a general solution to the nonlinear Schrödinger equation. Their solution for the stationary case with $s = 0$ reads

$$\phi(x) = \frac{\chi}{\sqrt{2\kappa|J|}} \text{sech}(\chi^2/(\kappa J)x) \qquad (9.56)$$

The wave function as written here is normalized and gives a self-trapping energy of

$$E_{ste} = \varepsilon_o - \frac{\chi^4}{\kappa^2|J|}$$

For a completely rigid helix, $\kappa \to \infty$ and the self-trapping vanishes and we have a band of delocalized excitations. It can also be seen from the sign of the trapping energy that the energy of the trapped exciton will always be below that of a delocalized exciton, ε_o, when the coupling between the exciton and the lattice is taken into account.

Problem 9.1 Demonstrate that $\phi(x) = \sinh(x/a)$ is a stationary solution of the nonlinear Schrödinger equation

$$J\psi'' + \kappa|\psi|^2\psi = E\psi$$

What is the normalization, a, and the ground-state energy in terms of J and κ?

TABLE 9.1

Parameter Values for Davydov Model from Lomdahl and Kerr

	κ	J	χ	$\chi^2/\kappa J$	t_o
	N/m	cm^{-1}	pN		ps
Discrete	5.0	20.0	75.0	2.83	19.5
Continuum	13.0	31.2	40.0	0.29	12.1
α helix	13.0	7.8	62.0	1.91	12.1

9.3.3 DISCUSSION: DO SOLITONS EXIST IN REAL α-HELIX SYSTEMS?

Although the formal analysis leading to Davydov's nonlinear Schrödinger equation in Equation 9.54 is quite compelling and has led to some interesting and beautiful analyses in the field of nonlinear dynamics, one must ask at some point whether such coherent structures actually exist in real biological proteins under physiological conditions.[123] In order to analyze this, we first need to know some specific parameters for the model. The dipole–dipole interaction is straightforward given the dipole moment of a C=O in a peptide chain. The nonlinear interaction χ is a bit more subtle and estimates place it between 30 and 62 pN. Table 9.1 lists a set of parameters from Lomdahl and Kerr's *Phys. Rev. Letter.*[123] from 1985.

In order to include the effects of a thermal environment, Lomdahl and Kerr add a damping force and noise term to the equations of motion for each amide-I site

$$F_i = -m\gamma\dot{u}_i + \eta_i(t)$$

where γ is the vibrational relaxation rate, and $\eta_i(t)$ is a noise term with

$$\overline{\eta_i(t)\eta_j(t')} = \delta_{ij}\delta(t-t')2m\gamma k_B T$$

and $\overline{\eta_i(t)} = 0$. In other words, each site is subject to the thermal noise and we assume there is no correlation between the thermal noise from site to site. This is certainly a reasonable assumption since it gives the average kinetic energy (per site) as

$$\frac{m}{2}\overline{\dot{u}_i^2(t)} = \frac{1}{2}k_B T$$

as expected for a classical oscillator. They conclude that at physiological temperatures, random fluctuations in the lattice are too strong and prevent self-trapping. Similar conclusions were reached by Schweitzer based upon quantum perturbation theory.[124] On the other hand, Förner concludes that Davydov solitons are stable at 300 K in reasonable parameter ranges, but only for special initial conditions close to the terminal sides of the chain.[125] Moreover, it is well known in the biophysical literature that polyamide in water undergoes a helix-to-coil transition at around $T = 280$ K. Consequently, for free polyamide chains, Davydov solitons do not exist in polyamide chains at physiologic temperatures; however, they may exist in other peptide chains and in proteins. Their existence remains a theoretical question.

9.4 VIBRONIC RELAXATION IN CONJUGATED POLYMERS

Our discussion of vibronic relaxation of an exciton in contact with a harmonic lattice concludes with a brief recapitulation of the theory of the electonic spectra in a molecular system. For the sake of discussion, consider the case of a two-level electronic system coupled with the nuclear motions of the molecule x. We shall work in a representation such that all of our energies are referenced to the electronic ground state of the molecule and that x represents the distortions of the molecule away from its minimum energy geometry. First, consider the variation of the ground-state energy with the distortions of the molecule away from its equilibrium position. The Hamiltonian describing the nuclear motions is, thus,

$$H_{vib} = \sum_n \frac{p_n^2}{2m_n} + \frac{1}{2}\sum_{nm} E_{gs}(x = 0) + \left(\frac{\partial^2 E_{gs}}{\partial x_m \partial x_n}\right)_{eq} x_m x_n + \cdots \quad (9.57)$$

where x_n represents the displacement of the nth atom with mass m_n from its equilibrium position. We can eliminate the explicit mass dependency by adopting mass-scaled coordinates, $\tilde{x}_n = m_n^{1/2} x_n$ and momenta. The phonons (or harmonic motions) of the molecule are obtained by diagonalizing the second derivative matrix (Hessian)

$$h_{nm} = \left(\frac{\partial^2 E_{gs}}{\partial \tilde{x}_m \partial \tilde{x}_n}\right)_{eq}$$

to obtain the phonon frequencies ω_n and normal coordinates q_n. Each q_n represents some collective motion in a harmonic well with frequency ω_n. As a result, we can write the vibrational Hamiltonian in terms of phonon creation and annihilation operators,

$$H_{vib} = \sum_n \hbar\omega_n\left(b_n^\dagger b_n + 1/2\right) + E_{gs}(x = 0) \quad (9.58)$$

where the $b_m^\dagger$ and b_m operators obey the familiar commutation relation for boson operators: $[b_n^\dagger, b_m] = \delta_{nm}$. The zero-point energy can be folded into the ground-state energy by defining the energy origin as

$$E_0 = E_{gs}(x = 0) + \hbar\frac{1}{2}\sum_n \omega_n$$

The electronic energy curve and corresponding harmonic levels for the ground state correspond to the lower parabola shown in Figure 9.5a [116].

Upon electronic excitation, the electronic density about the nuclei is changed. Consequently, a molecule in its ground-state equilibrium geometry will be subject to a potential force causing it to distort toward some new equilibrium geometry. Let $E_k(q)$ be the energy of the kth excited state as computed at nuclear coordinate q and expand this about the ground-state equilibrium geometry ($q = 0$):

$$E_k(q) = E_k(0) + \sum_n q_n \frac{\partial E_k(q)}{\partial q_n} + \sum_{nm} \frac{\partial E_k(q)}{\partial^2 q_n \partial q_m} q_n q_m + \cdots \quad (9.59)$$

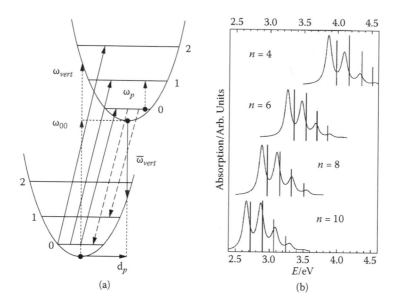

FIGURE 9.5 (a) Franck–Condon model in one normal dimension: 00 and vert: adiabatic and vertical transition frequencies, p: vibrational frequency, d_p: interstate distortion in the normal coordinate. Several of the v0 and 0v vibrational transitions in absorption and emission are given. (b) Theoretical 1Bu←1Ag absorption bands of all-*trans* polyenes with n double bonds. Vertical lines are the positions and intensities of the dominant vibrational features in absorption of *tert*-butyl-capped polyenes. Lines give the positions of the absorption and emission peaks from Ref. 126. (From S. Karabunarliev, M. Baumgarten, E. R. Bittner, and K. Müllen, *J. Chem. Phys.* **113**, 11372, 2000. Copyright (2000) American Institute of Physics.)

$E_k(0)$ is the excitation energy taken at the ground-state equilibrium geometry. We now make the assumption that the second derivative term is diagonal so that the ground-state normal modes q_n are also normal modes in the excited state. This assumption does not always hold and one can have the case where one needs to define new normal modes $q_n^{(k)}$ for each electronic state with frequencies $\omega_n^{(k)}$. Generally, the approximation is robust for larger conjugated polymer systems; however, for small molecules one should be careful in following this prescription. We also shall ignore at this point any coupling between the electronic states brought about by the geometric change in the molecule:

$$E_k(q) \approx E_k(0) + \sum_n \frac{1}{2}\omega^2 q_n^2 + f_n^{(k)} q_n$$

$$\approx E_k(0) + \frac{1}{2}\sum_n \omega_n^2 \left(q_n - d_n^{(k)}\right)^2 - \frac{1}{2}\sum_n \omega_n^2 \left(d_n^{(k)}\right)^2 \qquad (9.60)$$

Thus, within the assumptions here, a molecule in its kth excited state feels a linear force distorting it away from the ground-state geometry toward some new geometry

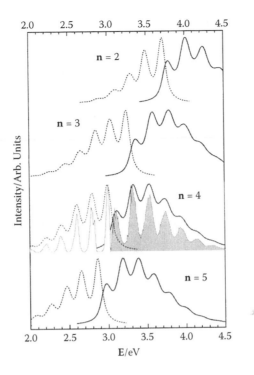

FIGURE 9.6 Theoretical absorption (solid) and emission (dotted) band shapes for the lowest electronic transitions in oligo(para-phenylenevinylenes) with n benzene rings. Computed transition probabilities without line-shape broadening are given for the tetramer as illustrated (from Ref. 116).

defined by the distortion coordinates:

$$d_n^{(k)} = -\frac{f_n^{(k)}}{\omega_n^2} \tag{9.61}$$

Finally, the quantity

$$E_{ad}^{(k)} = E_k(0) - \frac{1}{2}\sum_n \omega_n^2 \left(d_n^{(k)}\right)^2$$

is the relaxed (adiabatic) energy of the kth electronic state. Thus, our picture of the electronic energies of a molecule begins to look like what is shown in Figure 9.5a where we see a manifold of parabola each representing the potential energy surface for the various electronic states within the harmonic approximation. Once we know the harmonic frequencies of the phonons ω_k, the linear forces $f_n^{(k)}$, and the distortions $d_n^{(k)}$, we can compute the line-shape function for the absorption or emission of light.

Figures 9.5b and 9.6 give the predicted absorption and emission spectra for a series of conjugated oligomers. The absorpton peaks in Figure 9.5b are computed to the experimental positions and intensities for all trans-polyacetylene. Aside from a systematic shift in the position, the agreement between the theoretical Frank–Conden model and the experimental data generally improves with polymer chain length.

The line shape for transitions between initial and final electronic states $(k \rightarrow k')$ is given by the general expression

$$F(E) = \frac{1}{Z_k} \sum_{mn} e^{-E_m^{(k)}\beta} |\langle km_k|\mu|k'n_{k'}\rangle|^2 \delta\left(E_n^{(k')} - E_m^{(k)} - E\right) \qquad (9.62)$$

where the sum is over the Boltzmann-weighted initial vibrational state of electronic state k and all possible final vibrational states of electronic state k'. Z_k is the canonical partition function for the vibrations on state k. $E_m^{(k)}$ is the vibronic energy of the mth vibrational level in the kth electronic state. Implicit in this expression is that we are summing over all the vibrational levels of all the phonon modes. In order to simplify our notation considerably, we shall consider the case where there is only a single dominant phonon mode. The generalization to a multimode system is straightforward.[116,127] Notice that the line-shape function reduces to the Fermi golden rule rate in the limit

$$W = \frac{2\pi}{\hbar} \lim_{E \to 0} F(E)$$

To proceed, we make the Condon approximation that electronic transitions occur within a fixed nuclear framework. This allows us to approximate the transition matrix element as

$$\langle km_k|\mu|k'n_{k'}\rangle = \langle k|\mu|k'\rangle\langle m_k|n_{k'}\rangle$$

The first factor is simply the matrix element of the dipole operator between the initial and final electronic states. This determines the electronic selection rule and overall intensity of the transition. We take this to be independent of the vibrational state and thus pull it out of the summation. The second factor represents the overlap matrix element between two harmonic oscillator wave functions in displaced harmonic wells. For clarity and distinction, we label each vibrational state by the electronic state with which it is associated. Finally, since most spectra are taken from (or to) the ground electronic state, we assume from here on that one of our electronic states is the ground state $(k = 0)$. The modulus squared of the vibrational overlap between the m_k and $n_{k'}$ levels is given by

$$|\langle m_k|n_{k'}\rangle|^2 = e^{-X} \frac{m_k!}{n_{k'}!} X^{m_k - n_{k'}} \left(L_{m_k}^{n_{k'}-m_k}(X)\right)^2 \qquad (9.63)$$

where $X = d^2/2$, d is the dimensionless distortion between the minima of state k and the minima of state k', and $L_a^b(x)$ is an associated Laguerre polynomial.[128] At low temperatures—that is, where $kT < \hbar\omega_n$—the vibrational population of the initial state is concentrated in the lowest lying vibrational level; thus, we can simplify the Franck–Condon factor to

$$|\langle 0_k|n_{k'}\rangle|^2 = e^{-X} \frac{X^{n_{k'}}}{n_{k'}!} \qquad (9.64)$$

Finally, we recall the fact that the delta function can be represented as the limiting case of a Lorentzian

$$\delta(x - x_o) = \lim_{\varepsilon \to 0} \frac{1}{\pi} \frac{\varepsilon}{(x - x_o)^2 + \varepsilon^2} \qquad (9.65)$$

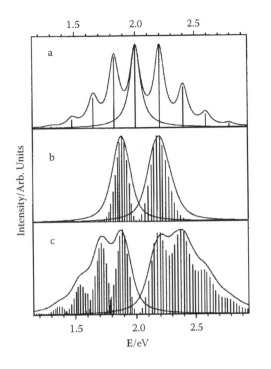

FIGURE 9.7 Vibrational transitions in absorption and emission within the harmonic Condon approximation. Models of (a) coupling in a double-bond stretching mode (0.2 eV), (b) coupling in a ring-torsional mode (0.01 eV), and (c) coupling in both modes. The curves are the convolutions for a Lorentzian linewidth of 0.04 eV for the vibrational transitions. The asymmetry between absorption and emission is a consequence of a minor stiffening of the accepting vibrational modes for absorption (from Ref. 127).

which is, of course, the line shape for damped oscillator. Consequently, we can associate a lifetime $\tau = 1/\gamma$ to each vibrational mode to account for the fact that the molecule is embedded in a continuum and, thus, write the absorption (or emission) line shape as

$$F(\omega) = |\langle k|\mu|k'\rangle|^2 \sum_{n_{k'}=0}^{\infty} e^{-X} \frac{X^{n_{k'}}}{n_{k'}!} \frac{\hbar\gamma}{\pi} \frac{1}{\left(E_0^{(k')} - \left(E_0^{(k)} - \hbar\omega_{nk'}n_{k'}\right)^2 + (\hbar\gamma)\right)^2} \quad (9.66)$$

Figure 9.7 shows the vibronic line shapes for a model conjugated polymer. In Figure 9.7a, only high frequency models are included in the model. This produces the familiar symmetric absorption and emission line structure. In Figure 9.7b, only low frequency phonon modes were included in our model. The absorption and emission line shapes are nearly symmetric; however, the absorption band is slightly broader than the emission. Finally, in Figure 9.7c, we include both high and low frequency modes in our model. Here, the absorption band is significantly broader than the absorption band and the high frequency vibronic fine structure is somewhat washed out. The emission, on the other hand, exhibits well resolved vibronic fine structure.

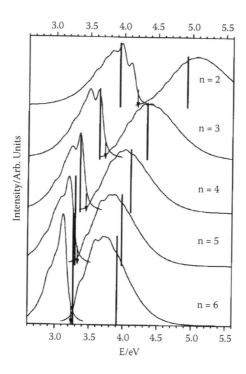

FIGURE 9.8 Computed S1←S0 absorption and emission bands of p-polyphenyls with **n** phenyl/phenylene rings. Arrows mark the theoretical electronic origins (from Ref. 127).

Figure 9.8 shows the theoretical absorption and emission spectra for various oligomers of polyphenylenevinylene (OPV) as computed using a semiempirical approach implemented within the MOPAC code.[116]* Here we can see a very clear vibronic progression indicative of the strong vibronic coupling between the π-electronic system and the skeletal vibrations. Notice that in the case for $n = 4$, OPV with four phenylene rings, we include the contribution from all the modes without imposing any additional vibronic line broadening. We can see that the main vibronic fine-structure features are composed of nearly a continuum of finer-grained lines corresponding to the contributions from the low-frequency modes of the molecule. In this case, the prominent vibronic progression in the emission and absorption spectra is due to the C=C stretching modes.

In the absorption spectra, the lines correspond to excitations from the lowest vibrational state to all possible vibrational states in the first electronic excited state. Consequently, all the fine structure is to the blue of the absorption origin (corresponding to the 0-0 vibronic transition). Thus, we can assign the main absorption peaks as 0-0, 0-1, 0-2, and so forth. The emission spectra, on the other hand, correspond to transitions from the 0 vibrational level in the upper electronic state to the nth vibrational level of the lower state. As n increases, the energy gap decreases so all emission peaks are to the red of the 0-0 line.

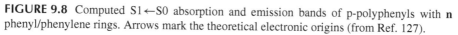

* Our modification of the MOPAC code is available upon request.

9.5   SUMMARY

Electron-phonon interactions play a significant role in the electronic states of conjugated systems. In this chapter, we have gone from an analytic treatment of elec-
tron/phonon coupling via the SSH model, used this approach to study exciton self-
trapping, and finally presented a linear-coupling model that when combined in con-
cert with a semiempirical technique (such as MOPAC) can reproduce the vibronic fine
structure of wide range of conjugated polymer systems.

9.6   PROBLEMS AND EXERCISES

Problem 9.2   Consider a problem for a mixed valency metal complex Ru^{II}-Ru^{III}.
Neglecting the electronic repulsion, we can write a simple two-state model as

$$H = H_{el} + H_{nuc}$$

where

$$H_{nuc} = \frac{p_{nuc}^2}{2M} + \frac{1}{2}kx^2$$

describes the nuclear motion and

$$H_{el} = \beta\left(a_1^{\dagger}a_2 + a_2^{\dagger}a_1\right) + g\hbar\omega\left(a_2^{\dagger}a_2 - a_1^{\dagger}a_1\right)X\left(\frac{2M\omega}{\hbar}\right)^{1/2}$$

describes the transfer of an electron from atom 1 to atom 2. M, k, β, ω, and g
are the nuclear mass, force constant, tunneling term, frequency $\omega = \sqrt{k/M}$, and
electron/phonon coupling (dimensionless). X represents the relative displacement
between the two atoms.

Define the Born–Oppenheimer (BO) potential surface as

$$V_{BO}(X) = V_{nucl}(X) + \langle\psi|H_{el}|\psi\rangle$$

for a given electronic state ψ.

1. Determine the Fock matrix elements for the electrons. This is a 2×2 matrix
   with
   $$f_{ij} = \langle\{[a_i, H], a_j^{\dagger}\}\rangle$$
   The elements of this matrix will depend upon X.
2. Determine the electronic eigenvalues from f. Add these to V_{nuc} to determine the two BO potential curves, $V_{BO}(X)$. Plot these curves using the
   parameters $\beta = -0.02$ eV, $g = 1$, $\omega = 500$ cm^{-1}, $M = 10,000m_e$. How
   do these curves change if $\beta = -2.0$ eV?
3. Suppose in describing a mixed valance Ru complex Ru^{II}-Ru^{III}, we take
   $a_1^{\dagger}$ as creating an electron in a d_{xy} orbital on the left atom analogously for
   the atom on the right. Assume at time zero the electronic configuration is
   Ru^{II}-Ru^{III} so that
   $$\psi_{el}(t = 0)a_1^{\dagger}|0\rangle$$

Using the Franck–Condon principle, calculate the electronic excitation energy for the parameters in part 2. This type of transition is termed the intervalence transfer band.

4. Using the Marcus theory of electron transfer, determine the driving force, the reorganization energy, and nonadiabatic coupling for the parameter cases you considered above. Comment on which "regime" each parameter set corresponds to and compute the electron transfer rate at $T = 300$ K.

5. Finally, using the Franck–Condon principle and Fermi's golden rule, plot the electronic absorption and emission spectra (at 300 K) as functions of photon frequency for the two parameter cases. [Hint: See Equation 9.66 and discussion in the text.]

Problem 9.3   Consider the time evolution of a spin state of an electron in a magnetic field. Take the unperturbed Hamiltonian to be the Zeeman Hamiltonian with static magnetic field B_o taken in the z direction:

$$H_o = \gamma B_o \hat{S}_z$$

where S_z is the usual spin operator $\hat{S}_z |m\rangle = m|m\rangle$ and γ is the gyromagnetic ratio. Consider the case where the coupling is a time-varying interaction with a magnetic field in the x direction so that the perturbing term is

$$V(t) = \gamma B_x \hat{S}_x \cos(\Omega t)$$

where B_x is the perturbing field and Ω the frequency.

1. At $t = 0$, the electron is prepared in the spin-down state α. What is the probability of ending up in the spin-up state β as a function of time? Take $\Omega = \gamma B_o$ and use first-order perturbation theory.

2. Now, let us solve the time-dependent Schrödinger equation exactly. For this, write the time-evolved α state as

$$\psi(t) = C_\alpha(t)|\alpha\rangle + C_\beta(t)|\beta\rangle$$

and write the equations of motion for the coefficients C_α and C_β. Next, take the limit that $\Omega = \gamma B_o$. Carefully consider the time-evolution of each term and neglect all terms that vary as $\exp(\pm 2i\Omega t)$ (Rotating Wave Approx). Show that these equations are the same as you derived in the first part.

3. Solve the equations of motion numerically for the case where the driving term is both on and off resonance with the Zeeman splitting. Make a plot of the survival probability of the initial spin-up state as a function of time for the resonant and nonresonant cases. How do your numerical results compare with the results you derived above?

10 Lattice Models for Transport and Structure

In Chapter 8 we largely cast our discussion of the electronic structure of π-conjugated systems upon the idea that the C $2p_z$ orbitals provided a good basis for the molecular orbitals. We also made the simplifying assumption that the atom-centered orbitals were orthogonal to each other. This approximation goes under the moniker of "neglect of differential overlap" (NDO) in the quantum chemical literature and "tight-binding approximation" in the solid-state physics literature. In this chapter, we shall continue along these lines in order to discuss transport and dynamics in extended systems.

10.1 REPRESENTATIONS

10.1.1 BLOCH FUNCTIONS

Consider the Schrödinger equation for a system with a periodic potential (and with Hamiltonian H_o) and some external potential U. For $U = 0$, the system corresponds to a perfectly periodic system while U represents the contributions from point defects or impurities within the lattice. Alternatively, U could represent an entirely external potential, say, from an electric or magnetic field, or the radiation field. While the physics is different in each case, the underlying mathematical technique used will be the same. We can expand the wave function in terms of a complete set of plane waves defined by the periodicity of the underlying system.

For an unbound, free electron in one dimension, $V(x) = 0$ at all points and we have the general solution of the Schrödinger equation

$$\psi(k, r) = ae^{ikr} + be^{-ikr}$$

with energy

$$E(k) = \frac{\hbar^2 k^2}{2m}$$

If we force the system to be confined to $x \in [0, L]$ as in the case of a particle in a box, then k can take only discrete values and the eigenstates read

$$\psi(k, r) = \sqrt{\frac{2}{L}} \sin(kr)$$

with $k = n\pi/L$ and $n = 1, 2, 3, \ldots$. Likewise, for the periodic system

$$\psi(k, r) = \sqrt{\frac{1}{L}} e^{\pm ikr}$$

with $k = n2\pi/L$. In Figure 10.1 we show $E(k)$ for the free particle, the bound particle, and the particle on a periodic lattice. For the free particle, all values of k give rise to

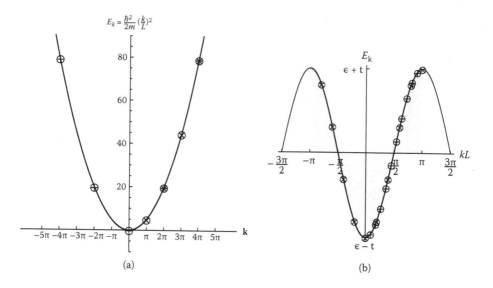

$$E_k = \frac{\hbar^2}{2m}\left(\frac{k}{L}\right)^2$$

(a)   (b)

FIGURE 10.1 (a) Energy $E(k)$ versus k for a free particle (solid line), a particle on a periodic lattice ($\oplus$) and a particle in a box $\otimes$. (b) Energy band for particle on a discrete lattice. The points indicated by $\oplus$ represent the eigenenergies for a linear chain ($\otimes$) and a ring ($\oplus$) of 10 atoms. The shaded region between $-\pi$ and π defines the first Brillouin zone.

stationary solutions of the Schrödinger equation. However, because of the imposed boundary conditions for the other two cases, only certain values of k give rise to stationary solutions.

Let us define for a periodic system the basis function ϕ_j as a localized atomic orbital centered about atom #j. As in the Hückel model, we consider ε_j to be the energy to put an electron in orbital #j and t_{ij} the energy to transfer an electron (of a given spin) from orbital #i to orbital #j. Taking all sites to be equivalent and $t_{ij} \neq 0$ only between neighboring sites ($i = j \pm 1$), the stationary Schrödinger equation reads

$$\phi_j \varepsilon + t(\phi_{j+1} + \phi_{j-1}) = E\phi_j \qquad (10.1)$$

Next, we take advantage of the fact that the system is periodic and write $\phi_{j+n} = e^{ikna}\phi_j$ where n is an integer and L is the spacing between adjacent atoms

$$\phi_j \varepsilon + t(e^{+iknL} + e^{iknL})\phi_j = E\phi_j \qquad (10.2)$$

This allows us to eliminate ϕ_j from both sides and obtain the energy band

$$E(k) = \varepsilon + 2t \cos(kL)$$

For a finite lattice of atoms, we have the Hückel results we have seen before. In Figure 10.1b we show the energies for both a chain and a ring of 10 atoms superimposed upon the energy curve for the infinite system. The expressions for the energies were given in Equations 8.18 and 8.45. For the case of a ring of N atoms, the boundary conditions give stationary solutions only when kL takes integer multiples of $2\pi/N$.

Extending this to an infinite ring, one can easily see that all points $kL \in [-\pi, \pi]$ would give rise to stationary solutions. Going beyond this range, the energies repeat themselves and thus we can limit our discussion to energies with wave vectors $k \in [-\pi/L, \pi/L]$. This defines the first Brillouin zone (BZ).

We can generalize this idea considerably by defining all basis functions in terms of complete sets of functions over the first BZ.

Bloch functions are eigenfunctions of H_o

$$H_o \psi_n(k, r) = E_n(k) \psi_n(k, r) \tag{10.3}$$

where n is a band index and k is the wave vector or crystal momentum. This representation is termed the *crystal momentum representation* (CMR) since it is based upon states with definite k. These functions are orthonormal:

$$\int \psi_n^*(k, r) \psi_m(k', r) dr = \delta_{nm} \delta(k - k') \tag{10.4}$$

Note that throughout our discussion here, the integral over r is taken over all space. One can also show that the Bloch functions are complete:

$$\sum_n \int \psi_n^*(k, r) \psi_n(k, r') dk = \delta(r - r') \tag{10.5}$$

where the integral is over a single BZ. Since the Bloch functions are a complete set, we can expand any general wave function in terms of the Bloch functions

$$\psi(r) = \sum_n \int \phi_n(k) \psi_n(k, r) \tag{10.6}$$

where the expansion coefficients describe the wave function in the CMR. Hence, the Bloch functions can be thought of as the transformation coefficient $\psi(k, r) = \langle r | nk \rangle =$ between the CMR and r.

$$|nk\rangle = \frac{1}{\sqrt{\Omega}} e^{ikr}$$

where Ω is the volume of a unit cell.

10.1.2 WANNIER FUNCTIONS

Wannier functions are essentially spatially localized basis functions that can be derived from the band structure of an extended system. Quantities such as the exchange interaction and Coulomb interaction can be easily computed within the atomic orbital basis; however, there are many known difficulties in computing these within the crystal momentum representation. Because of this, it is desirable to develop a set of orthonormal spatially localized functions that can be characterized by a band index and a lattice site vector, R_μ. These are the Wannier functions, which we shall denote by $a_n(r - R_\mu)$ and define in terms of the Bloch functions

$$a_n(r - R_\mu) = \frac{\Omega^{1/2}}{(2\pi)^{d/2}} \int e^{-ikR_\mu} \psi_{nk}(r) dk \tag{10.7}$$

The integral is over the Brillouin zone with volume $V = (2\pi)^d/\Omega$, and Ω is the volume of the unit cell (with d dimensions). A given Wannier function is defined for each band and for each unit cell. If the unit cell happens to contain multiple atoms, the Wannier function may be delocalized over multiple atoms. The functions are orthogonal and complete.

The Wannier functions are not energy eigenfunctions of the Hamiltonian. They are, however, linear combinations of the Bloch functions with different wave vectors and therefore different energies. For a perfect crystal, the matrix elements of H in terms of the Wannier functions are given by

$$\int a_l^*(r - R_\nu) H_o a_n(r - R_\mu) dr = \frac{\Omega}{(2\pi)^d} \int e^{i(qR_\nu - kR_\mu)} \psi_{lk}(r) H_o \psi_{nk}(r) dr dq dk$$

$$= \mathscr{E}_n(R_\nu - R_\mu)\delta_{nl} \qquad (10.8)$$

where

$$\mathscr{E}_n(R_\nu - R_\mu) = \frac{\Omega}{(2\pi)^d} \int e^{ik(R_\nu - R_\mu)} E_n(k) dk$$

Consequently, the Hamiltonian matrix elements in the Wannier representation are related to the Fourier components of the band structure, $E_n(k)$. Therefore, given a band structure, we can derive the Wannier functions and the single-particle matrix elements, F_{mn}°.

10.2   STATIONARY STATES ON A LATTICE

For convenience, let us restrict our attention to a single band and consider the Schrödinger for the stationary states in a general periodic potential

$$\left(-\frac{\hbar^2}{2m}\nabla^2 + V(r)\right)\psi(r) = E\psi(r) \qquad (10.9)$$

Let us now transform to the CMR using

$$\psi(r) = \frac{1}{\sqrt{\Omega}}\sum_k \phi_k e^{ikr} \qquad (10.10)$$

In the CMR, the Schrödinger equation becomes

$$\frac{\hbar^2 k^2}{2m}\phi_k + \sum_{k'}\langle k|V|k'\rangle\phi_{k'} = E\phi_k \qquad (10.11)$$

where we see that the kinetic energy term is diagonal in the CMR while the potential term couples components of the wave function with different values of the crystal momentum k.

Now, let us assume that the potential can be written as a superposition of core potentials centered about each atomic site. In other words, we expand

$$V(r) = \sum_j v(r - r_j)$$

where $v(r)$ is a weak pseudopotential specific to a given atom. Using this approximation, we can evaluate $\langle k|V|k'\rangle$ by taking advantage of the properties of the Fourier transformation

$$\langle k|V|k'\rangle = \frac{1}{\Omega} \int dr\, e^{i(k-k')r} \sum_j v(r - r_j) \tag{10.12}$$

Swapping the integral and sum,

$$\langle k|V|k'\rangle = \frac{1}{\Omega} \sum_j e^{i(k-k')r_j} \int dr\, e^{i(k-k')(r-r_j)} v(r - r_j) \tag{10.13}$$

Now we change the variable of integration from r to $r - r_j$ and factor the volume Ω into $\Omega = N\Omega_o$, where N is the number of atoms and Ω_o is the atomic volume

$$\langle k|V|k'\rangle = \frac{1}{N} \sum_j e^{i(k-k')r_j} \frac{1}{\Omega_o} \int dr\, e^{i(k-k')r} v(r)$$

$$= S(k' - k)v_{k'-k} \tag{10.14}$$

In this last step we factor the interaction into two terms: a structure factor $S(q)$ and a form factor $v(q)$, where $q = k' - k$. The structure factor is given as the sum over the atomic positions

$$S(q) = \frac{1}{N} \sum_j e^{-iqr_j} \tag{10.15}$$

and is equivalent to the structure factor obtained from diffraction theory. The second factor, termed the *form factor*, is given by

$$v(q) = \frac{1}{\Omega_o} \int d^3r\, e^{-iq\cdot r} v(r) \tag{10.16}$$

Here we have explicitly indicated that the integration is over a three-dimensional volume $dV = d^3r$. Since $v(r)$ is centered about an atomic site, we can expand it in terms of the spherical harmonics

$$v(r) = \sum_{l=0}^{\infty} \sum_{m=-l}^{+l} v_{lm}(r)Y_{lm}(\theta, \phi) \tag{10.17}$$

Likewise, the exponential can be expanded as

$$e^{iq\cdot r} = 4\pi \sum_{l=0}^{\infty} \sum_{m=-l}^{+l} i^l Y_{lm}^*(\theta_q, \phi_q) j_l(qr)Y_{lm}(\theta, \phi) \tag{10.18}$$

where (θ_q, ϕ_q) given the direction of q relative to some z axis and $j_l(qr)$ is a spherical Bessel function. If we choose the z axis to be along the direction of q, then $q_x = q_y = 0$ and $q_z = q$ and we can write

$$e^{iq\cdot r} = e^{iqr\cos\theta}$$

It is then straightforward to show that

$$e^{i\mathbf{q}\cdot\mathbf{r}} = \sum_{l=0}^{\infty} i^l (2l+1) j_l(qr) P_l(\cos\theta) \tag{10.19}$$

Pulling everything together, we obtain

$$v(q) = \frac{1}{\Omega_o} \sum_{l=0}^{\infty} i^l \sqrt{4\pi(2l+1)} \int_0^{\infty} v_l(r) j_l(qr) r^2 dr \tag{10.20}$$

For the case of a spherically symmetric pseudopotential, v_l has only one component and we arrive at

$$v(q) = \frac{4\pi}{\Omega_o} \int_0^{\infty} r^2 v(r) j_0(qr) dr$$

$$= \frac{4\pi}{\Omega_o} \int_0^{\infty} r^2 v(r) \frac{\sin(qr)}{qr} dr \tag{10.21}$$

Within the pseudopotential approximation, we simply need to know the pseudopotential for a given atom and the structure factor of the material to determine the band structure or other properties related to the electronic structure.

One form of the pseudopotential is the so-called empty-core potential, which takes the form[129]

$$v(r) = \begin{cases} 0 & \text{for } r < r_c \\ v_{free}(r) & \text{otherwise} \end{cases} \tag{10.22}$$

where v_{free} is the free-atom potential that may take the form of a screened Coulomb potential

$$v_{free}(r) = \frac{-Z_{eff} e^2}{r} e^{-\kappa r} \tag{10.23}$$

where Z_{eff} is the effective core charge and κ is a screening length. Hence, we can evaluate the integral for the form factor as

$$v(q) = -\frac{4\pi Z_{eff} e^2}{\Omega_o q} \int_{r_c}^{\infty} r^2 \frac{e^{-\kappa r}}{r} \frac{\sin(qr)}{r} dr \tag{10.24}$$

$$= -\frac{4\pi Z_{eff} e^2}{\Omega_o (q^2 + \kappa^2)} \cos(qr_c) \tag{10.25}$$

In order to determine the form factor for a given atomic species within the empty-core approximation, we need to adjust the cutoff radius r_c until the eigenenergy of the corepotential matches the energy of the actual atomic species. For example, if we consider the corepotential for the 3s state of the Na atom, we can use a $Z_{eff} = 1$ and adjust r_c until the lowest energy radial eigenstate of the pseudopotential is equal to $E(3s) = -4.96$ eV. Doing so, we find that $r_c = 1.037$ Å. Formally, we may

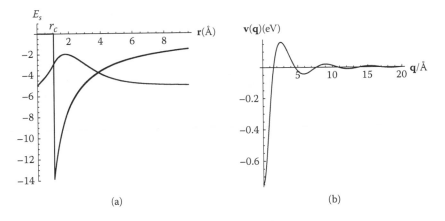

FIGURE 10.2 (a) Empty cone pseudo potential and radical wavefunction for sodium (Na) atom (b) Na pseudo potential from factor.

anticipate that the real purpose of κ is to ensure that the integral has a finite radius of convergence and that we should take $\kappa \to 0$. However, within the Thomas–Fermi statistical approximation, we find that

$$\kappa^2 = \frac{4me^2 k_F}{\pi \hbar^2} \qquad (10.26)$$

where k_F is the Fermi momentum. The empty-core pseudopotential, approximate 3s radial wave function, and pseudopotential form factor for a Na atom are given in Figure 10.2. For Na, we use $k_F = 0.92\text{Å}^{-1}$.

The structure factor $S(q)$ can be obtained from scattering data or from simulation. In general, the structure factor is related to the Fourier transform of the pair distribution function.[130] This gives us an independent way to incorporate the structure of a system into the calculation of the electronic band structure. The pair distribution is given by summing over all pairs of atoms in the sample. For an isotropic single-component system

$$g(r) = \frac{V}{N(N-1)} \left\langle \int d^3 r' \sum_{i \neq j} \delta(r' - r_i)\delta(r' + r - r_j) \right\rangle$$

The quantity $\rho g(r)dr$ gives the "probability" of finding a second atom (or molecule) at some distance r away, given that there is an atom or molecule at the origin and $\rho = N/V$ is the density. It is not really a probability since

$$4\pi \int_0^\infty \rho g(r) r^2 dr = N - 1 \approx N$$

To be precise, $\rho g(r)dr$ gives the number of atoms between r and $r + dr$ about a central atom. The relation between $g(r)$ and $S(q)$ is[130]

$$S(\mathbf{q}) = \rho \int e^{i \mathbf{q} \cdot \mathbf{r}}(g(r) - 1)d^3 \mathbf{r}$$

For a single chain with interatomic spacing d, the atoms on the chain are located at distances $r_n = nd$ from an atom located at the origin. Thus, it is easy to see that for a chain of N atoms

$$S(q) = \frac{1}{N} \sum_{n=0}^{N-1} e^{-iqnd} \tag{10.27}$$

$$= \frac{1}{N}(1 + x + x^2 + x^3 + \cdots + x^{N-1}) \tag{10.28}$$

$$= \frac{1}{N} \frac{x^{N+1} - 1}{x - 1} \tag{10.29}$$

with $x = \exp(iqd)$. For a periodic system, q takes integer multiples of $2\pi/(Nd)$. Consequently, the numerator in this last expression is always equal to zero. The denominator is zero when n/N is an integer. Thus, $S(q) = 1$ at values of $q = n2\pi/d$. These are termed the *lattice wave numbers* or reciprocal lattice vectors.

10.2.1 A SIMPLE BAND-STRUCTURE CALCULATION

We just found that for the one-dimensional chain the structure factor only takes nonvanishing values at integer multiples of $q = 2\pi/d$. This means that a given Bloch basis function $|k\rangle$ is coupled only to Bloch functions $|k \pm nq\rangle$. Since this only occurs once as we move across the BZ, we can write the Hamiltonian operator in the CMR as a 3×3 matrix where $|k\rangle$ is coupled to Bloch states $|k + q\rangle$ and $|k - q\rangle$,

$$H(k) = \begin{bmatrix} \dfrac{\hbar^2(k+q)^2}{2m} & v(q) & 0 \\[2ex] v(q) & \dfrac{\hbar^2 k^2}{2m} & v(q) \\[2ex] 0 & v(q) & \dfrac{\hbar^2(k-q)^2}{2m} \end{bmatrix} \tag{10.30}$$

These bands are shown in Figure 10.3 for the case where $v(q) = -2$ (with $m = 1$ and $\hbar = 1$). Note that bands are present in the final analysis, but the highest energy band is only (nearly) degenerate with the second band at $k = 0$. Consequently, we can approximate $H(k)$ as a 2×2 system when $k > 0$

$$H(k > 0) \approx \begin{bmatrix} \dfrac{\hbar^2 k^2}{2m} & v(q) \\[2ex] v(q) & \dfrac{\hbar^2(k-q)^2}{2m} \end{bmatrix} \tag{10.31}$$

Diagonalizing this yields a fairly good approximation for the two lower bands for $k > 0$,

$$E_k = \frac{1}{2}\left(\frac{\hbar^2 k^2}{2m} + \frac{\hbar^2(k-q)^2}{2m}\right) \pm \sqrt{\left(\frac{1}{2}\left(\frac{\hbar^2 k^2}{2m} - \frac{\hbar^2(k-q)^2}{2m}\right)\right)^2 + v(q)^2} \tag{10.32}$$

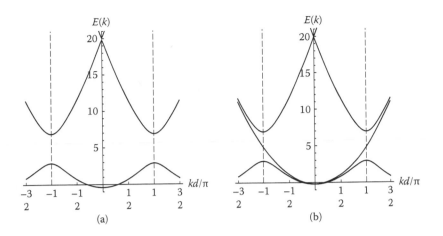

FIGURE 10.3 Bands from Eq. (10.32) for three 1D free-electron states coupled by a pseudopotential with form factor $v(q)$. (a) Three-state model. (b) Approximate two-state model.

For the left side ($k < 0$), we would write a similar 2×2 Hamiltonian except we would couple k to $k + q$. The band for the full BZ would the be approximated by joining these two cases as shown by the curves in Figure 10.3b. The approximation is robust across both respective half zones. However, once we move too far into the other half of the zone (that is, where $k < 0$ for the curve or $k > 0$ for the curve), the approximation breaks down considerably and we do not recover the second avoided crossings. Also, at $k = 0$ the curves cross, whereas in the three-state approximation we have a small gap.

10.3 KRONIG–PENNEY MODEL

A drastic simplification to the core-potential model is where we replace the Coulomb interaction between the electron and the core atom with a simple rectangular potential about each atom on the lattice. From Bloch's theorem, we only need to find a solution on a single period of the lattice and make sure that it is both continuous and smooth:

$$\psi(x) = e^{ikx} u(x)$$

where $u(x)$ is smooth and periodic. The function $u(x)$ satisfies

$$u(x + d) = u(x)$$

and

$$u'(x + d) = u'(x)$$

Since everything is periodic over d, we need only consider the solution in this range, making sure that the solution and its derivative are both continuous. We have two regions, region 1 where $0 < x < a$ and the second where $a < x < b$. We take $a + b = d$ to be the lattice spacing. In the first region,

$$\psi_1'' = -\alpha \psi_1$$

$(\alpha^2 = 2mE/\hbar^2)$, which gives

$$\psi_1 = A_1 e^{i\alpha x} + B_1 e^{-i\alpha x}$$

In region 2,

$$\psi_2'' = -\beta^2 \psi_2$$

$(\beta^2 = 2m(V_o - E)/\hbar^2)$ and the solution

$$\psi_2 = A_2 e^{i\beta x} + B_2 e^{-i\beta x}$$

To find $u(x)$ we need to do some manipulation of the wave function in each region:

$$\psi_1(x) = e^{ikx} u_1(x) = e^{ikx} \left(A_1 e^{i(\alpha-k)x} + B_1 e^{-i(\alpha+k)x} \right)$$

and

$$\psi_2(x) = e^{ikx} u_2(x) = e^{ikx} \left(A_2 e^{i(\beta-k)x} + B_2 e^{-i(\beta+k)x} \right)$$

Now we are in a position to determine the coefficients and $u_i(x)$ in each region. At the potential steps, the two solutions and their derivatives must match. Thus, at $x = 0$,

$$\psi_1(0) = \psi_2(0) \tag{10.33}$$
$$\psi_1'(0) = \psi_2'(0) \tag{10.34}$$

Likewise,

$$u_1(a) = u_2(-b) \tag{10.35}$$
$$u_1'(a) = u_2'(-b) \tag{10.36}$$

These last two conditions enforce the periodicity of the lattice. This leads to the following matrix equation:

$$\begin{pmatrix} 1 & 1 & -1 & -1 \\ \alpha & -\alpha & -\beta & \beta \\ e^{ia(\alpha-k)} & e^{-ia(k+\alpha)} & -e^{ib(\beta-k)} & e^{-ib(k+\beta)} \\ e^{ia(\alpha-k)}(\alpha-k) & -e^{-ia(k+\alpha)}(k+\alpha) & -e^{ib(\beta-k)}(\beta-k) & e^{-ib(k+\beta)}(k+\beta) \end{pmatrix} \begin{pmatrix} A_1 \\ B_1 \\ A_2 \\ B_2 \end{pmatrix}$$

$$\tag{10.37}$$

Since we have to ignore the trivial solution of $A_1 = A_2 = B_1 = B_2 = 0$, the determinant of the matrix must vanish. This leads to the condition

$$\cos(kd) = \cosh(\beta b)\cos(\alpha a) - \frac{\alpha^2 - \beta^2}{2\alpha\beta} \sinh(2\beta)\sin(\alpha a)$$

10.4   QUANTUM SCATTERING AND TRANSPORT

In order to study transport in materials, we need to discuss some basic principles of scattering theory. Our goal is to find an unbound stationary solution of the one-dimensional Schrödinger equation

$$\left(-\frac{\hbar^2}{2m}\frac{\partial^2}{\partial x^2} + V(x)\right)\psi(x) = E\psi(x) \tag{10.38}$$

where E is the energy of the scattering particle. Whereas in the bound-state problem, E is an eigenvalue of the Hamiltonian operator, here E can take any value. The basic idea is to consider how an incident wave function starting in the distant past is transformed into an outgoing wave function in the distant future. For convenience, we place the interaction close to the origin so that in regions to the left and right of the interaction region, the wave function behaves as a free particle. For a wave function moving in one dimension, we can write the wave function in the region to the left of the interaction as

$$\psi_L = A_L e^{ikx} + B_L e^{-ikx}$$

where A_1 and B_1 are the incident and reflected amplitudes and

$$k = \sqrt{2m(E - V(x))}/\hbar$$

is the wave vector. For regions to the right of the interaction we have

$$\psi_R = A_R e^{ikx} + B_R e^{-ikx}$$

where A_R is the amplitude traveling away from the interaction and B_R is the amplitude moving toward the interaction region. In most cases, we consider the case where the particle starts to the left and moves to the right. So, eventually we shall set $B_R = 0$. We shall keep it for now. The left and right amplitudes are related via the transfer matrix T:

$$\begin{pmatrix} A_R \\ B_R \end{pmatrix} = \begin{pmatrix} t_{11} & t_{12} \\ t_{21} & t_{22} \end{pmatrix} \cdot \begin{pmatrix} A_L \\ B_L \end{pmatrix} \tag{10.39}$$

We can also write the scattering matrix S as the transformation between the two incoming and the two outgoing amplitudes

$$\begin{pmatrix} B_L \\ B_R \end{pmatrix} = \begin{pmatrix} r & t \\ t^* & -r \end{pmatrix} \cdot \begin{pmatrix} A_L \\ A_R \end{pmatrix} \tag{10.40}$$

where r and t are the transmitted and reflected amplitudes. By normalization, $r^2 + t^2 = 1$. Typically, in scattering calculations, knowing the S or T matrix is sufficient for determining the scattering cross-section, quantum transport properties, or reaction rate for a particle of mass m and a potential $V(x)$.

As an example, let us take the case where the potential is taken to be a series of discrete steps. The simplest is where $V(x)$ is a step function where $V(x < x_s) = 0$

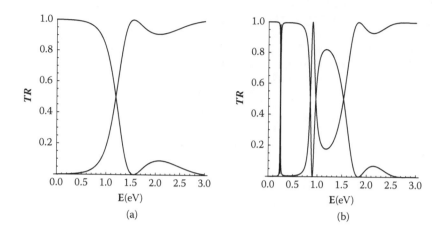

FIGURE 10.4 Transmission and reflection probabilities for a (a) single and (b) double-barrier problem.

and $V(x \geq x_s) = V_o$. The wave function and its first derivative must be continuous, so we need to join the left and right solutions at $x = x_s$:

$$\psi_L(x_s) = \psi_R(x_s) \tag{10.41}$$

$$\psi_L'(x_s) = \psi_R'(x_s) \tag{10.42}$$

Thus, we can relate the amplitude coefficients:

$$A_L e^{ik_L x_s} + B_L e^{-ik_L x_s} = A_R e^{ik_R x_s} + B_R e^{-ik_R x_s} \tag{10.43}$$

$$ik_L A_L e^{ik_L x_s} - ik_L B_L e^{-ik_L x_s} = ik_R A_R e^{ik_R x_s} - ik_R B_R e^{-ik_R x_s} \tag{10.44}$$

or in matrix form

$$\begin{pmatrix} e^{ik_L x_s} & e^{-ik_L x_s} \\ ik_L e^{ik_L x_s} & -ik_L e^{-ik_L x_s} \end{pmatrix} \begin{pmatrix} A_L \\ B_L \end{pmatrix} = \begin{pmatrix} e^{ik_R x_s} & e^{-ik_R x_s} \\ ik_R e^{ik_R x_s} & -ik_R e^{-ik_R x_s} \end{pmatrix} \begin{pmatrix} A_R \\ B_R \end{pmatrix} \tag{10.45}$$

Or, matrix form,

$$M[x_s, k_L] \cdot \begin{pmatrix} A_L \\ B_L \end{pmatrix} = M[x_s, k_R] \cdot \begin{pmatrix} A_R \\ B_R \end{pmatrix} \tag{10.46}$$

where $k_L = (2mE)^{1/2}/\hbar$ and $k_R = (2m(E - V_0))^{1/2}/\hbar$ are the wave vectors on either side of the step. Thus, we can arrive at the transmission matrix as

$$T[x_s, k_L, k_R] = M^{-1}[x_s, k_L] \cdot M[x_s, k_R] \tag{10.47}$$

To generalize this, let us assume that $V(x)$ can be represented as a series of discrete steps at $x = \{x_1, \ldots, x_N\}$ spanning the interaction region. As we move from left to right, we can relate the wave-function amplitudes such that after N steps,

$$\begin{pmatrix} A_R \\ B_R \end{pmatrix} = T[x_N, k_{N-1}, k_N] \cdot T[x_{N-1}, k_{N-1}, k_N] \cdots T[x_1, k_1, k_L] \begin{pmatrix} A_L \\ B_L \end{pmatrix}$$

$$= T_{tot} \begin{pmatrix} A_L \\ B_L \end{pmatrix}$$

$$= \begin{pmatrix} t_{11} & t_{12} \\ t_{21} & t_{22} \end{pmatrix} \begin{pmatrix} A_L \\ B_L \end{pmatrix} \qquad (10.48)$$

where we can construct the total transfer matrix as the product of the intermediate transfer matrices representing a series of transmitted and reflected amplitudes due to each change in the potential. Once we have the total transfer matrix, we can relate the transmitted and reflected wave-function coefficients (A_R and B_L) to the incident amplitude, A_L:

$$B_L = -\frac{t_{21}}{t_{22}} A_L \qquad (10.49)$$

$$A_R = \left(t_{11} - \frac{t_{12}t_{21}}{t_{22}} \right) A_L \qquad (10.50)$$

In Figure 10.4a and 10.4b we show the transmission and reflection probabilities for scattering past a single barrier (a) and a double barrier (b). In the single-barrier case, the potential "bump" of $V_o = 1$ a.u. extends from $x = -\pi/2$ to $x = +\pi/2$, whereas in the double barrier, each bump of $V_o = 1$ a.u. is of width $\pi/2$ with a space of π in between. These are purely toy problems and we have chosen the parameters to be as simple as possible to illustrate the problem. Moreover, we can easily do this problem by hand. First, notice that the transmission probability is not a step function at $E = 1$ a.u. as one would expect for a classically described particle scattering from left to right. This is due to quantum tunneling contributions. Second, notice that for $E > 1$ a.u. the transmission probability as slight dip at $E \approx 2$ a.u. This is a quasi resonance since it corresponds to the case where the width of the barrier is an integer number of de Broglie wavelengths of the scattering wave. Consequently, the particle is partially reflected.

The double-barrier case can also be handled analytically, again providing a convenient check of the transfer matrix calculation. Here we see much more complex features in the transmission/reflection spectra. First, we have two sharp resonance features, one at $E = 0.21$ a.u. and a second at $E = 0.84$ a.u.

10.5 DEFECTS ON LATTICES

Now, let us modify this description a bit by including a defect in the middle of the chain. Keeping the transfer term t the same for the chain, we can include the defect by changing the site energy of one or more sites. First, let us consider one defect, at $j = 0$, where the site energy is ε_0, and ε_s is the same everywhere else. For $j = 1, 2, 3, \ldots, u_j = T e^{ikdj}$ is the transmitted wave. For $j = -1, -2, -3, \ldots,$

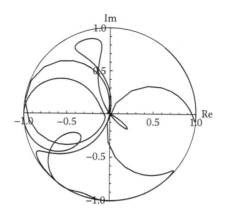

FIGURE 10.5 Argand diagram for the transmission and reflection coefficients for the model double-barrier problem. The loops indicate the presence of resonances.

we have both incident and reflected components: $u_j = Ie^{ikdj} + Re^{-ikdj}$. At $j = 0$, we have $u_0 = I + R$. Now, we match boundary conditions. At $j = 0$, we have the requirement that $I + R = T$. We also must satisfy the Schrödinger equation:

$$T(\varepsilon_s + \delta\varepsilon) + t(Te^{ikd} + Ie^{-ikd} + Re^{ikd}) = T\varepsilon \qquad (10.51)$$

where $\delta\varepsilon = \varepsilon_0 - \varepsilon_s$ is the barrier height. We can rearrange this last equation to

$$(-2t\cos(kd) + Te^{ikd} + Ie^{-ikd} + Re^{ikd}) = -T\delta\varepsilon \qquad (10.52)$$

We have too many unknowns: T, I, R. So, we can only specify the solution up to normalization: T/I and R/I. For example, we can derive

$$\frac{T}{I} = \frac{2it\sin(kd)}{2it\sin(kd) + \delta\varepsilon} \qquad (10.53)$$

We define the reflection and transmission probabilities as

$$\mathscr{R} = \left|\frac{R}{I}\right|^2 = \frac{\delta\varepsilon^2}{4t^2\sin^2(kd) + \delta\varepsilon^2} \qquad (10.54)$$

and

$$\mathscr{T} = 1 - \mathscr{R} = \frac{4t^2\sin^2(kd)}{4t^2\sin^2(kd) + \delta\varepsilon^2} \qquad (10.55)$$

In the limit that the barrier height is small, $\delta\varepsilon \approx 0$, the reflectivity $\mathscr{R} \to 0$ and the particle are transmitted through the system. We can define the velocity of the particle by taking the derivative of the band energy with respect to the momentum, k:

$$v = \frac{1}{\hbar}\frac{dE}{dk} = -2\frac{dt}{\hbar}\sin(kd) \qquad (10.56)$$

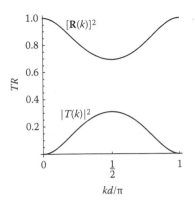

FIGURE 10.6 Transmission and reflection probablities for a chain with a single defect at the origin ($t = 1$, $\delta e = 3$).

Thus,

$$\mathscr{R} = \frac{\delta\varepsilon^2}{(\hbar v)^2/d^2 + \delta\varepsilon^2} \tag{10.57}$$

As $\delta\varepsilon^2 \to 0$, we can ignore the $\delta\varepsilon$ in the denominator and write

$$\mathscr{R} = \frac{\delta\varepsilon^2 d^2}{\hbar^2 v^2} \tag{10.58}$$

For this case, the particle is more or less delocalized over the entire chain and we can say that the probability of finding the particle on any given site is $1/Nd$ where N is the number of sites in the chain. So, the probability of the particle striking the barrier per unit time is v/Nd. In this limit the transmission rate is given by

$$k_T = \frac{v}{Nd}\frac{d^2\delta\varepsilon^2}{(\hbar v)^2} = \frac{d}{v}\frac{\delta\varepsilon^2}{N\hbar^2} \tag{10.59}$$

For larger barriers or smaller energies, we have the particle tunneling through the barrier and the transmission is given by

$$\mathscr{T} = 1 - \mathscr{R} = \frac{4t^2\sin^2(kd)}{4t^2\sin^2(kd) + \delta\varepsilon^2} \tag{10.60}$$

Substituting the definition of velocity from above and taking $\delta\varepsilon$ to be the dominant term in the denominator,

$$\mathscr{T} = \frac{\hbar^2 v^3}{\delta\varepsilon^2 d^2} \tag{10.61}$$

When the barrier is large compared to the energy, then the wave function is more or less a standing wave to the left of the barrier and the particle will strike the barrier at a rate $v/2L$. So, we can define the transmission rate as

$$k_T = \frac{v}{2L}\mathscr{T} = \frac{\hbar^2 v^3}{2L\delta\varepsilon^2 d^2} \tag{10.62}$$

To illustrate what is going on, let us consider an electron on a lattice with $\varepsilon_0 = -5$ eV, $\varepsilon_s = -10$ eV, and $t = -2.5$ eV. Rather than specifying the lattice spacing, we can plot (Figure 10.6) all of our results with respect to kd, which ranges from 0 to π. What we see is that the transmission probability peaks at $kd/\pi = 1/2$, giving a case where $\mathscr{R} = \mathscr{T}$.

10.6 MULTIPLE DEFECTS

Now, let us consider a more general case and examine what happens when the system has multiple defect sites. For the general case, we need to resort to numerical approaches and propagate a solution for u_j given some initial value. Rearranging the Schrödinger equation to solve for u_{j+1} gives

$$u_{j+1} = (u_j(E - \varepsilon_j) - tu_{j-1})/t \qquad (10.63)$$

If we have a particle transmitted to the right, then $u_j = Te^{ikdj}$. Since we need not specify the normalization, we can take $T = 1$ and iteratively determine u_j for the rest of the chain. This is a complex number and we can write it as $u_j = x_j + iy_j$. We are going to assume that the site energy to the left and right of the barrier is the same, so $E = \varepsilon_s + 2t\cos(k2)$ gives the dispersion relation between k and E for the two asymptotic regions. If this is the case, we can compute the transmission probability by comparing u_j and u_{j+1} on the opposite side of the chain (where we have incident and reflected components). The result is

$$\mathscr{T} = \frac{4\sin^2(kd)}{(x_{j+1} - x_j\cos(kd) + y_j\sin(kd))^2 + (y_{j+1} - x_j\sin(kd) - y_j\cos(kd))^2}$$
$$(10.64)$$

Consider the case in which we have two defect sites, one at $j = 0$ and another at $j = 3$ with $\varepsilon_j = \varepsilon_s + 3$ eV and $t = -1$ eV. We can set $\varepsilon_s = 0$ since it is an arbitrary zero of energy. The defects are spaced so that there are two nearly bound states in the

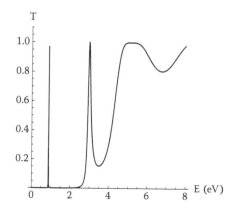

FIGURE 10.7 Transmission probability versus energy and kd.

one-dimensional well formed by defects. We consider here a particle incident from the left and transmitted to the right. In Figure 10.7 is the transmission probability vs energy and kd for this system. We see clearly two maxima in the transmission. These correspond to energies where the scattering energy closely matches the energy of a quasi-bound state within the well. Consequently, these are termed *tunneling resonances* and the width of the resonance is proportional to the rate of decay of a particle initially trapped in the well.

A Miscellaneous Results and Constants

A.1 PHYSICAL CONSTANTS AND CONVERSION FACTORS

TABLE A.1
Physical Constants

Constant	Symbol	SI Value
Speed of light	c	299792458 m/s (exact)
Charge of proton	e	$1.6021764 \times 10^{-19}$ C
Permitivity of vacuum	ε_o	$8.8541878 \, ^{-12}$ J^{-1}C^{2}m^{-1}
Avagadro's number	N_A	6.022142×10^{23} mol^{-1}
Rest mass of electron	m_e	9.109382×10^{-31} kg

TABLE A.2
Atomic Units: In Atomic Units, the Following Quantities Are Unitary: $\hbar$, e, m_e, a_o

Quantity	Symbol or Expression	CGS or SI Equivalent
Mass	m_e	9.109382×10^{-31} kg
Charge	e	$1.6021764 \times 10^{-19}$ C
Angular momentum	$\hbar$	1.05457×10^{-34} Js
Length (bohr)	$a_o = \hbar^2/(m_e e^2)$	$0.5291772 \, ^{-10}$ m
Energy (hartree)	$E_h = e^2/a_o$	4.35974×10^{-18} J
Time	$t_o = \hbar^3/(m_e e^4)$	2.41888×10^{-17} s
Velocity	$e^2/\hbar$	2.18770×10^6 m/s
Force	e^2/a_o^2	8.23872×10^{-8} N
Electric field	e/a_o^2	5.14221×10^{11} V/m
Electric potential	e/a_o	27.2114 V
Fine structure constant	$\alpha = \frac{e^2}{\hbar c}$	1/137.036
Magnetic moment	$\beta_e = e\hbar/(2m_e)$	9.27399×10^{-24} J/T
Permitivity of vacuum	$\varepsilon_o = 1/4\pi$	$8.8541878 \, ^{-12}$J^{-1}C^{2}m^{-1}
Hydrogen atom IP	$-\alpha^2 m_e c^2/2 = -E_h/2$	-13.60580 eV

TABLE A.3
Useful Orders of Magnitude

Quantity	Approximate Value	Exact Value
Electron rest mass	$m_e c^2 \approx 0.5$ MeV	0.511003 MeV
Proton rest mass	$m_p c^2 \approx 1000$ MeV	938.280 MeV
Neutron rest mass	$M_n c^2 \approx 1000$ MeV	939.573 MeV
Proton/electron mass ratio	$m_p / m_e \approx 2000$	1836.1515

One electron volt corresponds to a:

Quantity	Symbol/Relation	Exact Value
Frequency: $\nu = 2.4 \times 10^{14}$ Hz	$E = h\nu$	2.417970×10^{14} Hz
Wavelength: $\lambda = 12000$ Å	$\lambda = c/\nu$	12398.52 Å
Wave number: $1/\lambda = 8000$ cm^{-1}		8065.48 cm^{-1}
Temperature: $T = 12000$ K	$E = kT$	11604.5 K

A.2 THE DIRAC DELTA FUNCTION

A.2.1 DEFINITION

The Dirac delta function is not really a function, per se; it is really a generalized function defined by the relation

$$f(x_o) = \int_{-\infty}^{+\infty} dx \delta(x - x_o) f(x) \qquad (A.1)$$

The integral picks out the first term in the Taylor expansion of $f(x)$ about the point x_o and this relation must hold for any function of x. For example, let us take a function that is zero only at some arbitrary point, x_o. Then the integral becomes

$$\int_{-\infty}^{+\infty} dx \delta(x - x_o) f(x) = 0 \qquad (A.2)$$

For this to be true for any arbitrary function, we have to conclude that

$$\delta(x) = 0 \quad \text{for } x \neq 0 \qquad (A.3)$$

Furthermore, from the Reimann-Lebesque theory of integration

$$\int f(x)g(x)dx = \lim_{h \to 0} \left(a \sum_n f(x_n)g(x_n) \right) \qquad (A.4)$$

the only way for the defining relation to hold is for

$$\delta(0) = \infty \qquad (A.5)$$

This is a very odd function: it is zero everywhere *except* at one point, at which it is infinite. So it is not a function in the regular sense. In fact, it is more like a distrubution function that is infinitely narrow. If we set $f(x) = 1$, then we can see that the δ function is normalized to unity

$$\int_{-\infty}^{+\infty} dx \delta(x - x_o) = 1 \tag{A.6}$$

A.2.2 PROPERTIES

Some useful properties of the δ function are as follows:

1. It is real: $\delta^*(x) = \delta(x)$.
2. It is even: $\delta(x) = \delta(-x)$
3. $\delta(ax) = \delta(x)/a$ for $a > 0$
4. $\int \delta'(x) f(x) dx = f'(0)$
5. $\delta'(-x) = -\delta'(x)$
6. $x\delta(x) = 0$
7. $\delta(x^2 - a^2) = \dfrac{1}{2a}(\delta(x + a) + \delta(x - a))$
8. $f(x)\delta(x - a) = f(a)\delta(x - a)$
9. $\int \delta(x - a)\delta(x - b) dx = \delta(a - b)$

Exercise

Prove the above relations.

A.2.3 SPECTRAL REPRESENTATIONS

The δ function can be thought of as the limit of a sequence of regular functions. For example,

$$\delta(x) = \lim_{a \to \infty} \frac{1}{\pi} \frac{\sin(ax)}{x}$$

This is the "sinc" function or diffraction function with a width proportional to $1/a$. For any value of a, the function is regular. As we make a larger, the width increases and focuses about $x = 0$. This is shown in the Figure A.1 for increasing values of a. Notice that as a increases, the peak increases and the function itself becomes extremely oscillitory.

Another extremely useful representation is the Fourier representation

$$\delta(x) = \frac{1}{2\pi} \int e^{ikx} dk \tag{A.7}$$

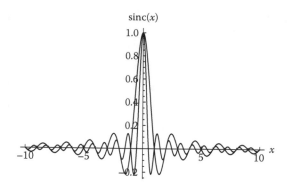

FIGURE A.1 $\sin(xa)/\pi x$ representation of the Dirac δ function.

Finally, another form is in terms of Gaussian functions as shown in Figure A.2,

$$\delta(x) = \lim_{a\to\infty} \sqrt{\frac{a}{\pi}} e^{-ax^2} \tag{A.8}$$

Here the height is proportional to a and the width to the standard deviation, $1/\sqrt{2a}$. Other representations include Lorentzian form,

$$\delta(x) = \lim_{a\to} \frac{1}{\pi} \frac{a}{x^2 + a^2}$$

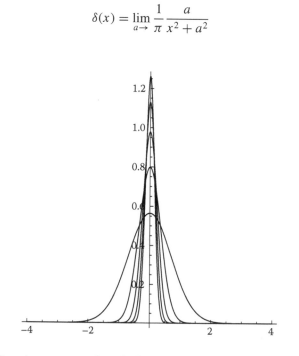

FIGURE A.2 Gaussian representation of δ function.

and derivative form

$$\delta(x) = \frac{d}{dx}\theta(x)$$

where $\theta(x)$ is the Heaviside step function

$$\theta(x) = \begin{cases} 0, & x \leq 0 \\ 1 & x \geq 0 \end{cases} \tag{A.9}$$

This can be understood as the cumulative distribution function

$$\theta(x) = \int_{-\infty}^{x} \delta(y)dy \tag{A.10}$$

A.3   SUMMARY OF ESSENTIAL EQUATIONS FROM QUANTUM MECHANICS

A.3.1   QUANTUM UNCERTAINTY RELATIONS

- De Broglie relation: $\lambda = h/p$ where p is the particle momentum and λ is the de Broglie wavelength.
- Planck–Einstein relation: $E = h\nu = \hbar\omega$
- Uncertainty relation (generalized)

$$(\delta A)^2(\delta B)^2 \geq \langle i[\hat{A}, \hat{B}]\rangle^2/4 \tag{A.11}$$

  where $\hat{A}$ and $\hat{B}$ are operators corresponding to physical observables with variance $(\delta A) = \langle A^2 \rangle - \langle A \rangle^2$ and $\langle \ldots \rangle$ denoting the quantum expectation value
- Position/momentum: $\delta x \cdot \delta p \geq \hbar/2$. This is also known as the "Heisenberg uncertainty relation"
- Energy/time: $\delta E \cdot \delta t \geq \hbar/2$.
- Occupation number/phase: $\delta n \cdot \delta \phi \geq 1/2$

A.3.2   STATES AND WAVE FUNCTIONS

If $\psi(x, t)$ is a quantum wave function,

- Probability density:

$$P(x, t) = |\psi(x, t)|^2 \tag{A.12}$$

- Probability density current:

$$\vec{j}(x, t) = \frac{\hbar}{2mi}(\psi^*(x, t)\vec{\nabla}\psi(x, t) - \psi(x, t)\vec{\nabla}\psi^*(x, t)$$
$$= Re(\psi^* \vec{p} \psi) \tag{A.13}$$

For particles in three dimensions, suitable units are $m^{-2}s^{-1}$.

- Schrödinger equation:

$$\left(i\hbar \frac{\partial}{\partial t} - \hat{H} \right) \psi = 0 \tag{A.14}$$

- Normalization:

$$\int \psi^* \hat{A} \psi \, dx = 1 \tag{A.15}$$

- Superposition principle: If ψ_n is a complete set of eigenfunctions over some finite or infinite range on x, then any other function on that range can be represented as

$$\phi = \sum_n c_n \psi_n \tag{A.16}$$

where

$$c_n = \int \psi_n^* \phi \, dx \tag{A.17}$$

Such sets of functions are orthogonal if

$$\int \psi_n^* \psi_m \, dx = \delta_{nm} \tag{A.18}$$

A.3.3 OPERATORS

If $\hat{A}$ is a quantum mechanical operator and ϕ and ψ are normalizable functions,

- Hermitian conjugate:

$$\int (\hat{A}\phi)^* \psi \, dx = \int \phi^* \hat{A} \psi \, dx \tag{A.19}$$

- Position operator:

$$\hat{x}^n = x^n \tag{A.20}$$

- Momentum operator:

$$\hat{p}^n = \left(\frac{\hbar}{i} \right)^n \frac{\partial^n}{\partial x^n} \tag{A.21}$$

- Kinetic energy operator:

$$\hat{T} = -\frac{\hbar^2}{2m} \nabla^2 \tag{A.22}$$

- Potential energy operator:

$$\hat{V} = V(\hat{x}) \tag{A.23}$$

- Hamiltonian operator:

$$\hat{H} = \hat{T} + \hat{V} \qquad (A.24)$$

for a one-dimensional system, this is written as

$$\hat{H} = -\frac{\hbar^2}{2m}\frac{\partial^2}{\partial x^2} + V(x) \qquad (A.25)$$

- Parity operator:

$$\hat{P} f(x) = f(-x) \qquad (A.26)$$

- Expectation value:

$$\langle \hat{A} \rangle = \langle \psi | \hat{A} | \psi \rangle = \int \psi^* \hat{A} \psi \, dx \qquad (A.27)$$

- Matrix element:

$$\langle \phi | A | \psi \rangle = \int \psi^* \hat{A} \psi \, dx \qquad (A.28)$$

- Time evolution (Heisenberg equations of motion):

$$\frac{d\langle A \rangle}{dt} = \frac{i}{\hbar}\langle [\hat{H}, \hat{A}] \rangle + \left\langle \frac{\partial A}{\partial t} \right\rangle \qquad (A.29)$$

- Ehrenfest theorem:

$$\frac{d\langle x \rangle}{dt} = \langle p \rangle \ \& \ \frac{d\langle p \rangle}{dt} = -\langle \nabla V \rangle \qquad (A.30)$$

- Expansion of expectation values in terms of eigenfunctions: If ψ_n is an eigenfunction of $\hat{A}$ such that $\hat{A}\psi_n = a_n \psi_n$, then

$$\langle \phi | \hat{A} | \phi \rangle = \sum_n |c_n|^2 a_n \qquad (A.31)$$

where c_n are given by

$$c_n = \int \psi_n^* \phi \, dx \qquad (A.32)$$

We interpret $|c_n|^2$ as the probability of finding the system in the nth eigenstate. More precisely, $|c_n|^2$ is the likelihood that making a physical observation described by the quantum mechanical operator $\hat{A}$ will result in a measurement of a_n.

- Dirac notation (bra-ket notation):
  - Matrix element

$$a_{nm} = \langle \psi_m | \hat{A} | \psi_n \rangle \qquad (A.33)$$

- Bra state vector: $|\psi_n\rangle$
- Ket state vector: $\langle\psi|$
- Wave function: $\psi_n(x) = \langle x|\psi_n\rangle$
- Scalar product:

$$\langle\psi_n|\psi_m\rangle = \int dx\,\psi_n(x)\psi_m(x) \tag{A.34}$$

- Resolution of the identity:

$$\mathbb{I} = \sum_n |\psi_n\rangle\langle\psi_n| \tag{A.35}$$

for a continuous set of states

$$\mathbb{I} = \int dx\,|x\rangle\langle x| \tag{A.36}$$

• Pauli matrices:

$$\sigma_x = \begin{pmatrix} 0 & 1 \\ 1 & 0 \end{pmatrix} \tag{A.37}$$

$$\sigma_y = \begin{pmatrix} 0 & -i \\ i & 0 \end{pmatrix} \tag{A.38}$$

$$\sigma_z = \begin{pmatrix} 1 & 0 \\ 0 & -1 \end{pmatrix} \tag{A.39}$$

$$\sigma_0 = \begin{pmatrix} 1 & 0 \\ 0 & 1 \end{pmatrix} \tag{A.40}$$

These matrices have the following properties:
- Anticommutation:

$$\sigma_i \cdot \sigma_j + \sigma_j \cdot \sigma_i = 2\delta_{ij}\sigma_o \tag{A.41}$$

- Commutation:

$$\sigma_i \cdot \sigma_j - \sigma_j \cdot \sigma_i = 2i\sigma_k \tag{A.42}$$

- Cyclic permutation of indices:

$$\sigma_i\sigma_j = i\sigma_k \tag{A.43}$$

$$\sigma_i\sigma_i = \sigma_o \tag{A.44}$$

- Boson operators (harmonic oscillator):
  - Properties of the creation/annihilation operators:

$$\hat{a}|0\rangle = 0 \tag{A.45}$$

$$\hat{a}^{\dagger}|0\rangle = |1\rangle \tag{A.46}$$

$$\hat{a}|n\rangle = \sqrt{n}|n\rangle \tag{A.47}$$

$$\hat{a}^{\dagger}|n\rangle = \sqrt{n+1}|n+1\rangle \tag{A.48}$$

$$\hat{n} = \hat{a}^{\dagger}\hat{a} \tag{A.49}$$

$$|n\rangle = \frac{1}{\sqrt{n!}}(\hat{a}^{\dagger})^{n}|n\rangle \tag{A.50}$$

$$\hat{n}|n\rangle = n|n\rangle \tag{A.51}$$

$$[\hat{a}, \hat{a}^{\dagger}] = 1 \tag{A.52}$$

$$\langle n-1|\hat{a}|n\rangle = \sqrt{n} \tag{A.53}$$

$$\langle n+1|\hat{a}^{\dagger}|n\rangle = \sqrt{n+1} \tag{A.54}$$

  - Relation to other operators:

$$\hat{a} = \hat{X} + i\hat{P} \tag{A.55}$$

$$\hat{a}^{\dagger} = \hat{X} - i\hat{P} \tag{A.56}$$

$$H = \hbar\omega(\hat{P}^2 + \hat{X}^2) = \hbar\omega(\hat{n} + 1/2) \tag{A.57}$$

where $\hat{X} = (m\omega/2\hbar)^2\hat{x}$ and $\hat{P} = (2m\hbar\omega)^{-1/2}\hat{p}$ are dimensionless position and momentum operators.
  - Harmonic oscillating matrix elements:

$$\langle m|\hat{H}|n\rangle = \hbar\omega(n + 1/2)\delta_{mn} \tag{A.58}$$

$$\langle n+1|\hat{x}|n\rangle = (n+1)^{1/2}\left(\frac{\hbar}{2m\omega}\right)^{1/2} \tag{A.59}$$

$$\langle n-1|\hat{x}|n\rangle = n^{1/2}\left(\frac{\hbar}{2m\omega}\right)^{1/2} \tag{A.60}$$

$$\langle n+1|\hat{p}|n\rangle = i(n+1)^{1/2}\left(\frac{\hbar m\omega}{2}\right)^{1/2} \tag{A.61}$$

$$\langle n-1|\hat{p}|n\rangle = -in^{1/2}\left(\frac{\hbar m\omega}{2}\right)^{1/2} \tag{A.62}$$

- Coordinate representation:

$$\hat{a} = \left(\frac{m\omega}{2\hbar}\right)^{1/2}x + \left(\frac{\hbar}{2m\omega}\right)^{1/2}\frac{d}{dx} \tag{A.63}$$

$$\hat{a}^{\dagger} = \left(\frac{m\omega}{2\hbar}\right)^{1/2}x - \left(\frac{\hbar}{2m\omega}\right)^{1/2}\frac{d}{dx} \tag{A.64}$$

- Wave functions: Since $\hat{a}|0\rangle = 0$, we have

$$\left[\left(\frac{m\omega}{2\hbar}\right)^{1/2} x + \left(\frac{\hbar}{2m\omega}\right)^{1/2} \frac{d}{dx}\right] \phi_0(x) = 0 \qquad \text{(A.65)}$$

which by simple integration leads to

$$\phi_0(x) = \left(\frac{m\omega}{\pi\hbar}\right)^{1/4} \exp\left[-\frac{m\omega}{\hbar}x^2\right] \qquad \text{(A.66)}$$

Thus, any other harmonic oscillating eigenfunction is given by

$$\phi_n = \frac{1}{(n!)^{1/2}} (\hat{a}^\dagger)^n \phi_0(x) \qquad \text{(A.67)}$$

When the $\hat{a}^\dagger$ acts upon a Gaussian, it generates the Hermite polynomials

$$H_n(x) = (-1)^n e^{x^2} \frac{d^n}{dx^n} e^{-x^2} \qquad \text{(A.68)}$$

Thus,

$$\phi_n(y) = \frac{1}{(2^n n!)^{1/2}} \left(\frac{\beta^2}{\pi}\right)^{1/4} H_n(y) e^{-y^2} \qquad \text{(A.69)}$$

where $y = \beta x$ and $\beta = (m\omega/\hbar)^{1/2}$
- Variances and uncertainty product:

$$\langle x^2 \rangle - \langle x \rangle^2 = (2n+1)\frac{\hbar}{2m\omega} \qquad \text{(A.70)}$$

$$\langle p^2 \rangle - \langle p \rangle^2 = (2n+1)\frac{\hbar m\omega}{2} \qquad \text{(A.71)}$$

$$\Delta x \Delta p = \hbar(n + 1/2) \qquad \text{(A.72)}$$

This last relation indicates that the harmonic oscillator ground state carries the minimal amount of uncertainty in accordance with the Heisenberg uncertainty principle.

A.4 MATHEMATICAL SERIES AND INTEGRAL TRANSFORMATIONS

- Fourier series (real forms): If $f(x)$ is a periodic function with period $2L$,

$$f(x) = \frac{a_o}{2} + \sum_{n=1}^{\infty} (a_n \cos(n\pi x/L) + b_n \sin(n\pi x/L)) \qquad \text{(A.73)}$$

with

$$a_n = \frac{1}{L} \int_{-L}^{+L} f(x) \cos(n\pi x/L) dx \qquad \text{(A.74)}$$

and

$$b_n = \frac{1}{L} \int_{-L}^{+L} f(x)\sin(n\pi x/L)dx \tag{A.75}$$

- Complex form:

$$f(x) = \sum_{n=-\infty}^{+\infty} c_n exp(in\pi x/L) \tag{A.76}$$

with

$$c_n = \frac{1}{2L} \int_{-L}^{+L} dx f(x)e^{in\pi x/L} \tag{A.77}$$

- Parseval's theorem:

$$\frac{1}{2L} \int_{-L}^{+L} dx |f(x)|^2 = \sum_{n=-\infty}^{\infty} |c_n|^2 \tag{A.78}$$

- Fourier transform: If $f(x)$ is a function of x, its Fourier transform is $\mathscr{F}(k)$. There are at least three ways one can define this transformation and its inverse.
  - Definition #1:

$$\mathscr{F}(k) = \int_{-\infty}^{\infty} f(x)e^{-2\pi ikx}dx \tag{A.79}$$

$$f(x) = \int_{-\infty}^{\infty} \mathscr{F}(k)e^{+2\pi ikx}dk \tag{A.80}$$

  - Definition #2:

$$\mathscr{F}(k) = \int_{-\infty}^{\infty} f(x)e^{-ikx}dx \tag{A.81}$$

$$f(x) = \frac{1}{2\pi} \int_{-\infty}^{\infty} \mathscr{F}(k)e^{+ikx}dk \tag{A.82}$$

  - Definition #3:

$$\mathscr{F}(k) = \frac{1}{\sqrt{2\pi}} \int_{-\infty}^{\infty} f(x)e^{-ikx}dx \tag{A.83}$$

$$f(x) = \frac{1}{\sqrt{2\pi}} \int_{-\infty}^{\infty} \mathscr{F}(k)e^{+ikx}dk \tag{A.84}$$

- Fourier transform theorems ($\leftrightharpoons$ denotes Fourier transform relation):
  - Convolution of two functions:

$$f(x) * g(x) = \int_{-\infty}^{\infty} f(u)g(x - u)du \qquad \text{(A.85)}$$

  - Convolution rules:

$$f * g = g * f \qquad \text{(A.86)}$$

$$f * (g * h) = (f * g) * h \qquad \text{(A.87)}$$

  - Convolution theorem:

$$f(x) * g(x) \leftrightharpoons \mathscr{F}(s) * \mathscr{G}(s) \qquad \text{(A.88)}$$

  - Autocorrelation (f^* denotes the complex conjugate of f):

$$f^*(x) \star f(x) = \int_{-\infty}^{\infty} du f^*(u - x) f(u) \qquad \text{(A.89)}$$

  - Wiener–Khintchine theorem:

$$f^*(x) \star f(x) \leftrightharpoons |\mathscr{F}(s)|^2 \qquad \text{(A.90)}$$

  - Cross-correlation:

$$f^*(x) \star g(x) = \int_{-\infty}^{\infty} du f^*(u - x)g(u) \qquad \text{(A.91)}$$

  - Correlation theorem:

$$h(x) \star g(x) \leftrightharpoons \mathscr{H}(s)\mathscr{G}(s) \qquad \text{(A.92)}$$

  - Parseval's relation (also called the power transform):

$$\int_{-\infty}^{\infty} f(x)g^*(x)dx = \int_{-\infty}^{\infty} \mathscr{F}(s)\mathscr{G}(s)ds \qquad \text{(A.93)}$$

  - Parseval's theorem (also called Rayleigh's theorem):

$$\int_{-\infty}^{\infty} |f(x)|^2 dx = \int_{-\infty}^{\infty} |\mathscr{F}(s)|^2 ds \qquad \text{(A.94)}$$

  - Derivatives of transform pairs:

$$\frac{df(x)}{dx} \leftrightharpoons 2\pi i k \mathscr{F}(k) \qquad \text{(A.95)}$$

$$\frac{d}{dx}(f(x) * g(x)) = \frac{df(x)}{dx} * g(x) + \frac{dg(x)}{dx} * f(x) \qquad \text{(A.96)}$$

- Symmetry relations:

$$f(x) \rightleftharpoons \mathcal{F}(s)$$
$$\text{even} \rightleftharpoons \text{even}$$
$$\text{odd} \rightleftharpoons \text{odd}$$
$$\text{real, even} \rightleftharpoons \text{real, even}$$
$$\text{real, odd} \rightleftharpoons \text{imaginary, odd}$$
$$\text{imaginary, even} \rightleftharpoons \text{imaginary, even}$$
$$\text{complex, even} \rightleftharpoons \text{complex, even}$$
$$\text{complex, odd} \rightleftharpoons \text{complex, odd}$$
$$\text{real, asymmetric} \rightleftharpoons \text{complex, Hermitian}$$
$$\text{imaginary, asymmetric} \rightleftharpoons \text{complex, anti-Hermitian}$$
$$\text{real}: f(x) = f^*(x)$$
$$\text{imaginary}: f(x) = -f^*(x)$$
$$\text{even}: f(x) = f(-x)$$
$$\text{odd}: f(x) = -f(-x)$$
$$\text{Hermitian}: f(x) = f^*(-x)$$
$$\text{anti-Hermitian}: f(x) = -f^*(-x)$$

- Miscellaneous Fourier transform pairs (using Definition #1 from above):

$$f(x) \rightleftharpoons \mathcal{F}(s) \tag{A.97}$$

$$f(ax) \rightleftharpoons \frac{1}{|a|}\mathcal{F}(s/a) \tag{A.98}$$

$$f(x-a) \rightleftharpoons e^{-2\pi i a s}\mathcal{F}(s) \tag{A.99}$$

$$\delta(x) \rightleftharpoons 1 \tag{A.100}$$

$$\delta(x-a) \rightleftharpoons e^{-2\pi i a s} \tag{A.101}$$

$$\frac{d^n f}{dx^n} \rightleftharpoons (2\pi i s)^n \mathcal{F}(s) \tag{A.102}$$

$$e^{-a|x|} \rightleftharpoons \frac{2a}{a^2 + 4\pi^2 s^2} \tag{A.103}$$

$$x e^{-a|x|} \rightleftharpoons \frac{8\pi i a s}{(a^2 + 4\pi^2 s^2)^2} \tag{A.104}$$

$$e^{-x^2/a^2} \rightleftharpoons a\sqrt{\pi}\, e^{-\pi^2 a^2 s^2} \tag{A.105}$$

$$\sin(ax) \rightleftharpoons \frac{1}{2i}\left(\delta\left(s - \frac{a}{2\pi}\right) - \delta\left(s + \frac{a}{2\pi}\right)\right) \tag{A.106}$$

$$\cos(ax) \rightleftharpoons \frac{1}{2}\left(\delta\left(s - \frac{a}{2\pi}\right) + \delta\left(s + \frac{a}{2\pi}\right)\right) \tag{A.107}$$

$$\sum_{n=-\infty}^{\infty} \delta(x - na) \rightleftharpoons \frac{1}{a}\sum_{n=-\infty}^{\infty} \delta(s - n/a) \tag{A.108}$$

A.5   NUMERICAL APPROXIMATIONS

- Numerical derivatives: The derivative of $f(x)$ at x can be approximated as (h = small interval in x)

$$\frac{df(x)}{dx} \approx \frac{1}{2h}(f(x+h) - f(x-h)) + \mathcal{O}(h^2) \tag{A.109}$$

$$\approx \frac{1}{12h}(-f(x+2h) + 8f(x+h)$$
$$- 8(f-h) + f(x+2h)) + \mathcal{O}(h^4) \tag{A.110}$$

$$\frac{d^2 f(x)}{dx^2} \approx \frac{1}{h^2}(f(x+h) - 2f(x) + f(x-h)) + \mathcal{O}(h^2) \tag{A.111}$$

$$\approx \frac{1}{12h^2}(-f(x+2h) + 16f(x+h) - 30f(x)$$
$$+ 16(x-h) - f(x-2h)) + \mathcal{O}(h^4) \tag{A.112}$$

$$\frac{d^3 f(x)}{dx^3} \approx \frac{1}{2h^3}(f(x+2h) - 2f(x+h)$$
$$+ 2f(x-h) - f(x-2h)) + \mathcal{O}(h^2) \tag{A.113}$$

- Finding roots $f(x) = 0$:
  - Secant method:

$$x_{n+1} = x_n - \frac{x_n - x_{n-1}}{f(x_n) - f(x_{n-1})} f(x_n) \tag{A.114}$$

  - Newton-Raphson method:

$$x_{n+1} = x_n - \frac{f(x_n)}{f'(x_n)} \tag{A.115}$$

  - The built-in **Mathematica** function `FindRoots[f[x]==0,x]` is very convenient for this and works in multiple dimensions.

References

Chapter 2

1. Louis de Broglie. *Recherches sur la théorie des quanta*. PhD thesis, Sorbonne, Paris, 1924.
2. Erwin Schrödinger. An undulatory theory of the mechanics of atoms and molecules. *Phys. Rev.*, 28(6):1049–1070, December 1926.
3. Erwin Schrödinger. Über das verhältnis der Heisenberg-Born-Jordanschen quanten-mechanik zu der meinen. *Ann. Phys. (Leipzig)*, 1926.

Chapter 4

4. E. Fermi. *Nuclear Physics*. University of Chicago Press, Chicago, 1950.
5. P. A. M. Dirac. The quantum theory of emission and absorption of radiation. *Proc. R. Soc. London, Ser. A*, 114:243–265, 1927.
6. Eric R. Bittner and Peter J. Rossky. Quantum decoherence in mixed quantum-classical systems: Nonadiabatic processes. *J. Chem. Phys.*, 103(18):8130–8143, Nov 1995.
7. Eric R. Bittner and Peter J. Rossky. Decoherent histories and nonadiabatic quantum molecular dynamics simulations. *J. Chem. Phys.*, 107(20):8611–8618, Nov 1997.
8. Robbie Grunwald and Raymond Kapral. Decoherence and quantum-classical master equation dynamics. *J. Chem. Phys.*, 126(11):114109, 2007.
9. Klaus Hornberger. Master equation for a quantum particle in a gas. *Phys. Rev. Lett.*, 97(6):060601, 2006.
10. Ahren W. Jasper and Donald G. Truhlar. Electronic decoherence time for non-Born-oOppenheimer trajectories. *J. Chem. Phys.*, 123(6):064103, 2005.
11. Gunter Kab. Fewest switches adiabatic surface hopping as applied to vibrational energy relaxation. *J. Phys. Chem. A*, 110(9):3197–3215, 2006.
12. Gil Katz, Mark A Ratner, and Ronnie Kosloff. Decoherence control by tracking a Hamiltonian reference molecule. *Phys. Rev. Lett.*, 98(20):203006, 2007.
13. M. Merkli, I. M. Sigal, and G. P. Berman. Decoherence and thermalization. *Phys Rev Lett*, 98(13):130401, 2007.
14. Maximilian A. Schlosshauer. *Decoherence and the quantum-to-classical transition*. Springer, Berlin, 2007.
15. L. D. Landau. *Phy. Z. Sowjetunion*, 1:89, 1932.
16. C. Zener. *Proc. R. Soc. London, Ser. A*, 137:696, 1933.
17. A. Nitzan. *Chemical Dynamics in Condensed Phases*. Oxford University Press, 2007.
18. R. A. Marcus. On the theory of electron-transfer reactions. Part VI. Unified treatment for homogeneous and electrode reactions. *J. Chem. Phys.*, 43(2):679–701, Jul 1965.
19. Rudolph A. Marcus. Nobel lecture: Electron transfer reactions in chemistry. Theory and experiment. *Rev. Mod. Phys.*, 65(3):599–610, 1993.
20. J. R. Miller, L Calcaterra, and G. L. Closs. "Intramolecular long-distance electron transfer in radical anions. The effects of free energy and solvent on the reaction rates." *J. Am. Chem. Soc.*, 106:3047, 1984.
21. Eyal Neria, Abraham Nitzan, R. N. Barnett, and Uzi Landman. Quantum dynamical simulations of nonadiabatic processes: Solvation dynamics of the hydrated electron. *Phys. Rev. Lett.*, 67(8):1011–1014, Aug 1991.

22. Eyal Neria and Abraham Nitzan. Semiclassical evaluation of nonadiabatic rates in condensed phases. *J. Chem. Phys.*, 99(2):1109–1123, 1993.
23. Louis de Broglie. *Ondes et mouvements*. Gauthiers-Villars, Paris, 1926.
24. Louis de Broglie. *La mécanique ondulatoire*. Gauthiers-Villars, Paris, 1928.
25. Peter Holland. *The Quantum Theory of Motion: An Account of the de Broglie-Bohm Causal Interpretation of Quantum Mechanics*. Cambridge, UK: Cambridge University Press, 1993.

CHAPTER 5

26. P. A. M. Dirac. *The Principles of Quantum Mechanics*. Oxford University Press, Oxford, NY, 4th edition, 1958.
27. Julian Schwinger. *Quantum Mechanics—Symbolism of Atomic Measurements*. Physics and Astonomy. Springer, Berlin, 2001.
28. Philip Pechukas. Time-dependent semiclassical scattering theory. Part I potential scattering. *Phys. Rev.*, 181(1):166–174, May 1969.
29. Philip Pechukas. Time-dependent semiclassical scattering theory. Part II atomic collisions. *Phys. Rev.*, 181(1):174–185, May 1969.
30. John C. Tully. Molecular dynamics with electronic transitions. *J. Chem. Phys.*, 93(2):1061–1071, 1990.
31. Frank J. Webster, Jurgen Schnitker, Mark S. Friedrichs, Richard A. Friesner, and Peter J. Rossky. Solvation dynamics of the hydrated electron: A nonadiabatic quantum simulation. *Phys. Rev. Lett.*, 66(24):3172–3175, Jun 1991.
32. B. Space and D. F. Coker. Nonadiabatic dynamics of excited excess electrons in simple fluids. *J. Chem. Phys.*, 94(3):1976–1984, 1991.

CHAPTER 6

33. L. Allen and J. H. Eberly. *Optical Resonance and Two-Level Atoms*. Dover Publications, 1987.
34. E. L. Hahn. Spin echoes. *Phys. Rev.*, 80(4):580–594, Nov 1950.
35. E. L. Hahn and D. E. Maxwell. Chemical shift and field independent frequency modulation of the spin echo envelope. *Phys. Rev.*, 84(6):1246–1247, Dec 1951.
36. N. A. Kurnit, I. D. Abella, and S. R. Hartmann. Observation of a photon echo. *Phys. Rev. Lett.*, 13(19):567–568, Nov 1964.
37. I. D. Abella, N. A. Kurnit, and S. R. Hartmann. Photon echoes. *Phys. Rev.*, 141(1): 391–406, Jan 1966.
38. A.A. Ovchinnikov and N.S. Erikhman. *Sov. Phys. JETP*, 40:733, 1975.
39. A. Madhukar and W. Post. *Phys. Rev. Lett.*, 39:1424, 1977.
40. A. M. Jayannavar and N. Kumar. *Phys. Rev. Lett*, 48:553, 1982.
41. S. M. Girvin and G. D. Mahan. *Phys. Rev. B*, 20:4896, 1979.
42. W. Fischer, H. Leschke, and P. Müller. *Phys. Rev. Lett.*, 73:1578, 1994.
43. I. Burghardt and L. S. Cederbaum. Hydrodynamic equation for mixed quantum states. i. general formulation. *J. Chem. Phys.*, 115(22):10303, December 2001.
44. I. Burghardt and L. S. Cederbaum. Hydrodynamic equations for mixed quantum states. ii. coupled electronic states. *J. Chem. Phys.*, 115(22):10312, December 2001.
45. Yasuteru Shigeta, Hideaki Miyachi, and Kimihiko Hirao. Quantal cumulant dynamics: General theory. *J. Chem. Phys.*, 125(24):244102, 2006.
46. Jeremy B. Maddox and Eric R. Bittner. Quantum relaxation dynamics using Bohmian trajectories. *J. Chem. Phys.*, 115(14):6309–6316, 2001.

47. Jeremy B. Maddox and Eric R. Bittner. Quantum dissipation in unbounded systems. *Phys. Rev. E*, 65(2):026143, Jan 2002.
48. J. B. Maddox and E. R. Bittner. Quantum dissipation in the hydrodynamic moment hierarchy: A semiclassical truncation strategy. *J. Phys. Chem. B*, 106(33):7981–7990, 2002.
49. L. A. Khalfin. *Sov. Phys. JETP*, 6:1053, 1958.
50. Wayne M. Itano, D. J. Heinzen, J. J. Bollinger, and D. J. Wineland. Quantum zeno effect. *Phys. Rev. A*, 41(5):2295–2300, Mar 1990.
51. R. J. Cook. What are quantum jumps? *Phys. Scr.*, T21:49, 1988.
52. E. P. Wigner. On the quantum correction for thermodynamic equilibrium. *Phys. Rev.*, 40:749–759, 1932.
53. H. Weyl. *The Theory of Groups and Quantum Mechanics*. Dover, New York, 1931.
54. H. Weyl. *Gruppentheorie und Quantenmechanik*. Hirzel, Leipzig, 1928.
55. H. Weyl. *Z. Phys.*, 46:1, 1927.
56. J. Ville. Théorie et applications de la notion de signal analytique. *Cables et Transmission*, 2A:61–74, 1948.
57. J. E. Moyal. Quantum mechanics as a statistical theory. *Proc. Cambridge Philos. Soc.*, 45:737–740, 1949.
58. H. J. Groenewold. *Physica*, 12(405), 1946.
59. William Frensley. Boundary conditions for open quantum systems driven far from equilibrium. *Rev. Mod. Phys.*, 62:745, 1990.
60. Jasper Koester and Shaul Mukamel. Transient gratings, four wave mixing, and polariton effects in non-linear optics. *Phys. Rep.*, 205(1):1–58, 1991.
61. Roger F. Loring, Daniel S. Franchi, and Shaul Mukamel. Anderson localization in Liouville space: The effective dephasing approximation. *Phys. Rev. B*, 37(4):1874–1883, Feb 1988.
62. Shaul Mukamel. *Principles of Nonlinear Optics and Spectroscopy*. Oxford University Press, 1995.
63. P. L. Bhatnagar, E. P. Gross, and M. Krook. A model for collision processes in gases. i. small amplitude processes in charged and neutral one-component systems. *Phys. Rev.*, 94(3):511–525, May 1954.

CHAPTER 7

64. T. Förster. Transfer mechanisms of electronic excitation. *Discuss. Faraday Soc. Aberdeen*, (7-18), 1960.
65. T. Förster. Zwischenmolekulare Energiewanderung und Fluoreszenz. *Ann. Phys. (Leipzig)*, 2(55–75), 1948.
66. T. Förster. Energiewanderung und Fluoreszenz. *Naturwissenschaften*, 6(166–175), 1946.
67. T. Förster. Transfer mechanisms of electronic excitation. *Discuss. Faraday Soc.*, 27:17, 1959.
68. E. Hass, E. Katchalski-Katzir, and I. Z. Steinberg. First demo of FRET in DNA loops. *Biopolymers*, 17(11–31), 1978.
69. Mark I. Wallace, Liming Ying, Shankar Balasubramanian, and David Klenerman. Non-arrhenius kinetics for the loop closure of a DNA hairpin. *Proc. Nat. Acad. Sci.*, 98(10):5584–5589, 2001.
70. E. Zazopoulos, E. Lalli, D. M. Stocco, and P. Sassone-Corsi. *Nature*, 390:311–315, 1997.
71. X. Dai, M. B. Greizerstein, K. Nadas-Chinni, and L. B. Rothman-Denes. *Proc. Nat. Acad. Sci.*, 94:2174, 1997.

72. S. Tyagi and F. R. Kramer. Molecular beacons. *Nature Biotechnol.*, 14:303–308, 1996.

73. Wichard J. D. Beenken and Tõnu Pullerits. Excitonic coupling in polythiophenes: Comparison of different calculation methods. *J. Chem. Phys.*, 120(5):2490–2495, 2004.

74. William Barford. Exciton transfer integrals between polymer chains. *J. Chem. Phys.*, 126(13):134905, 2007.

75. B. P. Krueger, G. D. Scholes, and G. R. Fleming. Calculation of couplings and energy-transfer pathways between the pigments of lh2 by the ab initio transition density cube method. *J. Phys. Chem. B*, 102(27):5378–5386, Jul 1998.

76. M. J. Frisch, G. W. Trucks, H. B. Schlegel, G. E. Scuseria, M. A. Robb, J. R. Cheeseman, J. A. Montgomery, Jr. et al. Gaussian 03, Revision C.02. Gaussian, Inc., Wallingford, CT, 2004.

77. Frank Neese. *ORCA—An ab initio Density Functional and Semiempirical Program Package, Version 2.4, Revision 45*, Jan 2006. Max Planck Institute for Bioinorganic Chemistry, Muelheim, Germany.

78. Arkadiusz Czader and Eric R. Bittner. Calculations of the exciton coupling elements between the DNA bases using the transition density cube method. *J. Chem. Phys.*, 128(3):035101, 2008.

79. B. Bouvier, T. Gustavsson, D. Markovitsi, and P. Millie. Dipolar coupling between electronic transitions of the DNA bases and its relevance to exciton states in double helices. *Chem. Phys.*, 275:75–92, 2002.

80. P. Claverie. *Intermolecular Interactions—From Diatomic to Biopolymers.*, Chapter Elaboration of Approximate Formulas for the Interaction between Large Molecules: Application in Organic Chemistry, pages 69–306. Wiley, 1978.

CHAPTER 8

81. J. C. Light, I. P. Hamilton, and J. V. Lill. Generalized discrete variable approximation in quantum mechanics. *J. Chem. Phys.*, 82(3):1400–1409, 1985.

82. J. C. Light. Discrete variable representations in quantum dynamics. In *Time-Dependent Quantum Molecular Dynamics*. Plenum-Press, 1992.

83. E. Hückel. *Zeitschrift für Physik*, 76:628, 1932.

84. J. E. Lennard-Jones. *Proc. R. Soc. London*, 158:280, 1937.

85. C. A. Coulson and A. Streitwieser. *Dictionary of π Electron Calculations*. Pergammon, New York, 1965.

86. W. Kutzelnigg. *Einführung in die Theoretische Cheme, Vol 2: Die Chemische Bindung*. Wiley-VCH, 1978.

87. J. Barriol and J. J. Metzger. *J. Chem. Phys.*, 47:433, 1950.

88. C. C. J. Roothaan. Self-consistent field theory for open shells of electronic systems. *Rev. Mod. Phys.*, 32(2):179–185, 1960.

89. C. C. J. Roothaan. New developments in molecular orbital theory. *Rev. Mod. Phys.*, 23(2):69–89, 1951.

90. Michael C. Zerner. Perspective on "New developments in molecular orbital theory." *Theor. Chim. Acta*, 103(3):217–218, 2000.

91. G. G. Hall. *Proc. R. Soc. London, Ser. A*, 205:541, 1951.

92. G. C. Schatz and M. A. Ratner. *Quantum Mechanics in Chemistry*. Prentice Hall, 1993.

93. Karl F. Freed. Is there a bridge between ab initio and semiempirical theories of valence? *Acc. Chem. Res.*, 16(4):137–144, 1983.

94. Charles H. Martin and Karl F. Freed. Ab initio computation of semiempirical π-electron methods. Part I, constrained, transferable valence spaces in $\mathcal{H}^v$ calculations. *J. Chem. Phys.*, 100(10):7454–7470, 1994.

95. Charles. H. Martin, R. L. Graham, and Karl. F. Freed. Ab initio study of cyclobutadiene using the effective valence shell Hamiltonian method. *J. Chem. Phys.*, 99(10):7833–7844, 1993.

96. J. Hubbard. Electron correlations in narrow energy bands. *Proc. R. Soc. London, Ser. A*, 276(1365):238–257, Nov 1963.

97. Elliott H. Lieb and F. Y. Wu. Absence of Mott transition in an exact solution of the short-range, one-band model in one dimension. *Phys. Rev. Lett.*, 20(25):1445–1448, Jun 1968.

98. Fabian H. L. Essler, Vladimir E. Korepin, and Kareljan Schoutens. Complete solution of the one-dimensional Hubbard model. *Phys. Rev. Lett.*, 67(27):3848–3851, Dec 1991.

99. J. C. Slater. Atomic shielding constants. *Phys. Rev.*, 36(1):57–64, Jul 1930.

100. S. F. Boys. *Proc. R. Soc. London, Ser. A*, 200:542, 1950.

101. Ira Levine. *Quantum Chemistry*. Prentice Hall, 4th edition, 1991.

CHAPTER 9

102. W. P. Su, J. R. Schrieffer, and A. J. Heeger. Soliton excitations in polyacetylene. *Phy. Rev. B*, 22(4):2099–2111, Aug 1980.

103. W. P. Su, J. R. Schrieffer, and A. J. Heeger. Solitons in polyacetylene. *Phys. Rev. Lett.*, 42(25):1698–1701, Jun 1979.

104. L. D. Landau. *Phys. Z. Sowjetunion*, 3:664, 1933.

105. E. I. Rashba and M. D. Sturge. *Excitons*. North-Holland, Amsterdam, Netherlands, 1982.

106. E. I. Rashba. *Opt. Spektrosk.*, 2:568, 1957.

107. Mark N. Kobrak and Eric R. Bittner. A dynamic model for exciton self-trapping in conjugated polymers. Part I. Theory. *J. Chem. Phys.*, 112(12):5399–5409, 2000.

108. Mark N. Kobrak and Eric R. Bittner. A dynamic model for exciton self-trapping in conjugated polymers. ii. implementation. *J. Chem. Phys.*, 112(12):5410–5419, 2000.

109. Mark N. Kobrak and Eric R. Bittner. A quantum molecular dynamics study of exciton self-trapping in conjugated polymers: Temperature dependence and spectroscopy. *J. Chem. Phys.*, 112(17):7684–7692, 2000.

110. Hitoshi Sumi and Atsuko Sumi. Dimensionality dependence in self-trapping of excitons. *J. Phys. Soc. Jp.* 63(2):637–657, 1994.

111. I. Franco and S. Tretiak. Electron-vibrational dynamics of photoexcited polyfluorenes. *J. Am. Chem. Soc.*, 126(38):12130–12140, Sep 2004.

112. Kirill I. Igumenshchev, Sergei Tretiak, and Vladimir Y. Chernyak. Excitonic effects in a time-dependent density functional theory. *J. Chem. Phys.*, 127(11):114902, 2007.

113. S. Tretiak, A. Saxena, R. L. Martin, and A. R. Bishop. Photoexcited breathers in conjugated polyenes: An excited-state molecular dynamics study. *Proc. Nat. Acad. Sci.*, 100(5):2185–2190, Mar 2003.

114. S. Tretiak, A. Saxena, R. L. Martin, and A. R. Bishop. Conformational dynamics of photoexcited conjugated molecules. *Phys. Rev. Lett.*, 89(9):097402, 2002.

115. Sergei Tretiak, Kirill Igumenshchev, and Vladimir Chernyak. Exciton sizes of conducting polymers predicted by time-dependent density functional theory. *Phys. Rev. B*, 71(3):033201–4, 2005.

116. Stoyan Karabunarliev, Martin Baumgarten, Eric R. Bittner, and Klaus Müllen. Rigorous Franck–Condon absorption and emission spectra of conjugated oligomers from quantum chemistry. *J. Chem. Phys.*, 113(24):11372–11381, 2000.

117. Stoyan Karabunarliev and Eric R. Bittner. Polaron–excitons and electron–vibrational band shapes in conjugated polymers. *J. Chem. Phys.*, 118(9):4291–4296, Mar 2003.

118. A. S. Davydov. Excitons and solitons in molecular systems. *Int. Rev. Cytology*, 106:183, 1987.

119. A. S. Davydov. Solitons and energy transfer along protein molecules. *J. Theor. Biol.*, 66(2):379–387, 1977.

120. A. S. Davydov. The theory of contraction of proteins under their excitation. *J. Theor. Biol.*, 38:559–569, 1973.

121. Alwyn Scott. Davydov's soliton. *Phys. Rep.*, 217(1):1–67, 1992.

122. V. E. Zakharov and A. B. Shabat. Exact theory of two dimensional self-focusing and one-dmensional modulations of waves in nonlinear media. *Sov. Phys. JETP*, 34:62–69, 1972.

123. P. S. Lomdahl and W. C. Kerr. Do Davydov solitons exist at 300 K? *Phys. Rev. Lett.*, 55(11):1235–1238, Sep 1985.

124. J. W. Schweitzer. Lifetime of the Davydov soliton. *Phys. Rev. A*, 45(12):8914–8923, Jun 1992.

125. W. Förner. Davydov solitons in proteins. *Int. J. Quantum Chem.*, 64(3):351, 1997.

126. Konrad Knoll and Richard R. Schrock. Preparation of tert-butyl-capped polyenes containing up to 15 double bonds. *J. Am. Chem. Soc.*, 111(20):7989–8004, 1989.

127. Stoyan Karabunarliev, Eric R. Bittner, and Martin Baumgarten. Franck–Condon spectra and electron-libration coupling in para-polyphenyls. *J. Chem. Phys.*, 114(13):5863–5870, 2001.

128. W. M. Gelbart, K. G. Speers, K. F. Freed, J. Jortner, and S. A. Rice. Boltzmann statistics and radiationless decay large molecules: optical selection rules. *Chem. Phys. Lett.*, 6(4):354, 1970.

CHAPTER 10

129. Walter Harrison. *Applied Quantum Mechanics*. World Scientific Publishing Company, Singapore, 2000.

130. Donald A. McQuarrie. *Statistical Mechanics*. Harper and Row, New York, 1976.

Index